INTRODUCTION TO DYNAMICS AND CONTROL IN MECHANICAL ENGINEERING SYSTEMS

Wiley-ASME Press Series List

Introduction to Dynamics and Control in Mechanical Engineering Systems	To	March 2016
Fundamentals of Mechanical Vibrations	Cai	May 2016
Nonlinear Regression Modeling for Engineering Applications: Modeling, Model Validation, and Enabling Design of Experiments	Rhinehart	August 2016

INTRODUCTION TO DYNAMICS AND CONTROL IN MECHANICAL ENGINEERING SYSTEMS

Cho W. S. To

Professor of Mechanical and Materials Engineering
University of Nebraska-Lincoln,
Lincoln, NE, USA

This Work is a co-publication between ASME Press and John Wiley & Sons, Ltd.

This edition first published 2016

Registered Office
John Wiley & Sons, Ltd, The Atrium, Southern Gate, Chichester, West Sussex, PO19 8SQ, United Kingdom

For details of our global editorial offices, for customer services and for information about how to apply for permission to reuse the copyright material in this book please see our website at www.wiley.com.

Library of Congress Cataloging-in-Publication data applied for.

ISBN: 9781118934920

A catalogue record for this book is available from the British Library.

Set in 10/12pt Times by SPi Global, Pondicherry, India
Printed and bound in Singapore by Markono Print Media Pte Ltd

1 2016

To my uncle
Mei Chang Cai (a.k.a. Muljanto Tjokro)

Contents

Series Preface

The Wiley-ASME Press Series in Mechanical Engineering brings together two established leaders in mechanical engineering publishing to deliver high-quality, peer-reviewed books covering topics of current interest to engineers and researchers worldwide.

The series publishes across the breadth of mechanical engineering, comprising research, design and development, and manufacturing. It includes monographs, references and course texts.

Prospective topics include emerging and advanced technologies in Engineering Design; Computer-Aided Design; Energy Conversion & Resources; Heat Transfer; Manufacturing & Processing; Systems & Devices; Renewable Energy; Robotics; and Biotechnology.

Preface

It is understood that there are many excellent books on system dynamics, control theory, and control engineering. However, the lengths of the majority of these books are of the order of six or seven hundred pages or more. There are, however, very few books that cover sufficient material and are limited to around 300 pages. The present book is aimed at addressing the balance. While it is more concise than those longer books, it does include many detailed steps in the example solutions. The author does believe that the detailed steps in the example solutions are essential in a first course textbook.

This book is based on lecture notes that have been developed and used by the author since 1986. These lecture notes have been employed in courses such as Mechanical Control and Process Control, as well as Dynamics and Control. The first two courses were taught by the author at the University of Western Ontario, London, Ontario, Canada while the third course has been given by the author at the University of Nebraska, Lincoln, Nebraska, USA, since 1996. All three courses have primarily been taken by junior undergraduates with majors in mechanical engineering and chemical engineering. Therefore, the subject matter dealt with in this book covers material for a first course of three credit hours per semester in system dynamics or control engineering. For a course in Mechanical Control or Process Control the material in the entire book, except the second half of Chapter 4, has been used. For a course in Dynamics and Control the material in the entire book except Chapter 11 has been covered. For a four credit hour course, the component of laboratory experiments has been omitted from the present book for two main reasons. First, the inclusion of the laboratory experiments is not feasible in the sense that its inclusion would increase drastically the length of the book. Second, nowadays many laboratory experiments are computer-aided in the sense that major software is required. Exclusion of laboratory experiments in the present book provides freedom for the instructors to select a particular software and allows them to tailor the design of their experiments to the availability of laboratory instrumentation in a particular department or engineering environment.

Under normal conditions, it is expected that the students using the present book have already taken courses in their sophomore year. These courses include linear algebra and matrix theory, a second course in mathematics with Laplace transformation, and engineering dynamics. In addition, students are expected to be able to use MATLAB, which is introduced during their first year or first semester of their sophomore year.

Acknowledgments

Many figures in Chapters 3–10 were drawn by Professor Jing Sun of Dalian University of Technology, Dalian, China. Professor Sun was a senior visiting scholar at the University of Nebraska, Lincoln, during the academic year 2012 to 2013. The author is grateful for Professor Sun's kindness in preparing these figures. Specifically, the latter are: Figures 3.2–3.4; Figures 4.4–4.7, 4.9, 4.11–4.13; Figures 5.1–5.4; Figures 6.1–6.3; Figures 7.2 and 7.3; Figures 8.1–8.9; Figures 9.1, 9.2, 9.4–9.8; and Figure 10.1.

Finally, the author would like to express his sincere thanks to Paul Petralia, Senior Editor, Clive Lawson, Project Editor, Annc Hunt, Associate Commissioning Editor, and their team members for their assistance and effort in the production of this book.

1

Introduction

This book is concerned with the introduction to the *dynamics* and *controls* of *engineering systems* in general. The emphasis, however, is on mechanical engineering system modeling and analysis.

- ***Dynamics*** is a branch of mechanics and is concerned with the studies of particles and bodies in motion.
- The term ***control*** refers to the process of *modifying* the *dynamic behavior* of a system in order to achieve some *desired outputs*.
- A ***system*** is a combination of components or elements so constructed to achieve an objective or multiple objectives.

1.1 Important Difference between Static and Dynamic Responses

The question of why one studies engineering dynamics as well as control, and not statics, is best answered by the fact that in control engineering it is the dynamic behavior of a system that is modified instead of the static one. Furthermore, the most important difference between statics and dynamics from the point of view of a mechanical engineering designer is in the responses of a system to an applied force.

Consider a lightly damped, simple, single degree-of-freedom (dof) system that is subjected to a unit step load. The dynamic response is shown in Figure 1.1. Note that the largest peak or overshoot is about 1.75 units, while the magnitude of the input is 1.0 unit. Owing to the positive damping in the system, the dynamic response approaches asymptotically to its steady-state (s.s.) value of unity. If one looks at the largest *mean square value* for the dynamic response, it is about 3.06 units squared. On the other hand, the mean square value for the s.s. or static

Introduction to Dynamics and Control in Mechanical Engineering Systems, First Edition. Cho W. S. To.

Companion website: www.wiley.com/go/to/dynamics

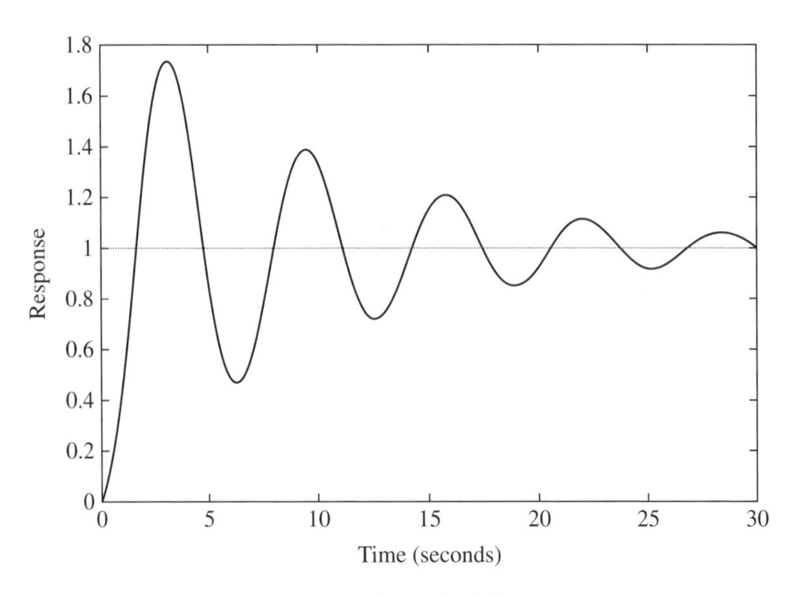

Figure 1.1 Dynamic response of a single dof system under unity input

response is 1.0 unit squared. Thus, the largest mean square value, which is the main design parameter, for the dynamic case is about 306% that of the static case, indicating the importance of dynamic response compared with that of the static case.

1.2 Classification of Dynamic Systems

This book deals with the study of dynamic and control systems in the engineering or physical world. In the latter many phenomena are *nonlinear* and *random* in nature, and therefore to describe, study, and understand such phenomena one has to formulate these phenomena in the conceptual or mathematical world as nonlinear differential equations. The latter, apart from some special cases, are generally very difficult to solve mathematically, and therefore in many situations these nonlinear differential equations are simplified to *linear* differential equations such that they may be solved analytically or numerically.

The meaning of a *linear* phenomenon may better be understood by considering a simple uniform cantilever beam of length L under a dynamic point load $f(t)$ applied transversely at the tip as shown in Figure 1.2. If the tip deflection $y(L,t)$, or simply written as y, satisfies the condition that

$$y \le \pm \frac{5}{100} L$$

then y is said to be linear, and therefore a linear differential equation can be used to describe the deflection y. If the deflection y is larger than 5% of the length L of the beam, a nonlinear differential equation has to be employed instead. The word *random* mentioned in the foregoing means that *statistical analysis* is required to study such phenomena, instead of the usual *deterministic* approaches that are employed throughout in this book.

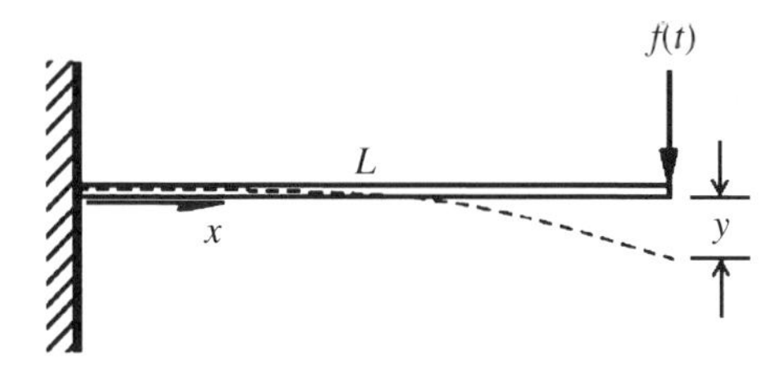

Figure 1.2 Cantilever beam with a point load

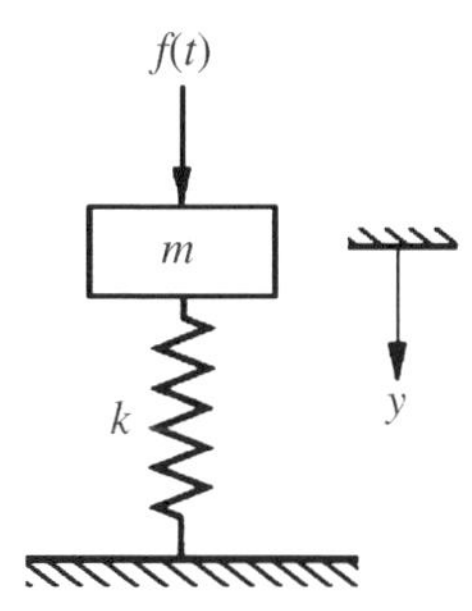

Figure 1.3 A lumped-parameter model of a massless cantilever beam

For the cantilever beam shown in Figure 1.2, the transverse deflection y at any point x along the length of the beam is a function of space x and time t, and therefore the differential equation required to describe the deflection is a *partial differential equation* (p.d.e.). Such a system is referred to as *continuous*. Continuous systems are also known as *distributed parameter* models and they possess an infinite number of dof.

On the other hand, for simplicity, if one approximates the uniform cantilever beam as massless such that the elasticity of the beam may be considered as a spring of constant coefficient $k = 3EI/L^3$, where E is the Young's modulus of elasticity of the material and I the second moment of cross-sectional area of the beam, and the mass of the beam m is considered concentrated at the tip of the beam, then the dynamic deflection of this *discrete* or *lumped-parameter* model, shown in Figure 1.3, can be described by an *ordinary differential equation* (o.d.e.).

1.3 Applications of Control Theory

It is believed that the first use of automatic control in Western civilization dated back to the period of 300 BC [1]. In the Far East the best-known automatic control in ancient China is the south-pointing chariot [1].

Fast forward to 1922, when Minorsky [2] introduced his three-term controller for the steering of ships, thereby becoming the first to use the proportional, integral, and derivative (PID) controller. In this publication [2] he also considered nonlinear effects in the closed-loop system (to be defined in Chapter 8). In modern times the theory of control has been applied in many fields. The following representative applications are important examples.

- The theory of control has been employed by economists, medical personnel, financial experts, political scientists, biologists, chemists, and engineers, to name but a few.
- In automobile engineering, many components of a car, such as the steering system, and the driverless car that has already appeared in the testing and refined design phase, employ many feedback control devices.
- Within the field of mechanical engineering, the speed control and maintenance of a turbine, and the heating system and water heater in a house, or the heating, ventilation and air conditioning (HVAC) system in a modern building, employ automatic control systems.
- In aerospace, the control of aircraft, helicopters, satellites, and missiles requires very sophisticated advanced control systems [3].
- In shipbuilding industries, control systems are often employed for steering and navigation [4].

1.4 Organization of Presentation

This book consists of 12 chapters. After this introduction, Chapter 2 is concerned with a brief review of Laplace transforms. The emphasis is on their applications in the analysis and design of dynamic and control systems. Use of the software MATLAB [5] provides several examples.

Chapter 3 presents the formulations and dynamic behaviors of hydraulic and pneumatic systems. A simple nonlinear system together with the linearization technique is included.

Chapter 4 deals with the formulations and dynamic behaviors of mechanical oscillatory systems. The focus in this chapter is on the formulation and analysis of linear single dof and many degree-of-freedom (mdof) vibration systems. Modal analysis of mdof systems is introduced in this chapter. Simple distributed-parameter models or continuous systems are included. Many solved problems are presented in this chapter.

The formulations and dynamic behaviors of thermal systems are introduced in Chapter 5. Dynamic equations of simple systems as well as the three-capacitance oven model are derived and investigated.

For completeness, the most basic electrical elements, laws, and networks, their corresponding dynamic equations, and derivations of transfer functions for various representative electromechanical systems are presented in Chapter 6.

The basic dynamic characteristics, theories, and operating principles of sensors or transducers are included in Chapter 7. The emphasis in this chapter is, however, on applications and derivations of dynamic equations of motion and their interpretations. Examples included in this chapter are accelerometers, microphones, and a piezoelectric hydrophone.

Chapter 8 is concerned with the fundamentals of engineering control systems. Transfer functions for open-loop and closed-loop feedback control systems are considered. System transfer functions of dynamic systems by block diagram reduction are illustrated with examples.

Modeling and analysis of engineering control systems are presented in Chapter 9. The time domain response of a unity feedback control system is developed and explained. Control types, such as the PID controls, s.s. error analysis, performance indices, and sensitivity functions are considered in this chapter.

The stability analysis of feedback control systems is introduced in Chapter 10. The focus in this chapter is the application of the Routh-Hurwitz stability criterion. For illustration, various examples are worked out in detail.

Chapter 11 is concerned with graphical methods in control systems. The methods introduced include the root locus method and root locus plots, polar and Bode plots, the Nyquist stability criterion and Nyquist diagrams, gain, phase margins in relative stability analysis, contours of magnitude, phase of system frequency response, the so-called *M* and *N* circles, and the Nichols chart. Various questions are solved by employing MATLAB at the end of this chapter. These questions are selected to show the powerful capability of MATLAB in the context of response computation.

The final chapter, Chapter 12, deals with modern control system analysis. The state space or vector space method is presented. The relationship between the Laplace transformed state equation and transfer function of a feedback control system is derived. The concepts of *controllability, observability, stabilizability, and detectability* are introduced, so as to provide a foundation for studies of multiple input and multiple outputs (MIMOs) feedback control systems. Various approximated system responses are obtained by employing MATLAB.

References

[1] Mayr, O. (1970). *The Origins of Feedback*. MIT Press, Cambridge, MA. (Published in German in 1969).
[2] Minorsky, N. (1922). Directional stability and automatically steered bodies. *Journal of the American Society for Naval Engineers*, **34**(2), 280–309.
[3] Siouris, G.M. (1993). *Aerospace Avionics Systems: A Modern Synthesis*. Academic Press, New York.
[4] Fossen, T.I. (1994). *Guidance and Control of Ocean Vehicles*. John Wiley & Sons, Ltd., Chichester.
[5] The Math Works, Inc. (2014). *MATLAB R2014a*, The Math Works, Inc., Natick, MA.

2

Review of Laplace Transforms

Laplace transformation [1–3] is one of several powerful transformations that can be applied to the analysis of signals and dynamic engineering problems. In the context of dynamic and control system analysis, Laplace transforms are applied to obtain the transfer functions and, in turn, the block diagram representation, and the solutions of linear differential equations. Of course, they can be applied to obtain the solutions of partial differential equations (p.d.e.).

While the method of Laplace transformation can be applied to the solution of ordinary differential equations (o.d.e.) and p.d.e., in this book it is applied to obtain the solutions of o.d.e. since application of Laplacc transformation to the solutions of p.d.e. is beyond the scope of this book. The process of solution by application of Laplace transforms has the following three stages:

- The given equation of motion is transformed into a *subsidiary* equation.
- The subsidiary equation is solved by *purely algebraic* manipulations.
- The solution of the subsidiary equation is transformed back (that is, taking the inverse Laplace transform) to provide the solution of the given problem.

The above solution of the o.d.e. by *algebraic operations* instead of *calculus operations* is referred to as *operational mathematics* [2].

This chapter begins with the definition of Laplace transforms and reviews of various important concepts and theorems. These topics are dealt with in Sections 2.1–2.6. The Laplace transforms of periodic functions and partial fraction method are considered in Sections 2.7 and 2.8, respectively. Section 2.9 is concerned with representative solved questions. These illustrative examples are included to demonstrate the solution of o.d.e. by the use of Laplace transforms. Applications of the software package MATLAB [4] for various problems are presented in Section 2.10.

Introduction to Dynamics and Control in Mechanical Engineering Systems, First Edition. Cho W. S. To.

Companion website: www.wiley.com/go/to/dynamics

2.1 Definition

The Laplace transform $F(s)$ of $f(t)$ is defined as [1,2]:

$$F(s) = \int_0^\infty f(t)e^{-st}\,dt \tag{2.1}$$

in which the function $f(t)$ is defined for all $t \geq 0$.

The inverse of $F(s)$ or inverse Laplace transform of $F(s)$ is represented as:

$$f(t) = \mathcal{L}^{-1}[F(s)],$$

where the symbol $\mathcal{L}^{-1}[.]$ is applied to denote the inverse Laplace transform of the enclosing quantity. In this book the uppercase represents the Laplace transform of the lower case function, unless stated otherwise.

- The function $f(t)$ in Equation (2.1) is linear. Thus, for example, one cannot operate:

$$\int_0^\infty [f(t)]^3 e^{-st}\,dt.$$

- If, however, $[f(t)]^3$ is convergent and has no *multiple values* then the Laplace transform of $[f(t)]^3$ can be evaluated by the so-called *multi-fold* or *multi-dimensional Laplace transform.* For the present course, this is not considered.

Some commonly applied Laplace transforms and properties of Laplace transforms are included, respectively in Tables 2.1 and 2.2. More Laplace transforms and properties may easily be found in a mathematical handbook [5,6].

Table 2.1 Functions and their Laplace transforms

$f(t)$, $t \geq 0$	$F(s)$
1. $\delta(t)$, unit impulse at $t = 0$	1
2. $u(t)$, unit step	$1/s$
3. t^n	$n!/s^{n+1}$
4. e^{-at}	$1/(s + a)$
5. $t^{n-1}\,e^{-at}/(n-1)!$	$1/(s + a)^n$
6. $1 - e^{-at}$	$a/[s(s + a)]$
7. $(e^{-at} - e^{-bt})/(b - a)$	$1/[(s + a)\,(s + b)]$
8. $[(c - a)e^{-at} - (c - b)e^{-bt}]/(b - a)$	$(s + c)/[(s + a)\,(s + b)]$
9. $\sin at$	$a/(s^2 + a^2)$
10. $\cos at$	$s/(s^2 + a^2)$
11. $e^{-at}\sin bt$	$b/[(s + a)^2 + b^2]$
12. $e^{-at}\cos bt$	$(s + a)/[(s + a)^2 + b^2]$

Table 2.2 Properties of Laplace transforms

$f(t)$	$F(s)=\int_0^\infty e^{-st}f(t)\,dt$
1. $af(t)+bg(t)$	$aF(s)+bG(s)$
2. $\frac{df}{dt}$	$sF(s)-f(0)$
3. $\frac{d^2f}{dt^2}$	$s^2F(s)-sf(0)-\dot{f}(0)$, $\dot{f}(0)=\left.\frac{df}{dt}\right\|_{t=0}$
4. $\frac{d^nf}{dt^n}$	$s^nF(s)-\sum_k^n s^{n-k}f_{k-1}^{(k-1)}(0)$, $f_{k-1}^{(k-1)}(0)=\left.\frac{d^{k-1}f(t)}{dt^{k-1}}\right\|_{t=0}$
5. $\int_0^t f(t)dt$	$\frac{1}{s}F(s)$
6. $g(t)=\begin{cases}0 & t<0\\ f(t-a) & t\ge a\end{cases}$	$G(s)=e^{-as}F(s)$
7. $e^{-at}f(t)$	$F(s+a)$
8. $f\left(\frac{t}{a}\right)$	$aF(as)$
9. $h(t)=\int_0^t f(\tau)g(t-\tau)d\tau=\int_0^t f(t-\tau)g(\tau)d\tau$	$H(s)=F(s)\,G(s)$

Before leaving this section it may be appropriate to mention that, strictly speaking, for a divergent function the integral defined by Equation (2.1) may not exist, so its Laplace transform cannot be established. For example, if the function $f(t)$ is the exponential e^{at} where a is a positive constant parameter such that Equation (2.1) becomes:

$$F(s)=\int_0^\infty e^{at}e^{-st}dt=\int_0^\infty e^{(a-s)t}dt$$

then $F(s)$ does not exist when $a>s$ since the integrand grows with time, and therefore the integral is not defined. Of course, one can use the shifting theorem to obtain the same result.

On the other hand, if one starts with the exponentially decaying function e^{-at} then the Laplace transform does exist. Having found the Laplace transform of the exponentially decaying function, one can then replace a in the resulting expression with $-a$. The result is then the Laplace transform of the exponentially rising function e^{at}. In other words, the Laplace transform of e^{at} is $\frac{1}{s-a}$.

Before leaving this section, it should be mentioned that there are many functions which do not have their Laplace transforms. An example of such a function frequently encountered in statistical analysis is $f(t) = e^{t^2}$.

2.2 First and Second Shifting Theorems

If f(t) has the transform F(s), where s > k, then $e^{at}f(t)$ has the transform F(s – a), where s – a > k with k being a constant. This is known as the *first shifting theorem*.

Symbolically,

$$\mathcal{L}[e^{at}f(t)] = F(s-a). \tag{2.2}$$

If one takes the inverse Laplace transform on both sides, one obtains

$$e^{at}f(t) = \mathcal{L}^{-1}[F(s-a)]. \tag{2.3}$$

The second shifting theorem can be stated as follow. *If f(t) has the transform F(s), then the "shifting function"*

$$\tilde{f}(t) = f(t-a)u(t-a) = \begin{cases} 0 & \text{if } t<0 \\ f(t-a) & \text{if } t>a \end{cases} \tag{2.4}$$

has the transform $e^{-as}F(s)$. That is,

$$\mathcal{L}[f(t-a)u(t-a)] = e^{-as}F(s). \tag{2.5}$$

If one takes the inverse Laplace transform on both sides of Equation (2.5), one has:

$$f(t-a)u(t-a) = \mathcal{L}^{-1}[e^{-as}F(s)]. \tag{2.6}$$

In the foregoing $u(t)$ is the *unit step function* which is also known as the *Heaviside function* such that $u(t-a)$ is defined as:

$$u(t-a) = \begin{cases} 0 & \text{if } t<a \\ 1 & \text{if } t>a \end{cases} \tag{2.7}$$

in which $a \geq 0$.

2.3 Dirac Delta Function (Unit Impulse Function)

The Laplace transform of the so-called generalized function,

$$\mathcal{L}[\delta(t-a)] = e^{-as}, \tag{2.8}$$

where the generalized function $\delta(t-a)$ is defined by

$$\delta(t-a)=\begin{cases}\infty & \text{if } t=a\\ 0 & \text{otherwise,}\end{cases}$$

and

$$\int_0^\infty \delta(t-a)dt=1.$$

The generalized function $\delta(t)$ is frequently called the unit impulse or Dirac delta function. Note that the units of Dirac delta function are s^{-1}.

2.4 Laplace Transforms of Derivatives and Integrals

Let $f(t)$ and its derivatives $f^{(1)}(t)$ or $f'(t)$, $f^{(2)}(t)$, or $f''(t),\ldots,f^{(n-1)}(t)$ be continuous functions for all $t\geq 0$, satisfying the existence condition, and the derivative $f^{(n)}(t)$ be piecewise continuous on every finite interval in the range $t\geq 0$. Then the Laplace transform of $f^{(n)}(t)$ is given by

$$\mathcal{L}\left[f^{(n)}(t)\right]=s^n\mathcal{L}[f(t)]-s^{n-1}f(0)-s^{n-2}f^{(1)}(0)-\cdots-s^0f^{(n-1)}(0). \tag{2.9}$$

Satisfying the existence condition, the Laplace transform of an integral is given by

$$\mathcal{L}\left[\int_0^t f(\tau)d\tau\right]=\frac{1}{s}F(s). \tag{2.10}$$

Therefore, the inverse transform, that is, the integral is

$$\int_0^t f(\tau)d\tau=\mathcal{L}^{-1}\left[\frac{1}{s}F(s)\right]. \tag{2.11}$$

2.5 Convolution Theorem

Let $f(t)$ and $g(t)$ satisfy the hypothesis of the existence condition. Then the product of their transforms $F(s)=\mathcal{L}[f(t)]$ and $G(s)=\mathcal{L}[g(t)]$ is the transform $H(s)=\mathcal{L}[h(t)]$ of the convolution $h(t)$ of $f(t)$ and $g(t)$, which is denoted by $f*g$ and defined by

$$h(t)=f*g=\int_0^t f(\tau)g(t-\tau)d\tau=\int_0^t f(t-\tau)g(\tau)d\tau, \tag{2.12}$$

in which τ is a dummy variable.

This theorem is very useful in the solution of complicated transfer functions and filter designs. The following example is a simple illustration of its use.

Example

Applying the convolution theorem, determine the inverse Laplace transform $h(t)$ of the transfer function $H(s)$ for a dynamic system, if the transfer function is given by

$$H(s) = \frac{2}{(s^2+1)^2}.$$

Solution:

Since

$$H(s) = \frac{2}{(s^2+1)^2} = \left(\frac{2}{s^2+1}\right)\left(\frac{1}{s^2+1}\right)$$

and

$$\mathcal{L}^{-1}\left(\frac{1}{s^2+1}\right) = \sin t,$$

therefore

$$h(t) = \mathcal{L}^{-1}[H(s)] = 2\sin t * \sin t = 2\int_0^t \sin\tau \sin(t-\tau)d\tau.$$

Note that the integrand of the above integral can be expressed as

$$\begin{aligned}\sin\tau\sin(t-\tau) &= \frac{1}{2}\{-\cos[\tau+(t-\tau)]+\cos[\tau-(t-\tau)]\}\\ &= \frac{1}{2}[-\cos t+\cos(2\tau-t)].\end{aligned}$$

Thus, substituting this result into the last integral, one obtains

$$\begin{aligned}h(t) &= 2\left(\frac{1}{2}\right)\int_0^t(-\cos t)d\tau + 2\left(\frac{1}{2}\right)\int_0^t\cos(2\tau-t)d\tau\\ &= [-\tau\cos t]_0^t + \left[\frac{1}{2}\sin(2\tau-t)\right]_0^t\\ h(t) &= -t\cos t + \frac{1}{2}\sin(t) - \frac{1}{2}\sin(-t)\\ &= -t\cos t + \sin t.\end{aligned}$$

2.6 Initial and Final Value Theorems

These theorems are particularly useful in the analysis and design of control systems. More specifically, the *final value theorem* is frequently applied in the evaluation of steady-state errors of feedback control systems. The *initial value theorem* can be applied to the impulse response analysis. The important assumption is that the functions $f(t)$ and their derivatives considered do have their Laplace transforms.

Mathematically, the final value theorem states that

$$\lim_{t\to\infty} f(t) = \lim_{s\to 0} sF(s), \tag{2.13}$$

where $F(s)$ is the Laplace transform of $f(t)$.

On the other hand, the initial value theorem states that

$$\lim_{t\to 0} f(t) = \lim_{s\to\infty} sF(s). \tag{2.14}$$

Before leaving this section it may be appropriate to note that the final value theorem cannot be applied to determine the stability of a dynamic system. The theorem is not valid if the denominator of $sF(s)$ contains any pole whose real part is zero or positive. An example is $F(s) = \dfrac{\omega}{s^2+\omega^2}$, which is the Laplace transform of $\sin \omega t$.

2.7 Laplace Transforms of Periodic Functions

Consider a periodic function

$$f(t) = f(t+\alpha) = f(t+2\alpha) = f(t+3\alpha) = \cdots \tag{2.15}$$

where α is the period.

One can write the Laplace transform of $f(t)$ as a series of integrals so that

$$\begin{aligned}\mathcal{L}[f(t)] &= \int_0^\infty f(t)e^{-st}dt \\ &= \int_0^\alpha f(t)e^{-st}dt + \int_\alpha^{2\alpha} f(t)e^{-st}dt + \int_{2\alpha}^{3\alpha} f(t)e^{-st}dt + \cdots\end{aligned}$$

Except the first integral on the right-hand side (rhs) of this relation, one can replace t with $t=\tau+\alpha$, $t=\tau+2\alpha$, $t=\tau+3\alpha$, and so on in the integrals, respectively, such that the integration limits for every integral are 0 and α. That is,

$$\begin{aligned}\mathcal{L}[f(t)] &= \int_0^{\alpha} f(t)e^{-st}dt + \int_0^{\alpha} f(\tau+\alpha)e^{-s(\tau+\alpha)}d\tau \\ &+ \int_0^{\alpha} f(\tau+2\alpha)e^{-s(\tau+2\alpha)}d\tau + \cdots \\ &= \int_0^{\alpha} f(t)e^{-st}dt + e^{-\alpha s}\int_0^{\alpha} f(\tau+\alpha)e^{-s\tau}d\tau \\ &+ e^{-2\alpha s}\int_0^{\alpha} f(\tau+2\alpha)e^{-s\tau}d\tau + \cdots\end{aligned}$$

According to Equation (2.15), the above equation becomes

$$\begin{aligned}\mathcal{L}[f(t)] = \int_0^{\alpha} f(t)e^{-st}dt + e^{-\alpha s}\int_0^{\alpha} f(\tau)e^{-s\tau}d\tau \\ + e^{-2\alpha s}\int_0^{\alpha} f(\tau)e^{-s\tau}d\tau + \cdots\end{aligned}$$

Without loss of generality the dummy variable of integration τ on the rhs of this equation can be replaced by t so that

$$\mathcal{L}[f(t)] = \int_0^{\alpha} f(t)e^{-st}dt + e^{-\alpha s}\int_0^{\alpha} f(t)e^{-st}dt + e^{-2\alpha s}\int_0^{\alpha} f(t)e^{-st}dt + \cdots$$

Upon factoring the integral on the rhs, this equation reduces to

$$\mathcal{L}[f(t)] = \left(1 + e^{-\alpha s} + e^{-2\alpha s} + \cdots\right)\int_0^{\alpha} f(t)e^{-st}dt.$$

Since $1 + x + x^2 + \cdots = \dfrac{1}{1-x}$, this equation can be written as

$$\mathcal{L}[f(t)] = \frac{1}{1-e^{-\alpha s}}\int_0^{\alpha} f(t)e^{-st}dt. \tag{2.16}$$

This is the Laplace transform of a periodic function defined by Equation (2.15). It should be noted that the integral on the rhs of Equation (2.16) is not the Laplace transform of $f(t)$ because the upper integration limit is not ∞.

2.8 Partial Fraction Method

In this section, the partial fraction method is applied to solve the following differential equation

$$4\frac{d^2x}{dt^2}+\frac{dx}{dt}+4x=1, \tag{2.17a}$$

$$x(0)=\dot{x}(0)=0. \tag{2.17b}$$

where the overdot in the initial conditions denotes the derivative with respect to t.

Taking the Laplace transform of Equation (2.17a), one has

$$4\left[s^2X(s)-s\,x(0)-\dot{x}(0)\right]+sX(s)-x(0)+4X(s)=\frac{1}{s}.$$

Substituting for the initial conditions, it leads to

$$4\left[s^2X(s)\right]+sX(s)+4X(s)=\frac{1}{s}.$$

Therefore,

$$X(s)=\frac{1}{s(4s^2+s+4)}.$$

By the partial fraction method, one can write the rhs of the last equation as

$$\frac{1}{s(4s^2+s+4)}=\frac{A}{s}+\frac{Bs+C}{4s^2+s+4} \tag{2.18}$$

Equating the numerators on both sides of Equation (2.18), one obtains

$$1=A\left(4s^2+s+4\right)+s(Bs+C)$$

Expanding the rhs and arranging terms in a polynomial of s such that

$$1=(4A+B)s^2+(A+C)s+4A.$$

and equating coefficients with equal degrees in s on both sides, one has

$$4A+B=0,\quad A+C=0,\quad 1=4A.$$

Solving the last three equations, it leads to

$$A=\frac{1}{4},\quad B=-1,\quad C=-A=-\frac{1}{4}.$$

Substituting the above into Equation (2.18), one obtains

$$\frac{1}{s(4s^2+s+4)}=\frac{1}{4}\left(\frac{1}{s}+\frac{-s-\frac{1}{4}}{s^2+\frac{s}{4}+1}\right)$$

The terms inside the brackets on the rhs can be written as

$$\frac{1}{s}+\frac{-s-\frac{1}{4}}{s^2+\frac{s}{4}+1}=\frac{1}{s}-\frac{s+\frac{1}{8}}{\left(s^2+\frac{1}{8}\right)^2+\left(\frac{\sqrt{63}}{8}\right)^2}-\frac{\frac{1}{8}}{\left(s^2+\frac{1}{8}\right)^2+\left(\frac{\sqrt{63}}{8}\right)^2}.$$

After taking the inverse Laplace transforms, one arrives at

$$x(t)=\frac{1}{4}\left(1-e^{-\frac{1}{8}t}\cos\frac{\sqrt{63}}{8}t-\frac{e^{-\frac{1}{8}t}}{\sqrt{63}}\sin\frac{\sqrt{63}}{8}t\right).$$

2.9 Questions and Solutions

The solved questions or illustrative examples included in this section are mainly concerned with uses of Laplace transforms and their inverses in the context of dynamics and transfer functions, and o.d.e. that arise from modeling of dynamic systems.

Example 1

a. The transfer function of a dynamic system is defined by

$$H(s)=\frac{X(s)}{R(s)}=\frac{3s+5}{3s^2+5s+4}.$$

Determine $h(t)$ by applying the Laplace transform method.

b. Suppose the above system is subjected to a unit step function, $r(t)=1$, such that $d[r(t)]/dt=0$ and therefore the above transfer function becomes

$$H(s)=\frac{X(s)}{R(s)}=\frac{5}{3s^2+5s+4}$$

Evaluate the response of the system in the time domain, $x(t)$.

Solution:

a. The transfer function is given by

$$H(s)=\frac{3s+5}{3s^2+5s+4}=\frac{s+\frac{5}{3}}{s^2+\left(\frac{5}{3}\right)s+\frac{4}{3}}$$

$$=\frac{s+\frac{5}{3}}{s^2+\left(\frac{5}{3}\right)s+\frac{25}{36}+\frac{23}{36}}.$$

$$H(s)=\frac{s+\frac{5}{6}}{\left(s+\frac{5}{6}\right)^2+\left(\frac{\sqrt{23}}{6}\right)^2}+\frac{\frac{\sqrt{23}}{6}\left(\frac{5}{6}\right)\left(\frac{6}{\sqrt{23}}\right)}{\left(s+\frac{5}{6}\right)^2+\left(\frac{\sqrt{23}}{6}\right)^2}.$$

Taking the inverse Laplace transforms, one obtains

$$h(t)=e^{-5t/6}\cos\left(\frac{\sqrt{23}}{6}t\right)+\frac{5}{\sqrt{23}}e^{-5t/6}\sin\left(\frac{\sqrt{23}}{6}t\right).$$

Hence

$$h(t)=e^{-5t/6}\left[\cos\left(\frac{\sqrt{23}}{6}t\right)+\frac{5}{\sqrt{23}}\sin\left(\frac{\sqrt{23}}{6}t\right)\right].$$

b. Since $R(s) = 1/s$,

$$X(s)=\frac{R(s)\,(3s+5)}{3s^2+5s+4}=\frac{5R(s)}{3s^2+5s+4},$$

because

$$R(s)\,(3s)=3\mathcal{L}\left\{\frac{dr(t)}{dt}\right\}=0.$$

Therefore, by the partial fraction method, one has

$$X(s)=\frac{5}{s(3s^2+5s+4)}$$

$$=-\frac{15s}{4(3s^2+5s+4)}-\frac{25}{4(3s^2+5s+4)}+\frac{5}{4s}$$

$$X(s)=-\left(\frac{5}{4}\right)\frac{3s+5}{(3s^2+5s+4)}+\frac{5}{4s}.$$

Note that the first term is similar to those in 1a above except for the negative factor $-\left(\frac{5}{4}\right)$ and therefore, following the above procedure, one obtains:

$$x(t)=\frac{5}{4}-\left(\frac{5}{4}\right)e^{-\frac{5t}{6}}\left[\cos\left(\frac{\sqrt{23}}{6}t\right)+\frac{5}{\sqrt{23}}\sin\left(\frac{\sqrt{23}}{6}t\right)\right].$$

Example 2

Solve the following second-order differential equations by the Laplace transform method:

a. $\frac{d^2x}{dt^2}+\frac{dx}{dt}+x=1, x(0)=\frac{dx(0)}{dt}=0.$

b. $\frac{d^2x}{dt^2}+2\frac{dx}{dt}+x=1, x(0)=\frac{dx(0)}{dt}=0.$

Solution:

a. Taking the Laplace transform of the given equation and applying the initial conditions,

$$\mathcal{L}\left\{\frac{d^2x}{dt^2}+\frac{dx}{dt}+x=1\right\}$$

giving

$$s^2X(s)+sX(s)+X(s)=\frac{1}{s}.$$

$$X(s)=\frac{1}{s(s^2+s+1)}=\frac{A}{s}+\frac{Bs+C}{s^2+s+1}.$$

$$\frac{1}{s(s^2+s+1)}=\frac{A}{s}+\frac{Bs+C}{s^2+s+1}.$$

Equating the numerator term on the left-hand side (lhs) to that on the rhs:

$$1=A(s^2+s+1)+Bs^2+Cs=(A+B)s^2+(A+C)s+A.$$

$$A+B=0,\quad A+C=0,\quad A=1.$$

Thus, $B=-1$; $C=-1$.

$$X(s)=\frac{1}{s}-\frac{s+1}{s^2+s+1}=\frac{1}{s}-\left[\frac{s+\frac{1}{2}}{\left(s+\frac{1}{2}\right)^2+\frac{3}{4}}+\frac{\frac{1}{2}}{\left(s+\frac{1}{2}\right)^2+\frac{3}{4}}\right].$$

Taking the inverse Laplace transform,

$$x(t) = 1 - e^{-\frac{1}{2}t}\cos\frac{\sqrt{3}}{2}t - \frac{1}{\sqrt{3}}e^{-\frac{1}{2}t}\sin\frac{\sqrt{3}}{2}t.$$

b. Taking the Laplace transform of the given equation and applying the initial conditions,

$$\mathcal{L}\left\{\frac{d^2x}{dt^2} + 2\frac{dx}{dt} + x = 1\right\}$$

giving

$$s^2X(s) + 2s\,X(s) + X(s) = \frac{1}{s}.$$

$$X(s) = \frac{1}{s(s^2+2s+1)} = \frac{1}{s(s+1)^2} = \frac{A}{s} + \frac{B}{s+1} + \frac{C}{(s+1)^2}.$$

Equating the numerator term on the lhs to that on the rhs:

$$1 = A(s+1)^2 + Bs(s+1) + Cs = (A+B)s^2 + (2A+B+C)s + A.$$

$$A+B=0,\quad 2A+B+C=0,\quad A=1.$$

$$B=-1,\quad C=-1.$$

$$X(s) = \frac{1}{s(s+1)^2} = \frac{1}{s} - \frac{1}{s+1} - \frac{1}{(s+1)^2}.$$

Taking the inverse Laplace transforms, it leads to

$$x(t) = 1 - e^{-t} - te^{-t}.$$

Example 3
Solve the following differential equation by the Laplace transform method:

$$\frac{d^4x}{dt^4} + \frac{d^3x}{dt^3} = \cos t,\quad x(0) = \frac{dx(0)}{dt} = \frac{d^3x(0)}{dt^3} = 0,\quad \frac{d^2x(0)}{dt^2} = 1.$$

Solution:
Taking the Laplace transform of the given equation and given initial conditions,

$$\mathcal{L}\left\{\frac{d^4x}{dt^4} + \frac{d^3x}{dt^3} = \cos t\right\}$$

giving

$$s^4X(s)-s(1)+s^3X(s)-1=\frac{s}{s^2+1}, \text{ since}$$

$$\mathcal{L}\left\{\frac{d^4x}{dt^4}\right\}=s^4X(s)-s^3x(0)-s^2\frac{dx(0)}{dt}-s\frac{d^2x(0)}{dt^2}-\frac{d^3x(0)}{dt^3}$$

$$=s^4X(s)-s, \text{ and}$$

$$\mathcal{L}\left\{\frac{d^3x}{dt^3}\right\}=s^3X(s)-s^2x(0)-s\frac{dx(0)}{dt}-\frac{d^2x(0)}{dt^2}=s^3X(s)-1.$$

Hence,

$$s^4X(s)+s^3X(s)=\frac{s}{s^2+1}+s+1.$$

$$X(s)=\frac{s}{s^3(s+1)(s^2+1)}+\frac{s+1}{s^3(s+1)}.$$

$$X(s)=\frac{1}{s^2(s+1)\,(s^2+1)}+\frac{1}{s^3}. \tag{i}$$

One can express the above equation as

$$X(s)=\frac{1}{s^3}+\frac{A}{s^2}+\frac{B}{s}+\frac{C}{s+1}+\frac{Ds+E}{s^2+1},$$

where the unknown constants are to be found as in the following.

The second through fifth terms on the rhs of the above equation are now considered. Thus,

$$\begin{aligned}\frac{1}{s^2(s+1)\,(s^2+1)} &= \frac{A}{s^2}+\frac{B}{s}+\frac{C}{s+1}+\frac{Ds+E}{s^2+1}\\ &= \frac{A(s+1)\,(s^2+1)}{s^2(s+1)\,(s^2+1)}+\frac{Bs(s+1)\,(s^2+1)}{s^2(s+1)\,(s^2+1)}\\ &\quad+\frac{Cs^2(s^2+1)}{s^2(s+1)\,(s^2+1)}+\frac{(Ds+E)s^2(s+1)}{s^2(s+1)\,(s^2+1)}.\end{aligned} \tag{ii}$$

Since the denominator terms on the rhs are identical, one only has to consider the numerator terms which become:

$$A(s^3+s^2+s+1)+B(s^4+s^3+s^2+s)$$
$$+C(s^4+s^2)+D(s^4+s^3)+E(s^3+s^2).$$

Writing the above expression in descending orders of s, one has

$$s^4(B+C+D)+s^3(A+B+D+E)+s^2(A+B+C+E) \\ +s(A+B)+A.$$

Equating the numerator term on the lhs of Equation (ii) to that on the rhs, one obtains

$$B+C+D=0, \quad A+B+D+E=0, \quad A+B+C+E=0, \\ A+B=0, \quad A=1.$$

Solving, it gives $B=-1$, $C=\frac{1}{2}$, $D=\frac{1}{2}$, $E=-\frac{1}{2}$. Hence, Equation (i) becomes:

$$X(s)=\frac{1}{s^3}+\frac{1}{s^2}-\frac{1}{s}+\left(\frac{1}{2}\right)\frac{1}{s+1}+\left(\frac{1}{2}\right)\frac{s}{s^2+1}-\left(\frac{1}{2}\right)\frac{1}{s^2+1}.$$

Taking the inverse Laplace transform, one has

$$x(t)=\left(\frac{1}{2}\right)t^2+t-1-\frac{1}{2}e^{-t}+\frac{1}{2}\cos t-\frac{1}{2}\sin t.$$

Example 4

A system has a transfer function given by

$$H(s)=\frac{3s}{(s^2+1)\ (s^2+4)}.$$

Find the time domain expression for the transfer function by taking the inverse Laplace transform.

Solution:

By the method of partial fractions,

$$H(s)=\frac{3s}{(s^2+1)\ (s^2+4)}=\frac{As+B}{s^2+1}+\frac{Cs+D}{s^2+4}.$$

Equating the numerator term on both sides, one obtains

$$3s=As^3+4As+Bs^2+4B+Cs^3+Cs+Ds^2+D.$$
$$3s=(A+C)s^3+(B+D)s^2+(4A+C)s+4B+D.$$

Thus, $A + C = 0$, $B + D = 0$, $4A + D = 3$, $4B + D = 0$.
Solving these equations gives $A = 1$, $B = 0$, $C = -1$, $D = 0$. Therefore,

$$H(s)=\frac{s}{s^2+1}-\frac{s}{s^2+4}.$$

Taking the inverse Laplace transforms, the time domain expression is

$$h(t) = \cos t - \cos 2t.$$

2.10 Applications of MATLAB

In this section several examples of applying MATLAB [4] are included. It is appropriate to mention that many similar problems can be solved. However, the following examples are included to demonstrate the steps required to obtain the solutions and to provide verifications of results found analytically.

Example 1

By applying MATLAB, find the partial fractions of the following transfer function of a system:

$$H(s) = \frac{1}{s^3 + s} = \frac{1}{s(s^2+1)}.$$

Solution:

The input to and output from MATLAB are presented in Program Listing 2.1. From the program listing one has

$$H(s) = \frac{1}{s(s^2+1)} = \frac{1}{s} - \frac{0.5}{s+i} - \frac{0.5}{s-i}.$$

Program Listing 2.1

```
>> num = [0 0 0 1];
>> den = [1 0 1 0];
>> [r,p,k] = residue(num,den)

 r =
    -0.5000
    -0.5000
     1.0000


 p =
     0.0000 + 1.0000i
     0.0000 - 1.0000i
     0.0000 + 0.0000i

 k =
      []
```

Example 2

By applying MATLAB, find the partial fractions of the following transfer function of a system:

$$H(s) = \frac{-7s^2 - 19s - 6}{s^3 + 6s^2 + 11s + 6}.$$

Solution:

The input to and output from MATLAB are presented in Program Listing 2.2. From the program listing one has

$$H(s) = \frac{-7s^2 - 19s - 6}{s^3 + 6s^2 + 11s + 6} = \frac{3}{s+1} - \frac{4}{s+2} - \frac{6}{s+3}.$$

Program Listing 2.2

```
>> num = [0 -7 -19 -6];
>> den = [1 6 11 6];
>> [r,p,k] = residue(num,den)

 r =
    -6.0000
    -4.0000
     3.0000

 p =
    -3.0000
    -2.0000
    -1.0000


 k =
     []
```

Example 3

By applying MATLAB, plot the response or time history of the dynamic system that has the transfer function

$$\frac{C(s)}{R(s)} = \frac{1}{s^2 + \frac{1}{5}s + 1}$$

and subjected to a unit step input.

Solution:

The input to MATLAB for the given system is presented in Program Listing 2.3. The time history plot from MATLAB is included in Figure 2.1.

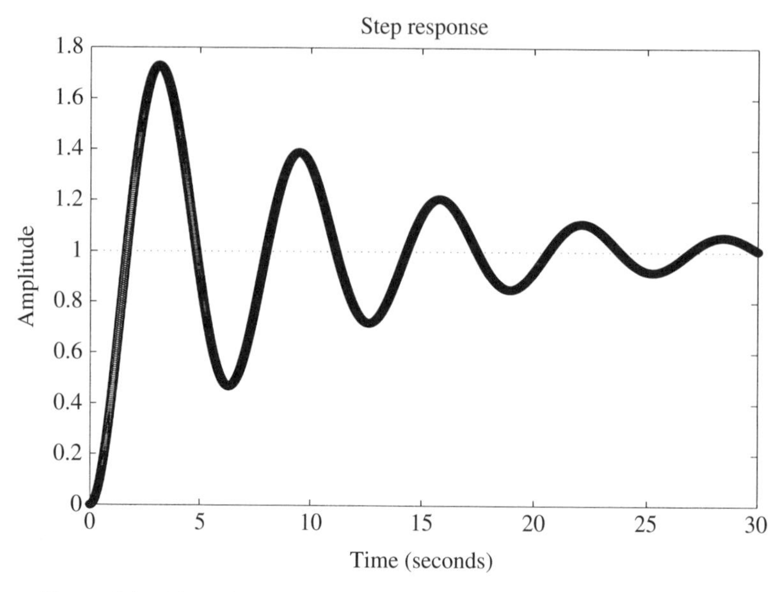

Figure 2.1 Time history of a second-order system under a unit step input

Program Listing 2.3

```
>> sys1 = tf([0,1],[1,0.2,1]);
>> t = (0:0.01:30);
>> step(sys1,'o',t)
```

Example 4
Consider two second-order systems whose transfer functions are

$$\frac{C(s)}{R(s)} = \frac{3s+5}{3s^2+5s+4}, \quad \text{and} \quad \frac{C(s)}{R(s)} = \frac{3s+5}{s^2+1s+4}.$$

Assuming the input is a unit step function, applying MATLAB plot the responses in the same figure.

Solution:
The input to MATLAB is presented in Program Listing 2.4.

Program Listing 2.4

```
>> sys1 = tf([3,5],[3,5,4]);
>> sys2 = tf([3,5],[3,1,4]);
>> t = (0:0.01:10);
>> step(sys1,' ',sys2,'*',t)
```

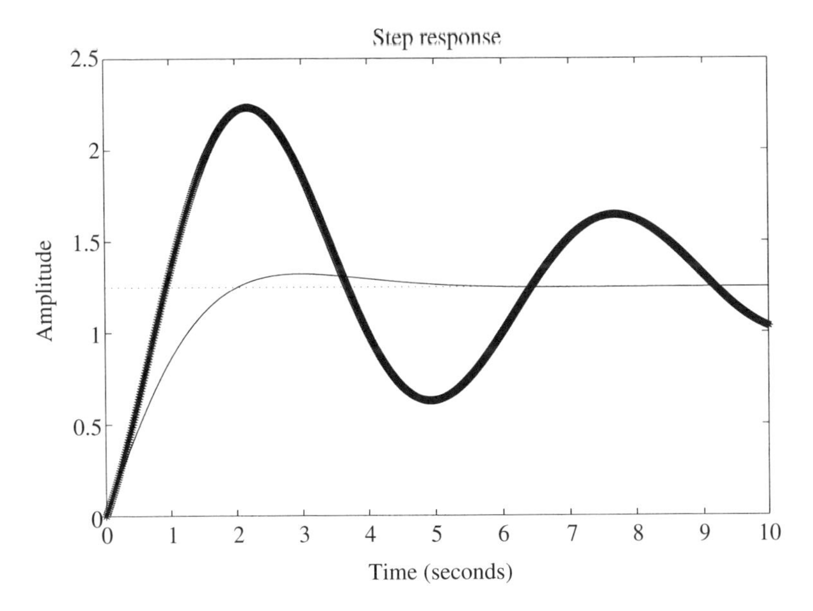

Figure 2.2 Time histories of two second-order systems

The responses from MATLAB are presented in Figure 2.2. It should be mentioned that the second system has considerably less damping and therefore the oscillation is very pronounced. The largest oscillation peak or overshoot is about 2.25, which is approximately 1.8 times the steady-state value of 1.25 (this steady-state value is obtained by setting $s = 0$ in the given transfer functions for the two systems).

Example 5

Repeat Example 3 for an impulse input instead of a unit step input. Plot the results of the impulse input for comparison.

Solution:

The input to MATLAB for the system in Example 3 is presented in Program Listing 2.5. The time history plot from MATLAB is included in Figure 2.3, which can be compared with those subjected to a unit input presented in Figure 2.1.

Program Listing 2.5

```
>> sys1 = tf([0,1],[1,0.2,1]);
>> t = (0:0.01:30);
>> impulse(sys1,'k',t)
```

In the last statement of Program Listing 2.5, the parameter k specifies the color of the plot being black in Figure 2.3.

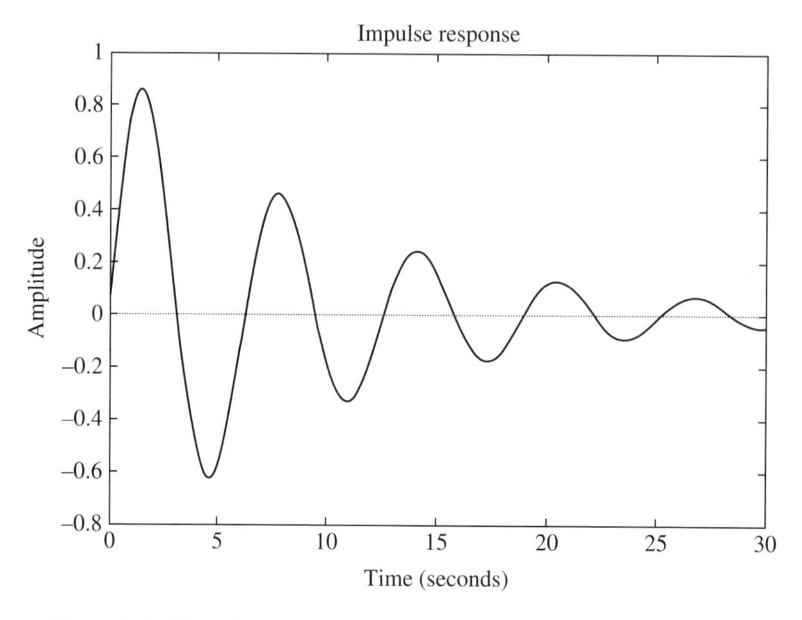

Figure 2.3 Time history of a second-order system under an impulse input

Exercise Questions

Q1. Solve the following differential equation by the Laplace transform method:

$$\frac{d^2x}{dt^2}+3\frac{dx}{dt}+x=1,\quad x(0)=\frac{dx(0)}{dt}=0.$$

Q2. Solve the following differential equation by the Laplace transform method:

$$\frac{d^2x}{dt^2}+\frac{dx}{dt}=t^2+2t,\quad x(0)=4,\quad \frac{dx(0)}{dt}=2.$$

Q3. By applying the inverse Laplace transform, find the time domain expression of the transfer function,

$$H(s)=\frac{1}{s(s^2-2s+5)}.$$

Q4. Determine the Laplace transform of the periodic triangular wave defined as

$$f(t)=\begin{cases}\dfrac{ht}{b}, & 0<t<b\\ \dfrac{h}{b}(2b-t), & a<t<2b.\end{cases}$$

Q5. Consider a simply-supported (SS) beam of constant cross-sectional area with uniformly distributed load μ and length L. The differential equation that describes the transverse deflection $w(x)$ can be shown to be

$$\frac{d^4w}{dx^4} = \frac{\mu}{EI},$$

where EI is known as the flexural rigidity of the beam. If EI and μ are assumed to be constant and the boundary conditions for this SS beam are given by $w(0) = \frac{d^2w(0)}{dx^2} = 0, w(L) = \frac{d^2w(L)}{dx^2} = 0$, find the transverse deflection $w(x)$ by the method of Laplace transforms.

Q6. By applying MATLAB, plot the response of the system with the following transfer function

$$\frac{C(s)}{R(s)} = \frac{1}{s^2 + \frac{s}{5} + 5}$$

and subjected to a unit impulse input.

Q7. By applying MATLAB, plot the response of the system with the following transfer function

$$\frac{C(s)}{R(s)} = \frac{s}{s^2 + \frac{1}{5}s + 1}$$

and subjected to a unit step input.

References

[1] Pipes, L.A. (1958). *Applied Mathematics for Engineers and Physicists*. McGraw-Hill, New York.
[2] Churchill, R.V. (1972). *Operational Mathematics*. McGraw-Hill, New York.
[3] Kreyszig, E. (1999). *Advanced Engineering Mathematics*, 8th edn. John Wiley & Sons, Inc., New York.
[4] The Math Works, Inc. (2014). *MATLAB R2014a*. The Math Works, Inc., Natick, MA.
[5] Spiegel, M.R. (1968). *Mathematical Handbook*. McGraw-Hill, New York.
[6] Abramowitz, M., and Segun, I.A. (eds) (1965). *Handbook of Mathematical Functions*. Dover Publications, Inc., New York.

3

Dynamic Behaviors of Hydraulic and Pneumatic Systems

Many physical dynamic systems can be represented and modeled by first-order differential equations. Two commonly encountered mechanical engineering systems modeled in this way are hydraulic tanks and pneumatic systems. In this chapter, different hydraulic tanks or liquid-level systems and pneumatic or gas systems are considered. Their corresponding transfer functions are derived.

3.1 Basic Elements of Liquid and Gas Systems

Before the consideration of hydraulic tanks and pneumatic systems, the typical basic elements of these systems are briefly presented in the following. The basic elements of liquid and gas systems are fluid resistance, inertia, and compressibility. These are sketched in Figure 3.1.

Fluid resistance R exists in flow valves, orifices, and fluid lines. In the simplest cases, which do not include turbulent flow, the flow rate q is proportional to the pressure difference between two stations or measurement points, that is:

$$q \propto p_1 - p_2 \quad \text{or} \quad q = R(p_1 - p_2) \tag{3.1}$$

where R is the fluid resistance or simply resistance, and p_i with $i = 1, 2$, are the pressures at stations 1 and 2.

Fluid inertia I or fluid mass in one-dimensional (1D) flow is defined by $I = \rho AL$ where ρ is the mass density of the fluid, and A the cross-sectional area of the line of length L.

Fluid compressibility C for liquids and gases is different. In this course, compressibilities of liquids and gases are defined in the following subsections.

Introduction to Dynamics and Control in Mechanical Engineering Systems, First Edition. Cho W. S. To.

Companion website: www.wiley.com/go/to/dynamics

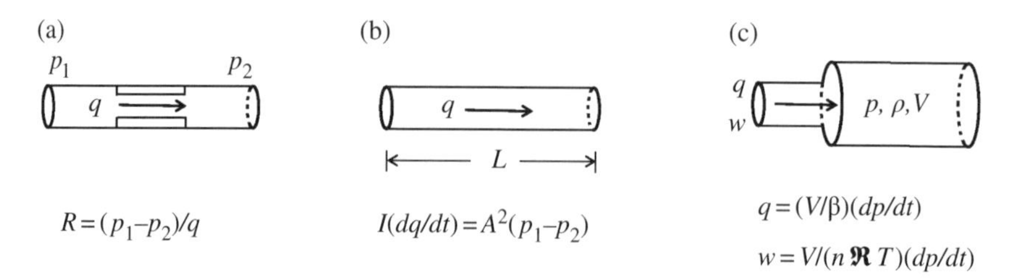

Figure 3.1 Basic elements of fluid systems: (a) resistance R; (b) inertia I; and (c) compressibility C

3.1.1 Liquids

In high-pressure hydraulic systems it is generally necessary to consider the effect of compressibility of the liquid of interest. Under the condition of high pressure the change of density is assumed to be proportional to the pressure difference. Thus one can write

$$\rho-\rho_o \propto p-p_o \quad \text{or} \quad \rho-\rho_o=\frac{\rho_o}{\beta}(p-p_o) \tag{3.2}$$

in which ρ_o and p_o are ρ and p at the reference or constant temperature, respectively, and β is the bulk modulus of the liquid. Typically, in oil pumping stations or pipelines, β is about 1000 psi, or in nuclear submarine model testing facilities or nuclear power stations it is in the range of 20,000 psi or higher.

Consider the mass flow rate w entering the constant volume V as shown in Figure 3.1c; that is:

$$w=\frac{d(\rho V)}{dt}=V\dot{\rho}. \tag{3.3}$$

From Equation (3.2), taking the time derivative on both sides leads to

$$\dot{\rho}=\left(\frac{\rho_o}{\beta}\right)\frac{dp}{dt}=\left(\frac{\rho_o}{\beta}\right)\dot{p} \tag{3.4}$$

Substituting Equation (3.4) into Equation (3.3), one has

$$w=V\left(\frac{\rho_o}{\beta}\right)\dot{p}$$

But $w = q\rho_o$, and therefore,

$$w=q\rho_o=V\left(\frac{\rho_o}{\beta}\right)\dot{p} \quad \text{or} \quad q=\left(\frac{V}{\beta}\right)\dot{p} \quad \text{or} \quad q=C_h\dot{p} \tag{3.5}$$

where $C_h=V/\beta$ is the compressibility of the liquid. Equation (3.5) states that the volume flow rate q is proportional to the rate of change of the pressure.

3.1.2 Gases

For gases, the relationship between pressure p and density ρ is given by

$$p = K\rho^n \tag{3.6}$$

in which K is a constant,

$$n = \begin{cases} 1 & \text{for isothermal processes} \\ \gamma & \text{for adiabatic frictionless processes} \end{cases}$$

and $\gamma = c_p/c_v$ is the ratio of specific heats, with c_p being the specific heat value at constant pressure and c_v the specific heat value at constant volume.

Differentiating Equation (3.6) with respect to time t and rearranging one can show that

$$\dot{\rho} = \left(\frac{\rho}{nK\rho^n}\right)\dot{p} \tag{3.7}$$

where in the denominator $K\rho^n$ is the pressure p according to Equation (3.6), so that this equation can be written as

$$\dot{\rho} = \left(\frac{\rho}{np}\right)\dot{p} \tag{3.8}$$

Substituting Equation (3.8) into Equation (3.3), one obtains

$$w = V\dot{\rho} = \left(\frac{V\rho}{np}\right)\dot{p} = C_g\dot{p} \tag{3.9}$$

where the compressibility of the gas, $C_g = \frac{V\rho}{np} = \frac{V\rho}{\beta} = \frac{V}{n\Re T}$, in which T is the absolute temperature and $\Re$ the gas constant of the ideal gas law $p = \rho\Re T$. It may be appropriate to note that in the foregoing $\frac{\rho}{np} = \frac{\rho}{\beta}$, where $\beta = np$ is the bulk modulus of gas, is considered constant since C_g in Equation (3.9) is assumed to be constant. Strictly speaking, neither ρ nor p is constant, and therefore $\frac{\rho}{np}$ is not a constant.

3.1.3 Remarks

With the foregoing elements, many liquid and gas systems can be modeled and analysed. However, for brevity, only hydraulic tanks and pneumatically actuated valves are studied in the remaining sections of this chapter.

3.2 Hydraulic Tank Systems

The principle of dynamic operation in hydraulic tanks is the *transient mass balance.* The latter is frequently referred to as mass balance and may be expressed as:

$$\textit{Mass flowing in} - \textit{mass flowing out} = \textit{rate of accumulation of mass in the tank}.$$

In sections 3.2.1 and 3.2.2 the non-interacting and interacting hydraulic tank systems are studied, and their differences and implications are emphasized.

3.2.1 Non-interacting Hydraulic Tank Systems

Consider the hydraulic two-tank system shown in Figure 3.2. Applying the transient mass balance principle to the first tank on the left-hand side (lhs) of Figure 3.2, one has:

$$q - q_1 = A_1 \frac{dh_1}{dt} \tag{3.10}$$

where q is the volumetric flow rate (in volume per unit time) into the first tank, q_1 is the volumetric flow rate out of the first tank, h_1 is the height of the liquid level in the first tank, and A_1 is the constant cross-section of the first tank.

Similarly, applying the principle of mass balance on the second tank gives:

$$q_1 - q_2 = A_2 \frac{dh_2}{dt}. \tag{3.11}$$

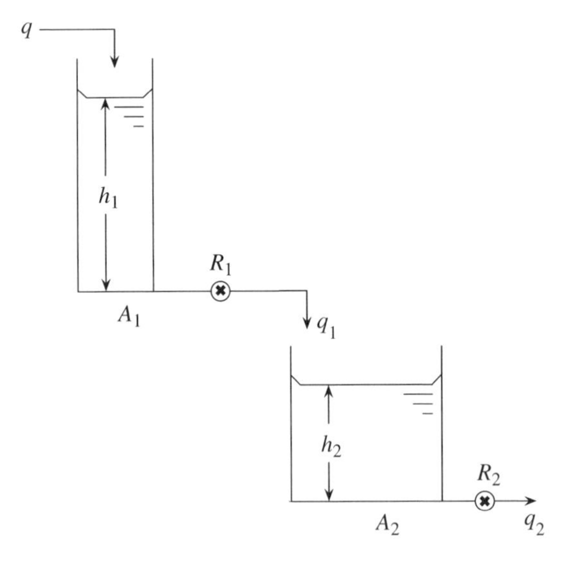

Figure 3.2 Non-interacting hydraulic tank system

For simplicity, it is assumed that the flow-head relationships for the two valves are linear; that is, the resistance for each valve is linear. This, in turn, implies that the flow through the valves is laminar. Thus, one can write the flow-head relationships as

$$q_1 = \frac{h_1}{R_1}, \tag{3.12}$$

$$q_2 = \frac{h_2}{R_2}, \tag{3.13}$$

where R_1 and R_2 are the linear resistances as indicated in Figure 3.2.

Combining Equations (3.10) and (3.12), one obtains

$$q - \frac{h_1}{R_1} = A_1 \frac{dh_1}{dt} \quad \text{or} \quad \frac{dh_1}{dt} + \frac{h_1}{\tau_1} = \frac{q}{A_1} \tag{3.14}$$

where $\tau_1 = A_1 R_1$ is the so-called *time constant* of the system.

Similarly, Equation (3.11) becomes

$$A_2 \frac{dh_2}{dt} + \frac{h_2}{R_2} - \frac{h_1}{R_1} = 0 \quad \text{or} \quad \frac{dh_2}{dt} + \frac{h_2}{A_2 R_2} - \frac{h_1}{A_2 R_1} = 0. \tag{3.15}$$

Introducing deviation or fluctuating variables

$$\tilde{q} = q - q_s \quad \tilde{h}_1 = h_1 - h_{1s}$$

into Equation (3.14), it becomes

$$\frac{d\tilde{h}_1}{dt} + \frac{\tilde{h}_1}{\tau_1} = \frac{\tilde{q}}{A_1}, \tag{3.16}$$

since at steady state the flow entering the tank equals that leaving the tank such that

$$q_s = q_{1s}, \quad \frac{dh_{1s}}{dt} = 0,$$

where the subscript s denotes the steady-state value.

After taking the Laplace transform and applying the initial conditions (assuming the system starts from rest), one can show that the transfer function for the system becomes

$$\frac{\tilde{H}_1(s)}{\tilde{Q}(s)} = \frac{R_1}{\tau_1 s + 1}, \tag{3.17}$$

in which

$$\tilde{H}_1(s) = \mathcal{L}\left[\tilde{h}_1\right], \quad \tilde{Q}(s) = \mathcal{L}[\tilde{q}], \quad \tau_1 = R_1 A_1.$$

Similarly, one can show that

$$\frac{\widetilde{H}_2(s)}{\widetilde{Q}(s)} = \left(\frac{1}{\tau_1 s+1}\right)\left(\frac{R_2}{\tau_2 s+1}\right), \tag{3.18}$$

where $\tau_2 = R_2 A_2$ is the time constant of the second hydraulic tank.

3.2.2 *Interacting Hydraulic Tank Systems*

For the system in Figure 3.3, upon application of mass balance, one can show that the governing equations of motion are identical to Equations (3.10) and (3.11) except that now the flow-head relationship of Equation (3.12) becomes

$$q_1 = \frac{h_1 - h_2}{R_1}. \tag{3.19}$$

Substituting Equations (3.19) and (3.13) into Equations (3.10) and (3.11), one has

$$A_1 \frac{dh_1}{dt} + \frac{h_1 - h_2}{R_1} = q \tag{3.20}$$

and

$$A_2 \frac{dh_2}{dt} + h_2\left(\frac{1}{R_2} - \frac{1}{R_1}\right) - \frac{h_1}{R_1} = 0. \tag{3.21}$$

- Note that Equations (3.20) and (3.21) are coupled because both equations have h_1 and h_2 on their lhs.
- In the non-interacting hydraulic tank system case, only one first-order differential equation has both h_1 and h_2 on the lhs.

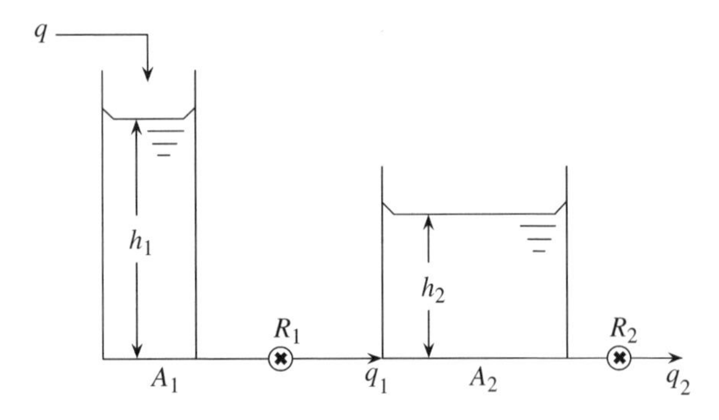

Figure 3.3 Interacting hydraulic tank system

Introducing the deviation or fluctuating variables again, and after taking the Laplace transform, one can show that the transfer function is (the detailed derivation is given in Appendix 3.A):

$$\frac{\widetilde{H}_2(s)}{\widetilde{Q}(s)}=\frac{R_2}{\tau_1\tau_2 s^2+(\tau_1+\tau_2+A_1R_2)s+1}. \tag{3.22}$$

The difference between Equations (3.18) and (3.22) is the additional term A_1R_2 in the denominator of Equation (3.22). In other words, in the interacting system an extra damping term is added to the system.

3.3 Nonlinear Hydraulic Tank and Linear Transfer Function

Most physical systems of practical significance are nonlinear. However, to provide a relatively simple means of analysing such systems, linearization techniques are often employed. The hydraulic tank shown in Figure 3.4 below is presented to illustrate the point, and to provide an appreciation of the steps involved and assumptions made in a commonly applied linearization technique.

By applying the principle of transient mass balance for the hydraulic tank in Figure 3.4, one has

$$q-q_1=A_1\frac{dh_1}{dt} \tag{3.23}$$

which is identical in form to Equation (3.10), but now the flow-head relationship has the following expression:

$$q_1=C\sqrt{h_1} \tag{3.24}$$

with C being a constant.

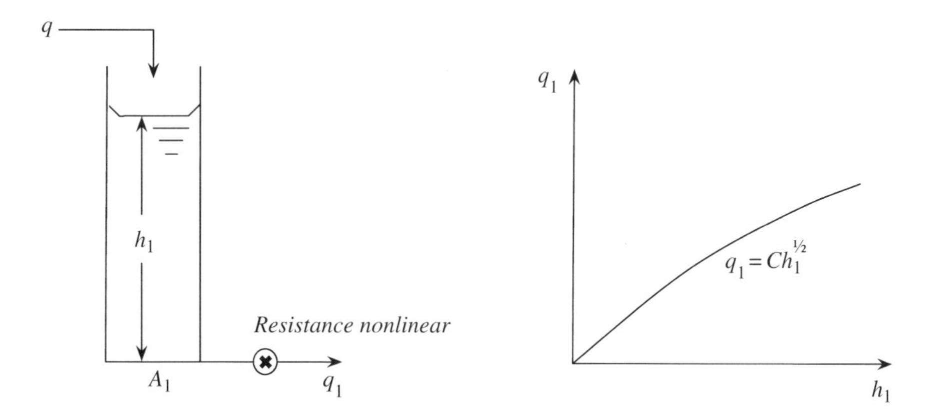

Figure 3.4 Nonlinear hydraulic tank system

Substituting Equation (3.24) into Equation (3.23), it results in:

$$A_1 \frac{dh_1}{dt} + C\sqrt{h_1} = q. \tag{3.25}$$

It should be pointed out that the method of Laplace transforms for solving differential equations reviewed in Chapter 2 cannot be applied to solve Equation (3.25) because it is a first-order nonlinear differential equation. However, by means of a Taylor series expansion, the function q_1 or with its argument included, $q_1(h_1)$ in Equation (3.24) may be expanded about the steady state value h_{1s} of h_1. Thus,

$$q_1(h_1) = q_1(h_{1s}) + (h_1 - h_{1s})\frac{dq_1(h_{1s})}{dh_1} + (h_1 - h_{1s})^2 \frac{d^2 q_1(h_{1s})}{dh_1^2} + \ldots \tag{3.26}$$

where $\frac{dq_1(h_{1s})}{dh_1}$ is the first derivative of $q_1(h_1)$ with respect to h_1 evaluated at h_{1s}, $\frac{d^2 q_1(h_{1s})}{dh_1^2}$ the second derivative with respect to h_1 evaluated at h_{1s}, and so on.

If one only keeps the linear term in Equation (3.26) it becomes:

$$q_1(h_1) = q_1(h_{1s}) + (h_1 - h_{1s})\frac{dq_1(h_{1s})}{dh_1}. \tag{3.27}$$

On the other hand, taking the derivative of q_1 with respect to h_1 in Equation (3.24) and evaluating the derivative at $h_1 = h_{1s}$ gives

$$\frac{dq_1(h_{1s})}{dh_1} = \frac{1}{2} C h_{1s}^{-\frac{1}{2}}. \tag{3.28}$$

Both sides of Equation (3.28) are constant since they are functions of the steady-state value h_{1s} so that one can write

$$\frac{1}{R_1} = \frac{1}{2} C h_{1s}^{-\frac{1}{2}} \tag{3.29}$$

and Equation (3.27) becomes

$$q_1(h_1) = q_1(h_{1s}) + (h_1 - h_{1s})\frac{1}{R_1} \tag{3.30}$$

Substituting $q_1(h_1)$ in Equation (3.30) into Equation (3.23), one has

$$q - q_1 = q - q_1(h_{1s}) - (h_1 - h_{1s})\frac{1}{R_1} = A_1 \frac{dh_1}{dt}$$

or

$$q - q_{1s} - (h_1 - h_{1s})\frac{1}{R_1} = A_1\frac{dh_1}{dt} \tag{3.31}$$

where $q_{1s} = q_1(h_{1s})$ has been used.

Now, introducing deviation variables, recall their definitions below Equation (3.15),

$$\widetilde{q} = q - q_s, \quad \widetilde{h}_1 = h_1 - h_{1s}$$

into Equation (3.31) so that it becomes

$$A_1\frac{d\widetilde{h}_1}{dt} + \frac{\widetilde{h}_1}{R_1} = \widetilde{q} \tag{3.32}$$

Note that $q_s = q_{1s}$ has been used in Equation (3.32) since at steady state the flow entering the tank equals the flow leaving the same tank.

Equation (3.32) has the same form as that for a single hydraulic tank system. Therefore, after taking the Laplace transform and applying the initial conditions (assuming the system starts from rest), one can show that the transfer function for the linearized system is given by Equation (3.17). This is not surprising since Equation (3.32) is identical to Equation (3.16).

Remark 3.3.1 The transfer function of a linear system is the ratio of the Laplace transformed output fluctuating variable to the Laplace transformed input fluctuating variable.

Remark 3.3.2 For the transfer function of a first-order system described by a first-order differential equation, the denominator is a first-degree polynomial in s.

Remark 3.3.3 Thus, for the transfer function of an output from n (the latter being an integer representing the number of hydraulic tanks connected in series) hydraulic tanks, the denominator is of a n-degree polynomial in s. For example, the transfer function of two interacting hydraulic tanks in Section 3.2.2, given by Equation (3.22), has a denominator of a second-degree polynomial in s.

3.4 Pneumatically Actuated Valves

A schematic sketch of a simple yet representative pneumatically actuated valve is shown in Figure 3.5a, in which k is the spring constant and x is the diaphragm and valve poppet displacement, being downward positive. It is assumed that the system starts from rest. The objective here is to find the transfer function of this valve. This simple model can be analysed in five steps, presented below.

a. **Step 1:**
   For simplicity, the mass and friction forces of the moving parts and fluid flow forces on the poppet are disregarded. It is assumed that x is small such that the valve is operated in the

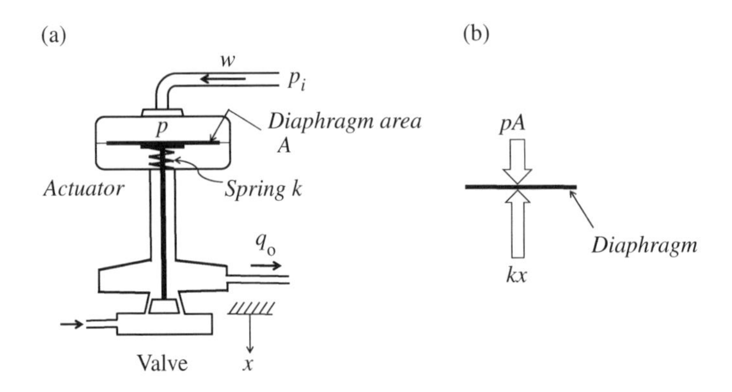

Figure 3.5 (a) Pneumatically actuated valve, and (b) free-body diagram of (a)

linear range and that in the pneumatic line the difference between the input pressure p_i and the pressure p above the diaphragm is proportional to the mass flow rate w. That is,

$$p_i - p \propto w \quad \text{or} \quad p_i - p = Rw \tag{3.33a}$$

where R is the constant resistance. Taking the Laplace transforms, one has

$$P_i(s) - P(s) = R\,W(s). \tag{3.33b}$$

b. **Step 2:**
For small displacement x, the compressibility C of the space above the diaphragm can be considered constant, so that the mass flow rate w is proportional to the rate of change of pressure p. Thus,

$$w = C\dot{p} \tag{3.34a}$$

with its Laplace transformation (recall that the system starts from rest) being

$$W(s) = CsP(s). \tag{3.34b}$$

c. **Step 3:**
With the free-body diagram (FBD) of the downward pressure force on the diaphragm and the restoring force in the spring, as indicated in Figure 3.5b, and by applying Newton's second law of motion, one has

$$Ap - kx = 0 \quad \text{or} \quad Ap = kx \tag{3.35a}$$

where A is the surface area of the diaphragm.
Taking the Laplace transform of Equation (3.35a), it becomes

$$AP(s) = kX(s). \tag{3.35b}$$

d. **Step 4:**
Now, consider the outflow rate q_o from the valve. With reference to Figure 3.5a, one can assume that

$$q_o = R_v x \tag{3.36a}$$

in which R_v is a proportionality constant. The Laplace transformation of Equation (3.36a) becomes

$$Q_o(s) = R_v X(s). \tag{3.36b}$$

e. **Step 5:**

To find the transfer function of the valve it is required to determine $\frac{Q_o(s)}{P_i(s)}$. To this end, one substitutes Equation (3.34b) into Equation (3.33b) to give

$$\frac{P(s)}{P_i(s)} = \frac{1}{RCs+1}. \tag{3.37}$$

Applying Equations (3.35b) and (3.36b), one has

$$\frac{Q_o(s)}{P(s)} = \frac{AR_v}{k}. \tag{3.38}$$

Applying Equations (3.37) and (3.38), the transfer function of the valve becomes

$$\frac{Q_o(s)}{P(s)}\frac{P(s)}{P_i(s)} = \frac{Q_o(s)}{P_i(s)} = \frac{\left(\frac{AR_v}{k}\right)}{RCs+1}. \tag{3.39}$$

3.5 Questions and Solutions

Four solved questions are included in this section. The first question deals with the derivation of the dynamic equations of two non-interacting hydraulic tanks, and the transfer functions of fluctuation variables of the water levels to input fluctuation variable. The second question is concerned with two non-interacting hydraulic tanks with two inputs to the first hydraulic tank. Time-dependent hydraulic levels are obtained by using the inverse Laplace transform. Derivation of the equation of motion of a hydraulic tank of non-uniform cross-sectional area is considered in the third question. The time-dependent liquid level is determined. Finally, the flapper-nozzle valve is considered in the fourth question.

Example 1

Two uniform cross-sectional area water tanks have inflow $q(t)$ into the first tank, and outflows $q_o(t)$ and $q_1(t)$, the latter of which is directed into the second tank, as shown in Figure 3E1.

a. From first principles, derive the dynamic equations in terms of deviation variables.
b. Determine the transfer functions $\frac{\widetilde{H}_2(s)}{\widetilde{Q}(s)}$ and $\frac{\widetilde{H}_1(s)}{\widetilde{Q}(s)}$, where $\widetilde{H}_i(s)$, $i=1, 2$, and $\widetilde{Q}(s)$ are the Laplace transforms of the deviation variables of the levels in the first and second tanks, and the inflow to the first tank, respectively.
c. If the valve resistances $R_1 = R_2 = 1$ unit, and $R_o = 2$ units, find the numerical values of the two transfer functions in b.

Solution:

a. Applying the transient mass balance principle, for the first tank, one has

$$q-(q_o+q_1)=A_1\frac{dh_1}{dt}.$$

Similarly, for the second tank

$$q_1-q_2=A_2\frac{dh_2}{dt}.$$

The flow-head relationships are

$$q_1=\frac{h_1}{R_1},\quad q_2=\frac{h_2}{R_2},\quad q_o=\frac{h_1}{R_o},\quad q_1+q_o=h_1\left(\frac{1}{R_1}+\frac{1}{R_o}\right)=\frac{h_1}{R_e},$$

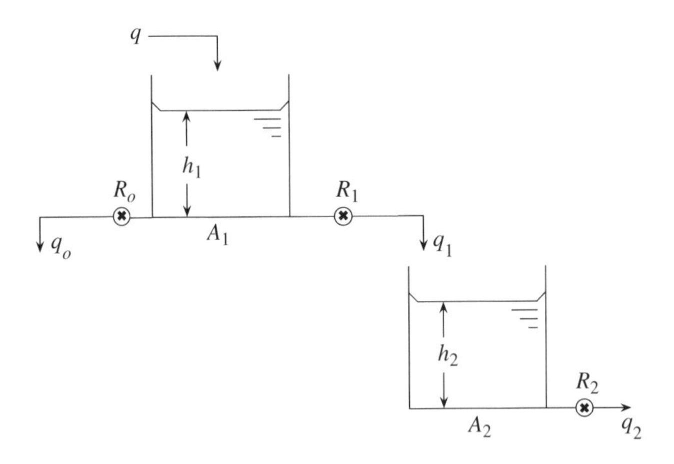

Figure 3E1 Two water tanks with one inflow and three outflows

where the equivalent resistance R_e is given by

$$\frac{1}{R_e} = \frac{1}{R_1} + \frac{1}{R_o}.$$

Therefore, the two equations are

$$A_1 \frac{dh_1}{dt} + \frac{h_1}{R_e} = q, \quad \text{and} \quad A_2 \frac{dh_2}{dt} + \frac{h_2}{R_2} - \frac{h_1}{R_1} = 0.$$

Thus, the two equations in terms of deviation variables are

$$A_1 \frac{d\widetilde{h}_1}{dt} + \frac{\widetilde{h}_1}{R_e} = \widetilde{q} \tag{i}$$

$$A_2 \frac{d\widetilde{h}_2}{dt} + \frac{\widetilde{h}_2}{R_2} - \frac{\widetilde{h}_1}{R_1} = 0. \tag{ii}$$

b. Taking the Laplace transform of Equation (i) and rearranging, one obtains the transfer function of the first tank as

$$\frac{\widetilde{H}_1(s)}{\widetilde{Q}(s)} = \frac{R_e}{A_1 R_e s + 1}. \tag{iii}$$

Since by making use of the flow-head relationship $q_1 = \frac{h_1}{R_1}$, deviation variables $\widetilde{h}_1 = h_1 - h_{1s}$, $\widetilde{q}_1 = q_1 - q_{1s}$, and $q_{1s} = \frac{h_{1s}}{R_1}$, one obtains $\widetilde{q}_1 = \frac{\widetilde{h}_1}{R_1}$.

Thus, upon taking a Laplace transform of the latter equation it results in

$$\widetilde{H}_1 = R_1 \widetilde{Q}_1.$$

Substituting into Equation (iii) one has

$$\frac{R_1 \widetilde{Q}_1}{\widetilde{Q}(s)} = \frac{R_e}{A_1 R_e s + 1}, \quad \text{or} \quad \frac{\widetilde{Q}_1}{\widetilde{Q}(s)} = \frac{(R_e/R_1)}{A_1 R_e s + 1}. \tag{iv}$$

With reference to Equations (ii) and (iii), and treating $\frac{\widetilde{h}_1}{R_1} = \widetilde{q}_1$ as input to Equation (ii), one obtains

$$\frac{\widetilde{H}_2(s)}{\widetilde{Q}_1(s)} = \frac{R_2}{A_2 R_2 s + 1}. \tag{v}$$

Multiplying Equations (v) and (iv) gives

$$\frac{\widetilde{H}_2(s)}{\widetilde{Q}_1(s)}\frac{\widetilde{Q}_1}{\widetilde{Q}(s)} = \frac{\widetilde{H}_2(s)}{\widetilde{Q}(s)} = \left[\frac{R_2}{A_2R_2s+1}\right]\left[\frac{(R_e/R_1)}{A_1R_es+1}\right].$$

Simplifying, the transfer function becomes

$$\frac{\widetilde{H}_2(s)}{\widetilde{Q}(s)} = \frac{(R_2R_e/R_1)}{(A_1R_eA_2R_2)s^2+(A_2R_2+A_1R_e)s+1}. \tag{vi}$$

The expressions in Equations (iii) and (iv) are the required transfer functions.

c. Now, substituting the numerical values of the given parameters for the water tanks into Equations (iii) and (vi), the transfer functions are

$$\frac{\widetilde{H}_1(s)}{\widetilde{Q}(s)} = \frac{2}{4s+3}, \quad \text{and} \quad \frac{\widetilde{H}_2(s)}{\widetilde{Q}(s)} = \frac{2}{4s^2+7s+3}.$$

Example 2

The two-tank system shown in Figure 3E2 is assumed to operate at steady state. At time $t = 0$, a volume of water of 10 m^3 is rapidly pumped through a large-diameter pipe to the first tank in 1.0 s. This rapid delivery of that volume of water can thus be represented as $q_a = 10\,\delta(t)$ m^3/s, where $\delta(t)$ is known as the unit impulse function.

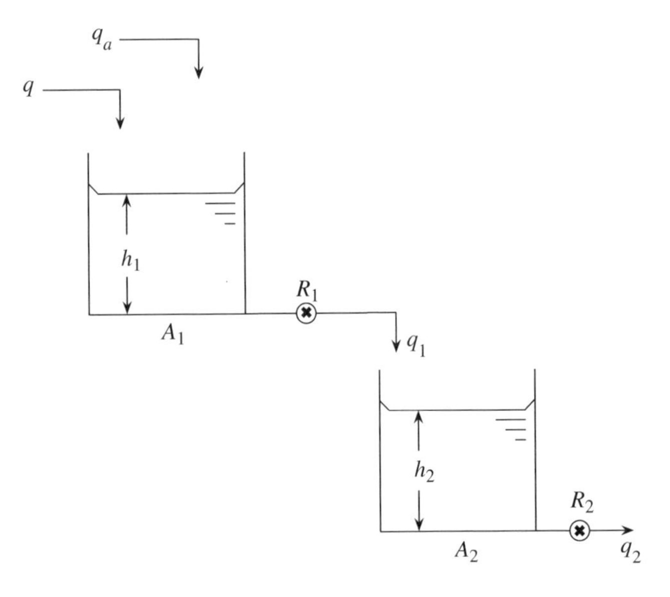

Figure 3E2 A two-tank system

a. Starting from first principles, derive the equations of motion in terms of the water levels h_1 and h_2 for the system shown; and

b. applying the equations of motion derived in (a), determine the water level in each tank by the method of Laplace transform if the instantaneous input to the first tank is $q(t) = 20\ \text{m}^3/\text{s}$, of uniform cross-sectional areas of tanks 1 and 2 are $A_1 = A_2 = 10\ \text{m}^2$, resistance of the valve associated with the first tank $R_1 = 0.10\ \text{s/m}^2$, and resistance of the valve associated with the second tank $R_2 = 0.35\ \text{s/m}^2$.

Solution:

Given	Cross-section areas of tanks $A_1 = A_2 = 10\ \text{m}^2$
	Resistances of valves $R_1 = 0.10\ \text{s/m}^2$ and $R_2 = 0.35\ \text{s/m}^2$
	Input to tank 1 at $t = 0$ is $q(t) = 20\ \text{m}^3/\text{s}$, and $q_a = 10\ \delta(t)\ \text{m}^3/\text{s}$
Find	(a) Equations of motion for tanks 1 and 2
	(b) Water level in each tank.
Approach	Method of Laplace transform and assume the system starts from rest. That is, all initial conditions are assumed to be zero.

Analysis:

a. Consider the two-tank system shown in Figure 3E2 and apply the transient mass balance principle to the first tank on the lhs. The equation of motion becomes

$$q + q_a - q_1 = A_1 \frac{dh_1}{dt}.$$

Similarly, the equation of motion of the second tank is

$$q_1 - q_2 = A_2 \frac{dh_2}{dt}.$$

The flow-head relationships are

$$q_1 = \frac{h_1}{R_1} \quad \text{and} \quad q_2 = \frac{h_2}{R_2}.$$

Substituting these flow-head relationships into the two equations of motion:

$$q + q_a - \frac{h_1}{R_1} = A_1 \frac{dh_1}{dt}, \tag{i}$$

$$\frac{h_1}{R_1} - \frac{h_2}{R_2} = A_2 \frac{dh_2}{dt}. \tag{ii}$$

Equations (i) and (ii) are the required equations of motion.

b. Substituting the given data into Equations (i) and (ii), and rearranging, one obtains

$$\frac{dh_1}{dt} + h_1 = 2 + \delta(t) \tag{iii}$$

$$\frac{dh_2}{dt} + \left(\frac{2}{7}\right)h_2 - h_1 = 0 \tag{iv}$$

Taking Laplace transform for Equations (iii) and (iv), one has

$$H_1(s) = \frac{\frac{2}{s} + 1}{s+1} = \frac{s+2}{s(s+1)} = \frac{2}{s} - \frac{1}{s+1}. \tag{v}$$

$$H_2(s) = \frac{H_1(s)}{s + \frac{2}{7}} = \frac{s+2}{s(s+1)\left(s+\frac{2}{7}\right)}. \tag{vi}$$

Writing Equation (vi) as in the following

$$H_2(s) = H_{21}(s) + H_{22}(s),$$

where

$$H_{21}(s) = \frac{2}{s(s+1)\left(s+\frac{2}{7}\right)} = \frac{\alpha}{s} + \frac{\beta}{s+1} + \frac{\gamma}{s+\frac{2}{7}},$$

$$H_{22}(s) = \frac{s}{s(s+1)\left(s+\frac{2}{7}\right)} = \frac{1}{(s+1)\left(s+\frac{2}{7}\right)} = \frac{\zeta}{s+1} + \frac{\eta}{s+\frac{2}{7}}.$$

Applying the partial fraction method, one obtains

$$\alpha = 7, \quad \beta = \frac{14}{5}, \quad \gamma = -\frac{49}{5}, \quad \zeta = -\frac{7}{5}, \quad \eta = \frac{7}{5}.$$

Therefore,

$$H_2(s) = \frac{7}{s} + \frac{7}{5(s+1)} - \frac{42}{5\left(s+\frac{2}{7}\right)}. \tag{vii}$$

Now, taking the inverse Laplace transform of Equations (v) and (vii), one has the water level in each tank as

$$h_1 = h_1(t) = 2 - e^{-t}\text{m}, \quad \text{and}$$

$$h_2 = h_2(t) = 7 + \left(\frac{7}{5}\right)e^{-t} - \left(\frac{42}{5}\right)e^{-\frac{2}{5}t}\text{m}.$$

Example 3

A non-uniform cross-section hydraulic tank is shown in Figure 3E3. It has three vertical walls and the fourth wall has a slope angle γ from the vertical. The distance separating the parallel walls (which are parallel to the plane of the figure) is 1 unit. The average operating level is h_o and the resistance of the valve is linear and of R units.

a. From first principles, derive the equation of motion of the hydraulic system shown in Figure 3E3.

b. With the result in (a), find the transfer function of the system and its time constant.

c. Derive an expression for liquid level $\tilde{h}(t)$ if the inflow to the tank can be represented by a unit step function, and comment on the effect of the slope angle γ on the liquid level.

Solution:

Given	Base cross-sectional area of tank $A = b$ (1) units
	Resistance of valve is R units
	Inflow to tank $q_i(t) = 1$ unit
	Angle of non-vertical wall of tank is γ
Find	(a) Equation of motion for the tank
	(b) Transfer function and time constant
	(c) Time history of liquid level
Approach	Obtain a formula for the volume of the tank
	Apply the principle of transient mass balance to obtain the equation of motion and express the equation in terms of deviation variables
	Use the method of Laplace transform and assume the system starts from rest

Analysis:

a. With reference to Figure 3E3, the flow-head relationship and volume V of the tank are, respectively,

$$q(t) = \frac{h}{R}, \quad V = b(1)h + \frac{1}{2}hx(1), \quad \text{where} \quad \frac{x}{h} = \tan\gamma.$$

Let $r = (\tan\gamma)/2$ so that the volume becomes $V = bh + rh^2$. Taking the differentiation with respect to time t, it gives

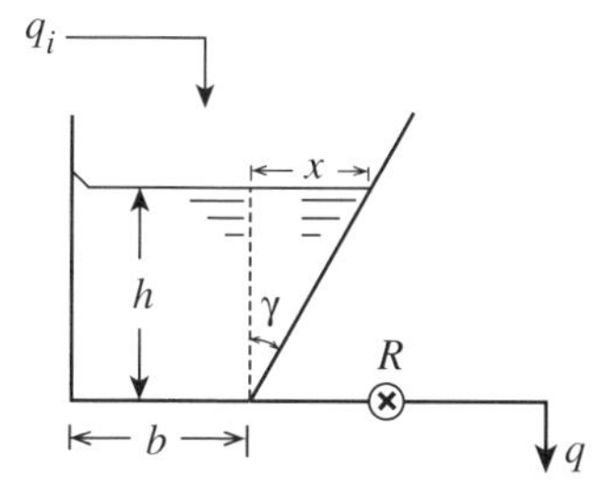

Figure 3E3 A hydraulic tank with a non-uniform cross-sectional area

$$\frac{dV}{dt}=(b+2rh)\frac{dh}{dt}.$$

Applying transient mass balance for the tank, one has

$$q_i-q=\frac{dV}{dt}.$$

Therefore, upon application of the flow-head relationship, one obtains

$$q_i-\frac{h}{R}=(b+2rh)\frac{dh}{dt}. \tag{i}$$

Since this equation does not have a constant coefficient for $\frac{dh}{dt}$, it means one cannot obtain the usual first-order differential equation. However, one can replace h in the coefficient $(b+2rh)$ by its average value h_o (this can be proved by applying the linearization technique introduced in Section 3.3) such that

$$q_i-\frac{h}{R}=(b+2rh_o)\frac{dh}{dt}.$$

Thus, the required equation of motion is

$$(b+2rh_o)\frac{dh}{dt}+\frac{h}{R}=q_i. \tag{ii}$$

b. At steady-state,

$$\frac{h_s}{R}=q_{is} \quad \text{since} \quad \frac{dh_s}{dt}=0. \tag{iii}$$

Introducing deviation variables and substituting Equation (iii) into Equation (ii), it results in

$$(b+2rh_o)\frac{d\tilde{h}}{dt}+\frac{\tilde{h}}{R}=\tilde{q}_i,\quad \tilde{h}=h-h_s,\quad \tilde{q}_i=q_i-q_{is}$$

Taking Laplace transform, one has

$$(b+2rh_o)\,s\,\tilde{H}(s)+\frac{\tilde{H}(s)}{R}=\tilde{Q}_i(s).$$

Therefore, the transfer function of the system becomes

$$\frac{\tilde{H}(s)}{\tilde{Q}_i(s)}=\frac{R}{\tau s+1} \tag{iv}$$

where the time constant of the system is defined by

$$\tau = R(b + 2rh_o) = R(b + h_o \tan\gamma).$$

c. The inflow is a unit step function so that $\widetilde{Q}_i(s) = \frac{1}{s}$. Substituting this into Equation (iv) gives

$$\widetilde{H}(s) = \frac{R}{s(\tau s + 1)}.$$

By the partial fraction method,

$$\widetilde{H}(s) = R\left(\frac{C_1}{s} + \frac{C_2}{\tau s + 1}\right).$$

This gives $C_1 = R$, and $C_2 = -R\tau$, so that

$$\widetilde{H}(s) = \frac{R}{S} - \frac{R\tau}{\tau s + 1}.$$

Taking the inverse Laplace transform,

$$\widetilde{h} = \widetilde{h}(t) = R\left(1 - e^{-\frac{t}{\tau}}\right). \tag{v}$$

This is the required solution. The latter indicates that when the angle γ increases, the time constant increases such that it takes a longer time to reach the steady-state level. At the critical value of $\gamma = 90°$ the time constant becomes infinite and $\widetilde{h}(t) = 0$. This is the case where the hydraulic tank is no longer a tank.

Example 4

Flapper-nozzle valves are frequently applied in pneumatic and hydraulic systems. A flapper-nozzle valve, shown in Figure 3E4, is used to control piston displacement y. The input pressure p_i is constant. The piston mass m, damping coefficient c, and spring coefficient k are constant. The piston cross-sectional area is A.

a. For small piston and flapper displacements, derive the transfer function $Y(s)/X(s)$, given that the constant volume under pressure p is V with constant density of air inside the volume being ρ and bulk modulus β. Assume the system starts from rest.

b. Generally, the mass m is small and therefore it may be disregarded. Determine this reduced transfer function and comment upon it.

(*Hint*: Equation of motion for the piston, from Chapter 4, is $m\ddot{y} + c\dot{y} + ky = Ap$.)

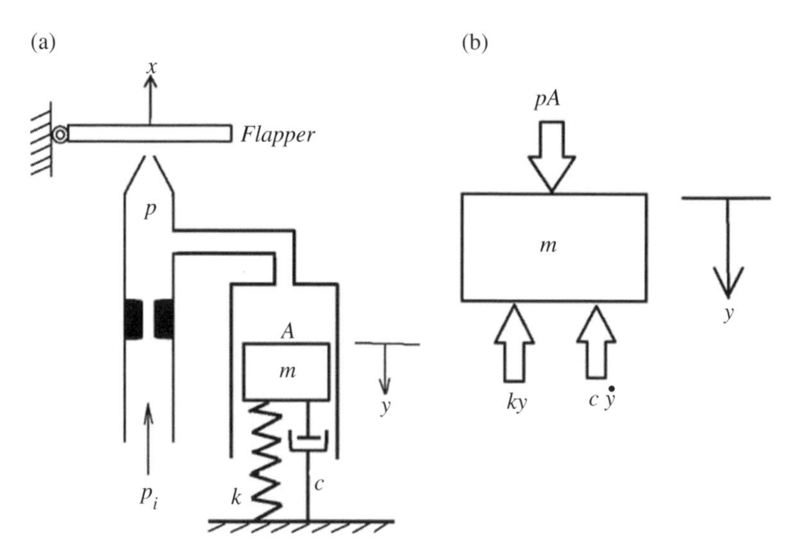

Figure 3E4 (a) Pneumatically actuated valve and (b) free-body diagram of (a)

Solution:

a. With reference to Figure 3E4, one has the supply flow $q = -k_i p$, where k_i is a constant. The nozzle flow is governed by:

$$q_n = k_1 x + k_2 p, \quad k_1 \text{ and } k_2 \text{ are constant.}$$

The flow difference is

$$q - q_n = -k_i p - (k_1 x + k_2 p) = -k_1 x - (k_i + k_2) p.$$

This gives rise to the compressibility flow of $\left(\frac{V}{\beta}\right)\dot{p}$ and piston flow of $A\dot{y}$. Therefore,

$$-k_1 x - (k_i + k_2) p = \left(\frac{V}{\beta}\right)\dot{p} + A\dot{y}$$

Taking the Laplace transform and recalling the system starts from rest, one has

$$-k_1 X(s) - (k_i + k_2) P(s) = \left(\frac{V}{\beta}\right) sP(s) + AsY(s). \tag{i}$$

By applying Newton's second law of motion and the FBD of the piston in Figure 3E4b, one obtains

$$m\ddot{y} + c\dot{y} + ky = Ap.$$

Taking the Laplace transform and again recalling the system starts from rest, then

$$ms^2Y(s)+csY(s)+kY(s)=AP \quad \text{or} \qquad \text{(ii)}$$
$$(ms^2+cs+k)Y(s)=AP.$$

Multiply Equation (i) by A and substitute Equation (ii) into the resulting equation so that, upon rearranging, one has

$$\frac{Y(s)}{X(s)}=\frac{-Ak_1}{\left[\left(\frac{V}{\beta}\right)s+k_i+k_2\right](ms^2+cs+k+A^2s)}.$$

Simplifying, one obtains the required transfer function,

$$\frac{Y(s)}{X(s)}=\frac{-Ak_1}{a_1s^3+a_2s^2+a_3s+a_4}, \qquad \text{(iii)}$$

where

$$a_1=\frac{V}{\beta}m, \quad a_2=m(k_i+k_2)+\frac{V}{\beta}(c+A^2),$$
$$a_3=\frac{V}{\beta}k(c+A^2)(k_i+k_2), \quad \text{and} \quad a_4=k(k_i+k_2).$$

b. Disregarding terms associated with mass m, one obtains the transfer function

$$\frac{Y(s)}{X(s)}=\frac{-Ak_1}{a_2s^2+a_3s+a_4}. \qquad \text{(iv)}$$

Note that now $a_2=\frac{V}{\beta}(c+A^2)$.

Equation (iv) indicates that the degree of the polynomial in the determinator is reduced from 3 in Equation (iii) to 2. The piston thus acts as a second-order system. The negative sign on the rhs of Equation (iv) indicates that the displacement of the flapper moves in the opposite direction to that of the piston.

Further simplification of the model is possible since, in general, $\frac{V}{\beta}$ is small. But it is not pursued here.

Appendix 3A: Transfer Function of Two Interacting Hydraulic Tanks

Consider Equations (3.20) and (3.21) which will be identified here as Equations (A1) and (A2), respectively:

$$A_1\frac{dh_1}{dt}+\frac{h_1-h_2}{R_1}=q \qquad \text{(A1)}$$

$$A_2\frac{dh_2}{dt}+h_2\left(\frac{1}{R_2}-\frac{1}{R_1}\right)-\frac{h_1}{R_1}=0. \tag{A2}$$

Now, consider Equation (A1) in terms of the fluctuation or deviation variables:

$$h_1=h_{1s}+\widetilde{h}_1,\quad h_2=h_{2s}+\widetilde{h}_2,\quad q=q_s+\widetilde{q},$$

so that

$$A_1\frac{d\left(h_{1s}+\widetilde{h}_1\right)}{dt}+\frac{h_{1s}+\widetilde{h}_1-\left(h_{2s}+\widetilde{h}_2\right)}{R_1}=q_s+\widetilde{q}. \tag{A3}$$

But the steady-state equation for the first tank is

$$A_1\frac{dh_{1s}}{dt}+\frac{h_{1s}-h_{2s}}{R_1}=q_s.$$

Subtracting this equation from Equation (A3), one has the equation in terms of the deviation variables

$$A_1\frac{d\widetilde{h}_1}{dt}+\frac{\widetilde{h}_1-\widetilde{h}_2}{R_1}=\widetilde{q}. \tag{A4}$$

Similarly, by making use of Equation (A2) for the second tank, the equation in terms of the deviation variables for the second tank becomes

$$A_2\frac{d\widetilde{h}_2}{dt}+\widetilde{h}_2\left(\frac{1}{R_2}+\frac{1}{R_1}\right)-\frac{\widetilde{h}_1}{R_1}=0. \tag{A5}$$

In order to derive the transfer functions one may use the Laplace transform method. Consider taking the Laplace transform of Equation (A4). Thus,

$$\mathcal{L}\left[A_1\frac{d\widetilde{h}_1}{dt}+\frac{\widetilde{h}_1-\widetilde{h}_2}{R_1}=\widetilde{q}\right],\quad A_1\mathcal{L}\left[\frac{d\widetilde{h}_1}{dt}\right]+\mathcal{L}\left[\frac{\widetilde{h}_1-\widetilde{h}_2}{R_1}\right]=\mathcal{L}[\widetilde{q}].$$

That is,

$$A_1\left[s\widetilde{H}_1(s)-\widetilde{h}_1(0)\right]+\left[\frac{\widetilde{H}_1(s)-\widetilde{H}_2(s)}{R_1}\right]=\widetilde{Q}(s).$$

Therefore, after applying the initial condition which is zero, or the system starts from rest, the above equation reduces to

$$[(A_1R_1)s+1]\widetilde{H}_1(s)-\widetilde{H}_2(s)=R_1\widetilde{Q}(s)$$

or

$$(\tau_1 s+1)\widetilde{H}_1(s)=R_1\widetilde{Q}(s)+\widetilde{H}_2(s),$$

where $\tau_1 = A_1R_1$. This gives

$$\widetilde{H}_1(s)=\frac{R_1\widetilde{Q}(s)+\widetilde{H}_2(s)}{\tau_1 s+1}. \tag{A6}$$

Now, take the Laplace transform of Equation (A5) so that

$$\mathcal{L}\left[A_2\frac{d\widetilde{h}_2}{dt}+\widetilde{h}_2\left(\frac{1}{R_2}+\frac{1}{R_1}\right)-\frac{\widetilde{h}_1}{R_1}=0\right].$$

This gives

$$A_2\left[s\widetilde{H}_2(s)-\widetilde{h}_2(0)\right]+\widetilde{H}_2\left(\frac{1}{R_2}+\frac{1}{R_1}\right)-\frac{\widetilde{H}_1(s)}{R_1}=0.$$

Substituting Equation (A6) and recalling the system starts from rest, the last equation reduces to

$$A_2\left[s\widetilde{H}_2(s)\right]+\widetilde{H}_2\left(\frac{1}{R_2}+\frac{1}{R_1}\right)-\frac{R_1\widetilde{Q}(s)+\widetilde{H}_2(s)}{(\tau_1 s+1)R_1}=0.$$

Multiplying by R_2 and re-arranging terms, it becomes

$$\widetilde{H}_2(s)(A_2R_2s+1)+\frac{R_2}{R_1}\widetilde{H}_2-\frac{R_2R_1\widetilde{Q}(s)+R_2\widetilde{H}_2(s)}{(\tau_1 s+1)R_1}=0,$$

$$\widetilde{H}_2(s)(A_2R_2s+1)+\frac{R_2}{R_1}\widetilde{H}_2-\frac{R_2\widetilde{H}_2(s)}{(\tau_1 s+1)R_1}=\frac{R_2R_1\widetilde{Q}(s)}{(\tau_1 s+1)R_1}$$

Further simplifying, one has

$$\widetilde{H}_2(s)(\tau_2 s+1)(\tau_1 s+1)+A_1R_2\widetilde{H}_2 s=R_2\widetilde{Q}(s),\quad \tau_2=A_2R_2.$$

The transfer function is

$$\frac{\widetilde{H}_2(s)}{\widetilde{Q}(s)}=\frac{R_2}{(\tau_1\tau_2)s^2+(\tau_1+\tau_2+A_1R_2)s+1}. \tag{A7}$$

This is Equation (3.22).

Exercise Questions

Q1. Three identical uniform cross-sectional area water tanks have inflow $q(t)$ and outflows $q_o(t)$ as shown in Figure 3Q1.

a. From first principles, derive the dynamic equations in terms of deviation variables.

b. Determine the transfer functions $\frac{\widetilde{H}_3(s)}{\widetilde{Q}(s)}$, $\frac{\widetilde{H}_2(s)}{\widetilde{Q}(s)}$, and $\frac{\widetilde{H}_1(s)}{\widetilde{Q}(s)}$, where $\widetilde{H}_i(s)$, $i = 1, 2, 3$, and $\widetilde{Q}(s)$ are the Laplace transforms of the deviation variables of the levels in the first, second, and third tanks, and the inflow to the first tank, respectively.

c. If the valve resistances $R_1 = R_2 = R_3 = 1$ unit, find the numerical values of the three transfer functions in (b).

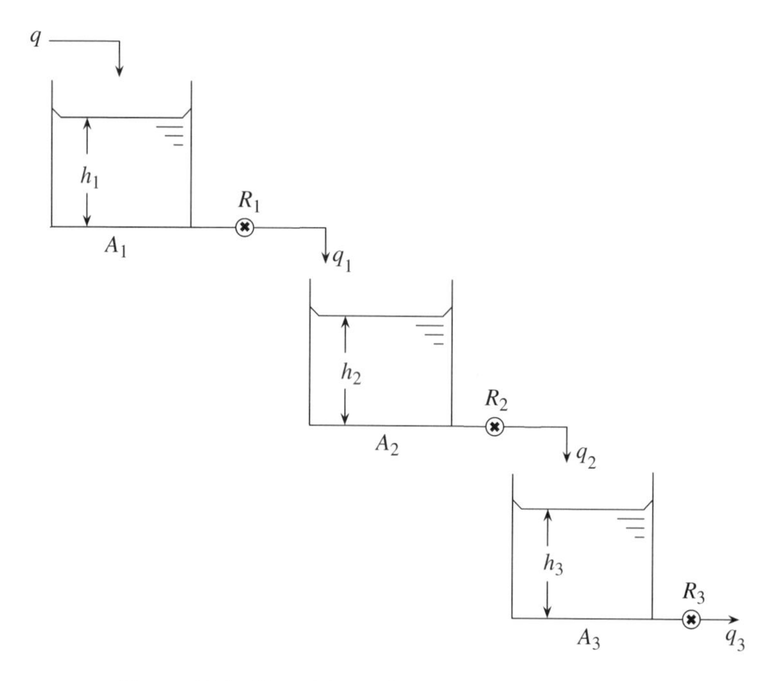

Figure 3Q1 Three identical non-interacting water tanks

Q2. Three identical, uniform cross-sectional area, interacting water tanks have inflow $q(t)$ and outflows $q_3(t)$ as shown in Figure 3Q2.

a. Derive the dynamic equations in terms of deviation variables.

b. Determine the transfer functions $\frac{\widetilde{H}_3(s)}{\widetilde{Q}(s)}$, $\frac{\widetilde{H}_2(s)}{\widetilde{Q}(s)}$, and $\frac{\widetilde{H}_1(s)}{\widetilde{Q}(s)}$, where $\widetilde{H}_i(s)$, $i = 1, 2, 3$, and $\widetilde{Q}(s)$ are the Laplace transforms of the deviation variables of the levels in the first, second, and third tanks, and the inflow to the first tank, respectively.

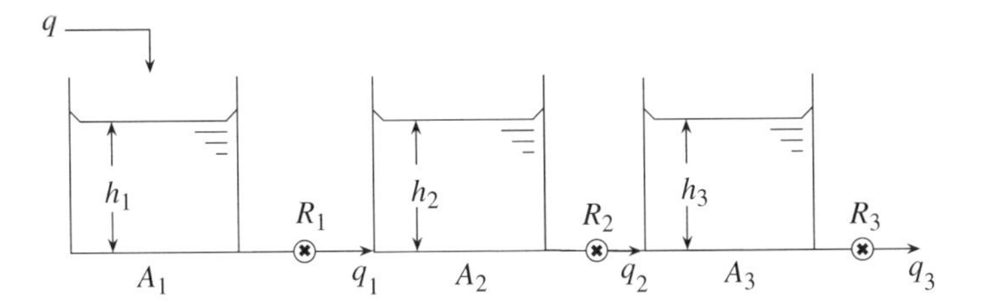

Figure 3Q2 Three interacting water tanks

c. If the valve resistances $R_1 = R_2 = R_3 = 1$ unit, find the numerical values of the three transfer functions in (b).

Q3. Three identical, uniform cross-sectional area, interacting and non-interacting water tanks have two identical inflows $q(t)$ to the two interacting water tanks and one outflow $q_3(t)$ as shown in Figure 3Q3.

a. Derive the dynamic equations in terms of deviation variables.

b. Determine the transfer functions $\frac{\widetilde{H}_3(s)}{\widetilde{Q}(s)}$, $\frac{\widetilde{H}_2(s)}{\widetilde{Q}(s)}$, and $\frac{\widetilde{H}_1(s)}{\widetilde{Q}(s)}$, where $\widetilde{H}_i(s)$, $i = 1, 2, 3$, and $\widetilde{Q}(s)$ are the Laplace transforms of the deviation variables of the levels in the first, second, and third tanks, and the inflow to the first tank, respectively.

c. If the valve resistances $R_1 = R_2 = R_3 = 1$ unit, find the numerical values of the three transfer functions in (b).

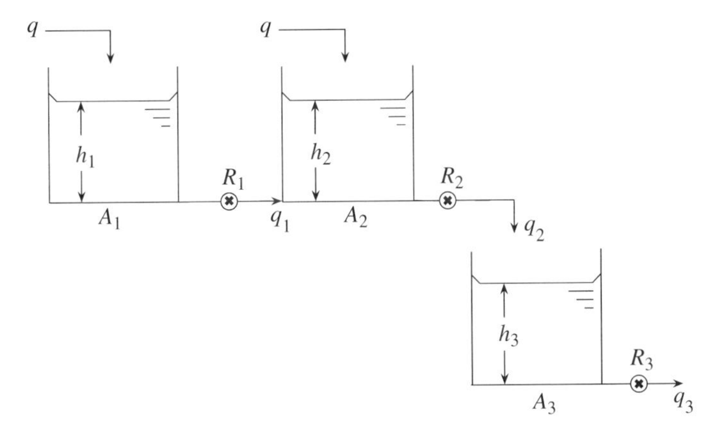

Figure 3Q3 Three interacting and non-interacting water tanks

Q4. The hydraulic system shown in Figure 3Q4 consists of two cylinders of uniform cross-sectional areas A_1 and A_2 connected by a channel which is modeled as an orifice. The pressures at the piston faces are denoted by p_1 and p_2. Suppose the masses of the pistons and rigid rods in the cylinders are considered zero. The friction between the sliding rods and supporting bearings is negligibly small and the region between the pistons is filled with an incompressible fluid. The flow resistance R through the orifice (representing by a pipe in the figure) is considered constant and there is no leakage in the system.

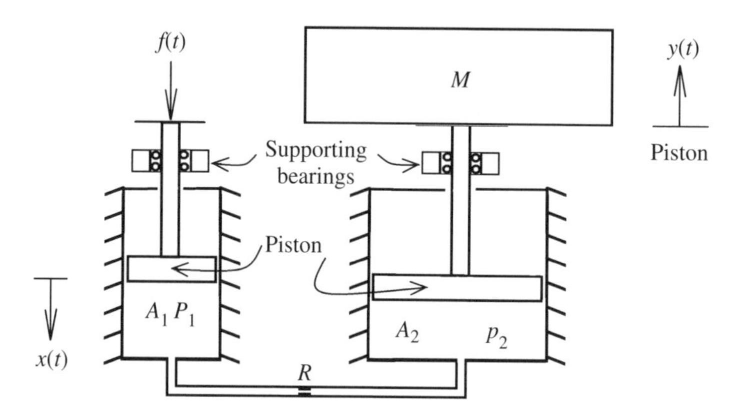

Figure 3Q4 A hydraulic system consisting of two cylinders

a. From first principles, derive the dynamic equation relating the inertia force of the mass M to the input force $f(t)$.
b. Determine the transfer function $\frac{Y(s)}{F(s)}$, assuming the system starts from rest (that is, the initial conditions are zero).

Q5. The section view of a commonly used hydraulic actuator is shown in Figure 3Q5. The main piston rod and the rods among the input pistons are assumed to be rigid and the piston masses are negligible. It is assumed that the main piston or output rod is connected to a mass m (which is not shown in the figure). The flow due to compressibility effects,

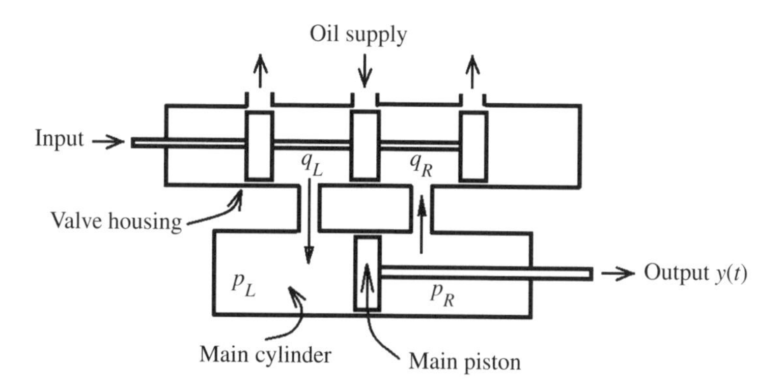

Figure 3Q5 Section view of a hydraulic valve actuator

according to Equation (3.5), is $(V/\beta)(dp/dt)$ where β is the bulk modulus of the fluid, V is the volume, and p is the pressure. The flow due to motion across the actuator is $A(dy/dt)$ in which A is the uniform cross-section of the main cylinder and $y(t)$ or y is the displacement of the actuator rod. Flow due to leakage across the actuator piston is $\mu(p_L - p_R)$, where μ is the coefficient of leakage and assumed to be constant, and p_L and p_R are the pressures in the left and right chambers as indicated in the figure.

a. Derive the equation of motion that includes the effects of compressibility.
b. Determine the transfer function $\dfrac{Y(s)}{Q(s)}$ where $Y(s)$ is the Laplace transform of $y(t)$, and $Q(s)$ is the Laplace transform of the average flow $q(t) = (q_L + q_R)/2$.

4

Dynamic Behaviors of Oscillatory Systems

In translating the oscillatory or oscillatory mechanical systems in the physical world into mathematical models in the conceptual world, there is always a question of how accurate and realistic the representation is of the physical system. It is generally logical to approach the problem by choosing the simplest model, and if one desires, a more realistic model can be formulated at a later stage. The simplest model is generally governed by the economy of solution, and can often provide information and insight to the problem at hand.

The present chapter therefore begins with the presentation of translational and rotational elements of oscillatory systems, formulation, and analysis of simplest dynamic systems. The translational and rotational elements are included in Section 4.1 while the single degree-of-freedom (dof) systems without applied forces are dealt with in Section 4.2. Section 4.3 deals with single dof linear systems under harmonic forces. Section 4.4 is concerned with non-harmonically forced single dof linear systems. In this particular system the excitation is a manifestation of the initial condition. For more realistic and accurate representations of the physical systems, more refined models are required. This leads to the linear multi-degrees-of-freedom (mdof) systems presented in Section 4.5. Free and forced mdof systems are considered. In order to limit the scope of the present chapter, only harmonically excited mdof systems are dealt with. The focus is, however, on a particular passive vibration control, the dynamic absorber. While the formulations in this part of the chapter are general and can be applied to linear systems of many dof, for illustration purpose, only systems having two dof are presented in details. Linear continuous systems which are also known as distributed parameter models are dealt with in Section 4.6. This chapter concludes with Section 4.7 in which eight questions are solved.

4.1 Elements of Oscillatory Systems

Some basic translational and rotational elements of oscillatory systems are presented in this section and illustrated in Figure 4.1. In Figure 4.1a, the direction of the arrow in the

Introduction to Dynamics and Control in Mechanical Engineering Systems, First Edition. Cho W. S. To.

Companion website: www.wiley.com/go/to/dynamics

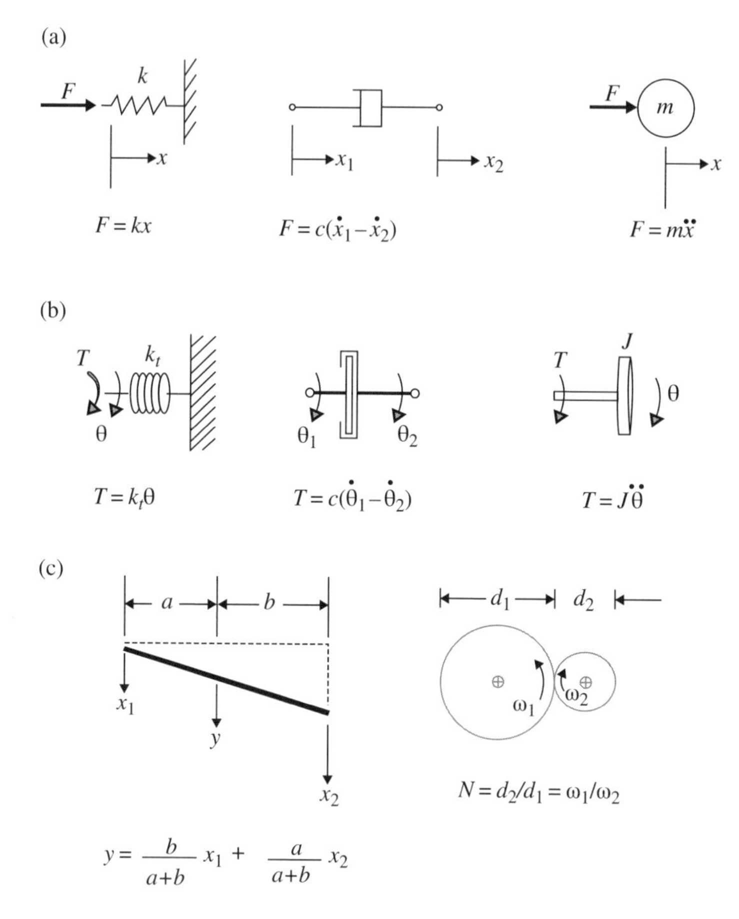

Figure 4.1 (a) Basic translational elements of oscillatory systems; (b) basic rotational elements of oscillatory systems; (c) lever mechanism (rigid bar with pinned joints) and simple gear mechanism

translational motion x denotes the positive direction. The counter-clockwise arrow, viewed from the rhs of Figure 4.1b, indicates positive rotational or angular displacement θ. In oscillatory systems of translational motion the three basic elements are the spring represented by stiffness coefficient k, dash-port by damping coefficient c, and mass m. In oscillatory systems with rotational displacements the three basic elements corresponding to the translational motion are the torsional spring constant k_t, torsional damping constant c_t, and rotary mass moment of inertia J. It may be appropriate to note that the translational and rotational elements have different units. For example, k is in N/m, while k_t is in N/rad. Of course, both translational and rotational motions can occur in a single mechanical system, and therefore care has to be taken with their combination in the derivation and solution of equations of motion.

Aside from the six basic elements presented in Figure 4.1a and b, two more important basic mechanisms are commonly encountered in dynamic and control system analysis. These are the lever mechanism (rigid bar with pinned joints) and simple gear mechanism (spur gears with rotating shafts in parallel which are supported by bearings not shown in the figure). These

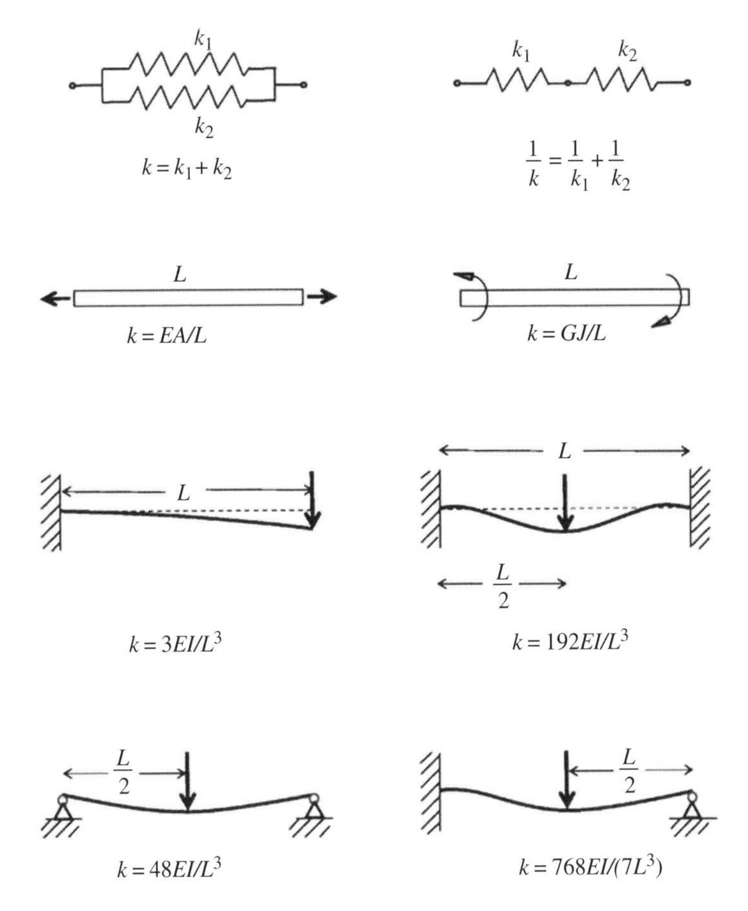

Figure 4.2 Elementary spring constants

two mechanisms are shown in Figure 4.1c. Note that for simplicity only the pitch circles (circles in the figure) of the gears are shown.

Before leaving this section, several elementary spring constants are included in Figure 4.2. These spring constants are useful in the design and analysis of oscillatory mechanical systems to be considered in this chapter.

4.2 Free Vibration of Single Degree-of-Freedom Systems

The presentation in this section is as follows. First, the systems in the physical world as shown in Figure 4.3 are given. Second, their corresponding models in the conceptual world are presented in Figure 4.4. Third, the formulation and analysis of one of the models in the conceptual world is selected to illustrate the steps.

Stage 1: Systems in the Physical World

There are many systems in the engineering physical world. It suffices to give two common examples for the purpose of illustrating the process of modeling. These two systems in the physical world are the automobile and the television tower as sketched in Figure 4.3.

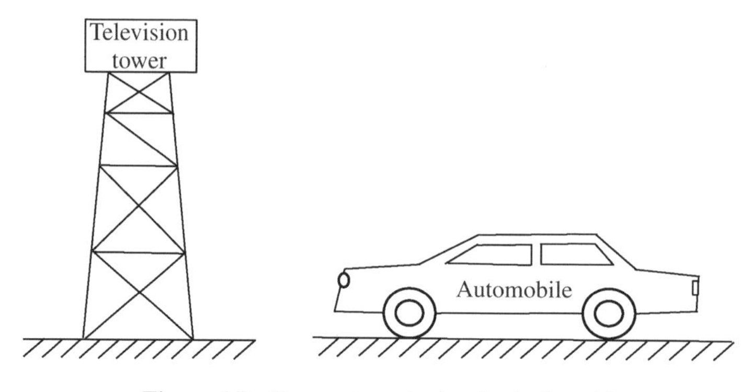

Figure 4.3 Two systems in the physical world

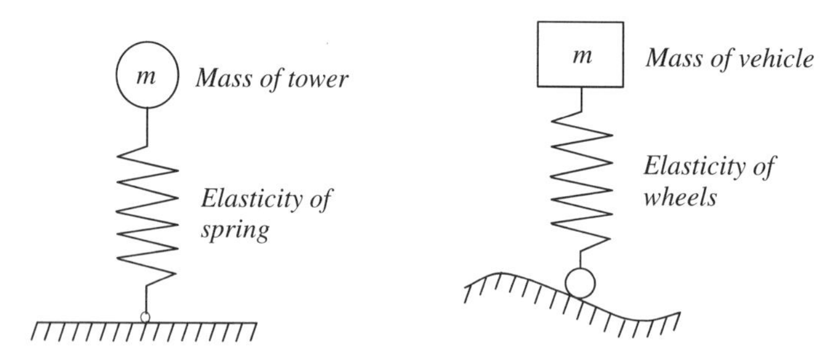

Figure 4.4 Single dof models of the systems in Figure 4.3

Stage 2: Systems in the Conceptual World

The systems in Figure 4.3 may be represented by the simplest mathematical models, known as lumped-parameter models or discrete systems, in the conceptual world. They are illustrated in Figure 4.4.

Stage 3: Formulation and Solution

The equation of motion governing each of the conceptual world models sketched in Figure 4.3 may now be derived. In general, the equation of free vibration for a single dof system may be obtained by one of the following approaches:

- definition of simple harmonic motion;
- knowledge of oscillatory motion;
- law of conservation of energy;
- Newton's second law of motion; and
- method of virtual work or virtual power.

Consider the discrete system on the left-hand side (lhs) in Figure 4.4. The latter and its free body diagram (FBD) are included in Figure 4.5.

By Newton's second law of motion for the lumped mass,

$$m\ddot{x} = \sum F = mg - k(\Delta + x) - c\dot{x},$$

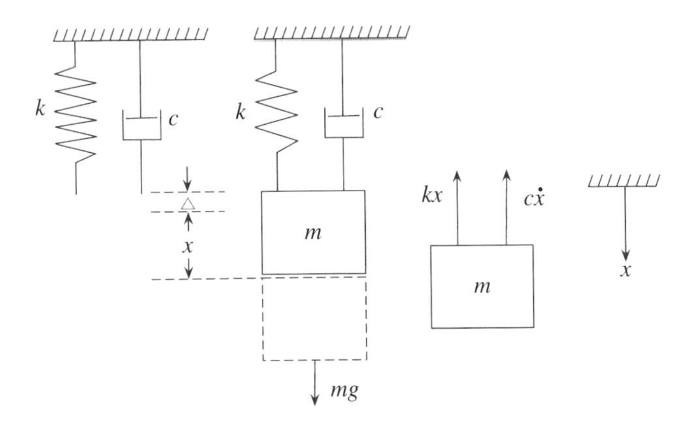

Figure 4.5 Single dof system and free body diagram

in which m is the mass, k the spring constant, and c the damping coefficient of the system. The symbol $\sum F$ denotes summing of the forces. The spring force $k\Delta$ is equal to the gravitational force mg acting on the mass m; that is, $mg = k\Delta$.

Therefore, the equation of motion for the above system is

$$m\ddot{x} + c\dot{x} + kx = 0. \tag{4.1}$$

The first term, second term, and third terms on the lhs of Equation (4.1) are, respectively, the inertia, damping, and restoring forces.

One can solve Equation (4.1) by the Laplace transform method, but in the following another independent approach is employed to provide one more alternative.

Let the solution of the above system be $x = e^{st}$, where s is a constant parameter. The basis for such an assumption is that physically one is dealing with oscillatory motion, and mathematically the governing equation of motion is a second-order linear differential equation.

Performing the differentiations of x with respect to time t and substituting the displacement, velocity, and acceleration terms into Equation (4.1), one obtains:

$$\left(ms^2 + cs + k\right)e^{st} = 0$$

which is satisfied for all values of t (physically, it implies that if the system undergoes vibrating motion) when

$$\begin{aligned} s^2 + \frac{c}{m}s + \frac{k}{m} = 0 \quad \text{or} \\ s^2 + 2\zeta\omega_n s + \omega_n^2 = 0. \end{aligned} \tag{4.2}$$

where $c/m = 2\zeta\omega_n$ and $k/m = \omega_n^2$. In the foregoing, ζ is known as the damping ratio and ω_n is the undamped natural frequency or simply referred to as the natural frequency of the system.

Equation (4.2) is known as the *characteristic equation* (c.e.) and has two roots since it is a quadratic equation. The two roots are

$$s_1, s_2 = -\zeta\omega_n \pm i\omega_n\sqrt{1-\zeta^2}. \tag{4.3}$$

- For a *lightly damped* system, that is $0 < \zeta < 1$,

$$s_1, s_2 = -\alpha \pm i\omega_d \tag{4.4}$$

where $\alpha = \zeta\omega_n$, and $\omega_d = \omega_n\sqrt{1-\zeta^2}$ is known as the *damped natural frequency* or *resonant frequency* of the system.
The general solution to Equation (4.1) can be written as

$$x = A_1 e^{(-\alpha + i\omega_d)t} + A_2 e^{(-\alpha - i\omega_d)t} \tag{4.5}$$

or factoring the exponential term such that it becomes

$$x = e^{-\alpha t}\left(A_1 e^{i\omega_d t} + A_2 e^{-i\omega_d t}\right)$$

where $e^{i\omega_d t} = \cos\omega_d t + i\sin\omega_d t$, $e^{-i\omega_d t} = \cos\omega_d t - i\sin\omega_d t$, which can be rewritten as

$$x = e^{-\alpha t}\left[(A_1 + A_2)\cos\omega_d t + i(A_1 - A_2)\sin\omega_d t\right]. \tag{4.6}$$

Introducing the following notation,

$$A_1 + A_2 = A\cos\phi, \quad i(A_1 - A_2) = A\sin\phi$$

Equation (4.6) becomes

$$x = Ae^{-\alpha t}(\cos\phi\cos\omega_d t + \sin\phi\sin\omega_d t).$$

But by using the trigonometric identity,

$$\cos(\phi - \omega_d t) = \cos(\omega_d t - \phi) = \cos\phi\cos\omega_d t + \sin\phi\sin\omega_d t,$$

the solution can be simplified to

$$x = Ae^{-\alpha t}\cos(\omega_d t - \phi) \tag{4.7}$$

where A and ϕ are constants of integration and are dependent on the initial displacement and initial velocity because A and ϕ are defined in terms of A_1 and A_2 in the foregoing.

Remark 4.2.1 In the special case in which *damping is critical*, that is, $\zeta = 1$, the solution to Equation (4.1) can be shown to be

$$x = (A_1 + tA_2)e^{-\alpha t}.$$

Remark 4.2.2 For the case in which $\zeta > 1$, that is *over-damped* or *heavily damped* free vibration, Equation (4.3) gives two real roots and the solution becomes

$$x = e^{-\alpha t}\left(A_1 e^{\omega_n \sqrt{\zeta^2 - 1}t} + A_2 e^{-\omega_n \sqrt{\zeta^2 - 1}t}\right).$$

Remark 4.2.3 In general, the lightly damped free vibration is the most common type of free vibration that can occur in many mechanical dynamic and control systems.

4.3 Single Degree-of-Freedom Systems under Harmonic Forces

Having derived the equation of free vibration for a single dof oscillatory system, it is logical to study the vibration of a forced single dof system. In practice, machines and structural systems are often accompanied by various dynamic forces. One type of dynamic force is the harmonic force.

- Common sources of harmonic forces in single dof systems are:
  a. reciprocating machines;
  b. unbalance in rotating machines; and
  c. support motions.
- One of the main objectives is to obtain the amplitude of vibration developed in the system.
- To prevent machine failure or catastrophic consequences, *resonance* (that is, when the applied frequency is equal to the natural frequency of the system) in the system is to be avoided.

Consider the single dof system shown in Figure 4.6. The equation of motion may be derived as

$$m\ddot{x} + c\dot{x} + kx = f(t) = F \sin \omega t \tag{4.8}$$

where F is the amplitude of the harmonic force $f(t)$, for simplicity written as f.

- The solution to this equation consists of two parts, the *complementary function* or *transient response,* which is the solution of the homogeneous equation, and the *particular integral.*

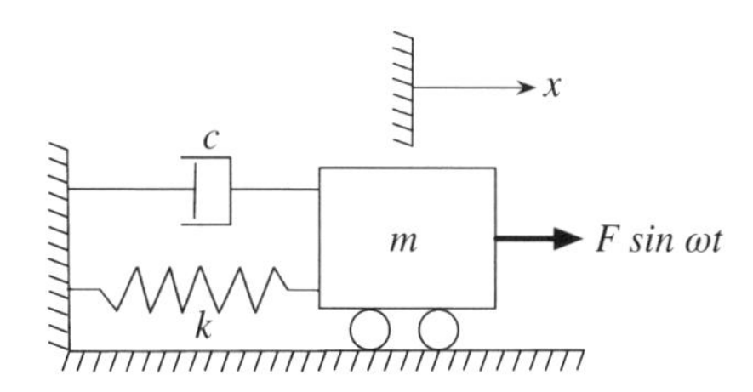

Figure 4.6 Harmonically forced single dof system

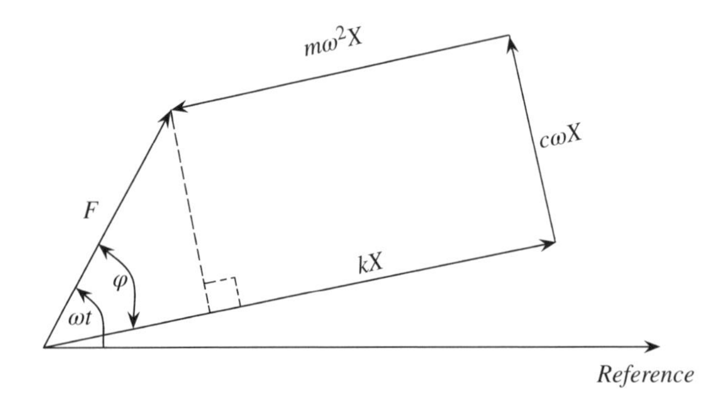

Figure 4.7 Vector diagram of a forced single dof system

The transient response has been discussed in Section 4.2. The present objective is to obtain the particular integral. Since it is assumed that the system has positive damping and therefore the response occurs at a later time, or the response is *delayed* by a phase angle φ with respect to the applied force f, so that the solution may be written as

$$x = X \sin(\omega t - \varphi) \tag{4.9}$$

where X is the amplitude of the displacement and not to be confused with the Laplace transform of the instantaneous displacement x.

The vector relationship for forced vibration with viscous damping is shown in Figure 4.7 below.

With reference to the vector diagram, one has

$$F^2 = \left(kX - m\omega^2 X\right)^2 + \left(c\omega X\right)^2$$

and therefore

$$X = \frac{F}{\sqrt{(k - m\omega^2)^2 + (c\omega)^2}} \tag{4.10}$$

$$\tan\varphi = \frac{c\omega}{k - m\omega^2} \tag{4.11}$$

The dimensionless form of Equation (4.10) can be obtained by dividing the numerator and denominator on the rhs by k as

$$X = \frac{F/k}{\sqrt{\left(1 - \frac{m\omega^2}{k}\right)^2 + \left(\frac{c\omega}{k}\right)^2}}$$

so that upon moving the resulting numerator on the rhs to the denominator on the lhs, and making use of the definitions $c/m = 2\zeta\omega_n$ and $k/m = \omega_n^2$, one can show that

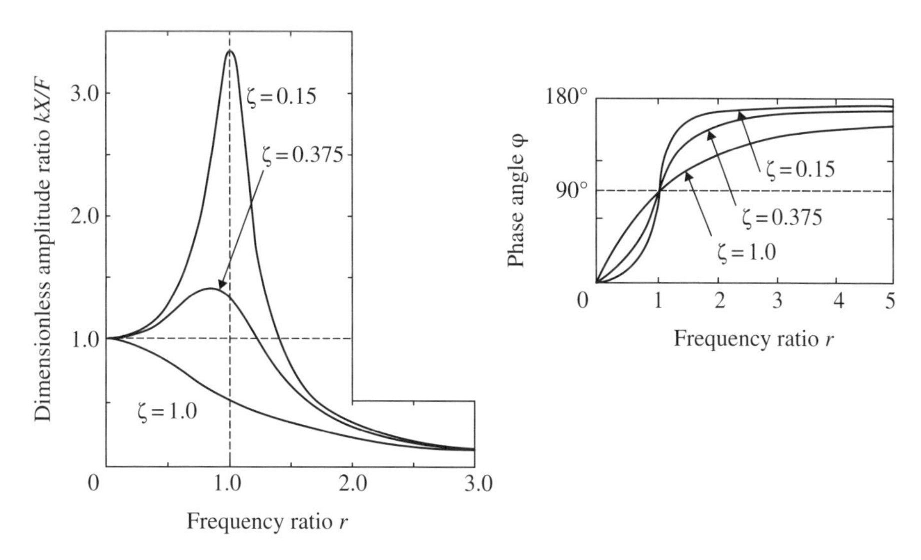

Figure 4.8 Response and phase relationships of single dof system

$$\frac{Xk}{F} = \frac{1}{\sqrt{\left[1-\left(\frac{\omega}{\omega_n}\right)^2\right]^2 + \left[2\zeta\left(\frac{\omega}{\omega_n}\right)\right]^2}}.$$

Writing the ratio of frequencies $r = \omega/\omega_n$, it reduces to

$$\frac{Xk}{F} = \frac{1}{\sqrt{(1-r^2)^2 + (2\zeta r)^2}}. \tag{4.12}$$

Similarly, Equation (4.11) becomes

$$\tan\varphi = \frac{2\zeta r}{1-r^2}. \tag{4.13}$$

The magnification factor defined by Equation (4.12) and phase angle given by Equation (4.13) are included in Figure 4.8 for three different damping ratios.

- Note that in practice, in order to avoid resonance or significantly reduce the vibration of the system, the applied frequency should be two or more times the natural frequency.

4.4 Single Degree-of-Freedom Systems under Non-Harmonic Forces

For completeness, the case of a spring-mass dropping through a vertical distance of h, as shown in Figure 4.9 is considered in this section. This is a simple case of non-harmonically forced vibration. Use of such a model can be found in analysis of the free fall of a landing helicopter or aircraft, and packaging design.

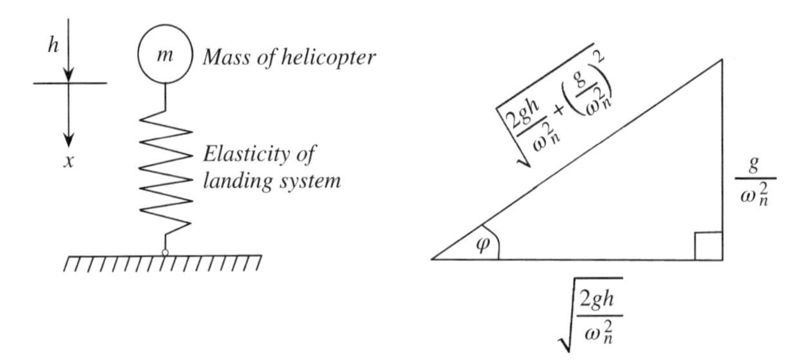

Figure 4.9 Dropping of a spring-mass system

Assume t = 0 at the instant when the spring just touches the ground. With reference to Figure 4.9, one has the equation of motion as

$$m\ddot{x} + kx = mg, \tag{4.14}$$

where g is the acceleration due to gravity.

In the above system the initial conditions are: $x(0)$ = 0 and $\dot{x}(0) = \sqrt{2gh}$. One may use the Laplace transform method for the solution of the response since the rhs of Equation (4.14) is a constant. Thus, taking the Laplace transform of Equation (4.14), one has

$$m\left[s^2X(s) - sx(0) - \dot{x}(0)\right] + kX(s) = \frac{mg}{s}.$$

Substituting for the initial conditions, one obtains

$$s^2X(s) - \sqrt{2gh} + \frac{k}{m}X(s) = \frac{g}{s}.$$

Re-arranging terms to give

$$X(s)\left(s^2 + \omega_n^2\right) = \sqrt{2gh} + \frac{g}{s}.$$

the subsidiary equation becomes

$$X(s) = \frac{\sqrt{2gh}}{s^2 + \omega_n^2} + \frac{g}{s\left(s^2 + \omega_n^2\right)}. \tag{4.15}$$

Taking the inverse Laplace transform of $X(s)$, one has

$$x(t) = \frac{\sqrt{2gh}}{\omega_n}\sin\omega_n t + \frac{g}{\omega_n^2}(1 - \cos\omega_n t).$$

By making use of the right-angle triangle relationship shown in Figure 4.9 and the trigonometric identity

$$\sin(\omega_n t-\varphi) = \sin\omega_n t\cos\varphi - \cos\omega_n t\sin\varphi,$$

one can show that the displacement becomes

$$x(t) = \frac{g}{\omega_n^2} + \sqrt{\frac{2gh}{\omega_n^2} + \left(\frac{g}{\omega_n^2}\right)^2}\sin(\omega_n t-\varphi),\quad x(t)>0\cdot \tag{4.16}$$

This equation agrees with that given in the book by Thomson [1].

4.5 Vibration Analysis of Multi-Degrees-of-Freedom Systems

For many situations in which a first insight is provided by a simple model such as a single dof system, it may have to be refined in order to obtain a more accurate solution, or the physical system is so complicated that a single dof model is not adequate. Therefore, a mdof system has to be considered. This, of course, will increase the amount of algebraic manipulation in the solution. In Section 4.5.1 the formulation and solution of two mdof systems are presented. Section 4.5.2 is concerned with the passive vibration control in which the so-called vibration absorber is employed. For completeness, the normal mode analysis is included in Section 4.5.3.

4.5.1 Formulation and Solution for Two-Degrees-of-Freedom Systems

In order to provide approaches in dealing with mdof systems while limiting the amount of algebraic manipulation, in this subsection only two 2-dof systems are included for illustration. These two physical systems are the coal sizing machine and the expensive electronic equipment installed inside a rocket. They are sufficiently representative to provide an appreciation of how to translate them from the physical world into the models of the conceptual world. The three stages in the analysis are outlined below.

Stage 1: Systems in Physical World

The coal sizing machine essentially consists of a jig and an absorber mass connected by springs, as shown in Figure 4.10a in which the screen reciprocates with a frequency. This reciprocating action is the source of harmonic force applied to the jig. The second physical world example is the electronic equipment installed inside the rocket. The electronic equipment is supported by a frame structure that can be modeled as massless beams, as sketched in Figure 4.10b.

Stage 2: Systems in Conceptual World

The two 2-dof systems in the physical world may be analysed as the lumped-parameter models in the conceptual or mathematical world. They are presented in Figure 4.11.

Stage 3: Formulation and Solution

Once the conceptual model is established one may proceed with the derivation of the governing equation of motion for the individual system. For brevity, only the coal sizing machine is

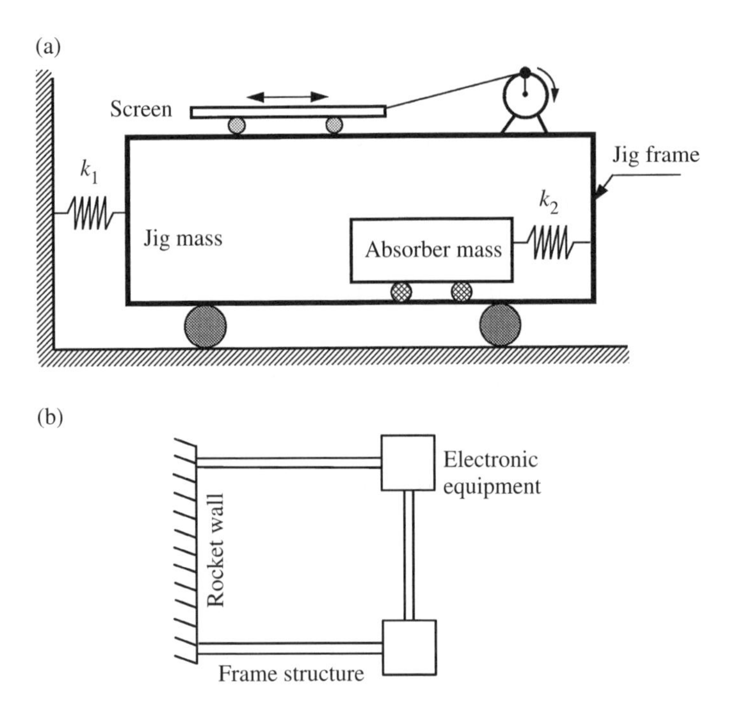

Figure 4.10 Two 2-dof systems in the physical world: (a) the coal sizing machine; (b) electronic equipment in a rocket

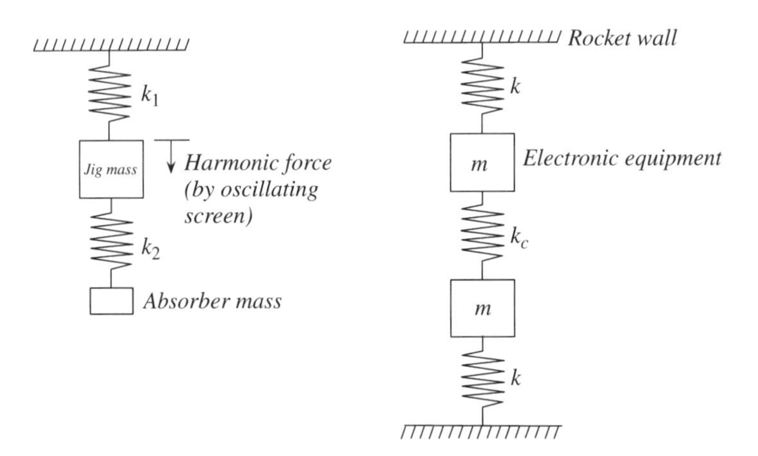

Figure 4.11 Lumped-parameter models of systems in Figure 4.10

analysed in detail in the following. Consider the lumped-parameter model and the FBD of this 2-dof system, as shown in Figure 4.12.

For simplicity, damping in the system is disregarded so that by applying Newton's law of motion to the jig mass, one has

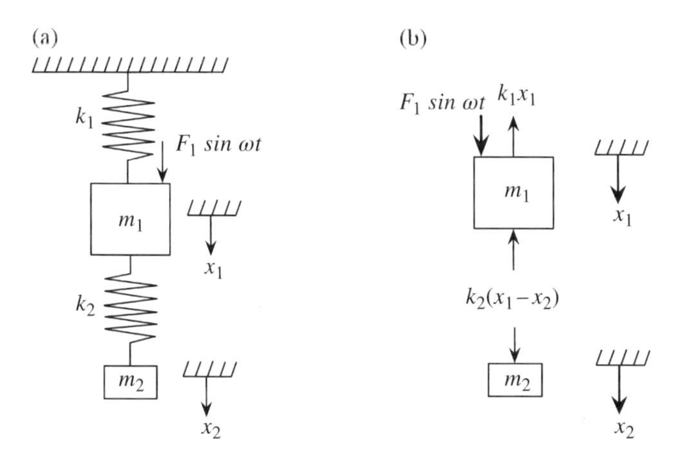

Figure 4.12 (a) Coal sizing machine model and (b) FBD of (a)

$$m_1\ddot{x}_1 = \sum F = -k_1x_1 - k_2(x_1 - x_2) + F_1 \sin \omega t.$$

Re-arranging terms, it becomes

$$m_1\ddot{x}_1 + (k_1 + k_2)x_1 - k_2x_2 = F_1 \sin \omega t.$$

Similarly, applying Newton's second law of motion for the absorber or second mass and the FBD, the equation of motion is given by

$$m_2\ddot{x}_2 = -k_2(x_2 - x_1)$$

or after re-arranging terms

$$m_2\ddot{x}_2 + k_2(x_2 - x_1) = 0.$$

For subsequent analysis, one can express the above two equations of motion in a more compact form. To this end, a matrix is applied. Thus, the above two equations of motion can be written in matrix form as

$$\begin{bmatrix} m_1 & 0 \\ 0 & m_2 \end{bmatrix} \begin{pmatrix} \ddot{x}_1 \\ \ddot{x}_2 \end{pmatrix} + \begin{bmatrix} (k_1 + k_2) & -k_2 \\ -k_2 & k_2 \end{bmatrix} \begin{pmatrix} x_1 \\ x_2 \end{pmatrix} = \begin{pmatrix} F_1 \\ 0 \end{pmatrix} \sin \omega t.$$

This equation can be written in a more conventional matrix notation as

$$\begin{bmatrix} m_{11} & 0 \\ 0 & m_{22} \end{bmatrix} \begin{pmatrix} \ddot{x}_1 \\ \ddot{x}_2 \end{pmatrix} + \begin{bmatrix} k_{11} & k_{12} \\ k_{21} & k_{22} \end{bmatrix} \begin{pmatrix} x_1 \\ x_2 \end{pmatrix} = \begin{pmatrix} F_1 \\ 0 \end{pmatrix} \sin \omega t, \tag{4.17}$$

where $m_{11} = m_1$, $m_{22} = m_2, k_{11} = k_1 + k_2, k_{12} = k_{21} = -k_2, k_{22} = k_2$.

Note that the diagonal elements of the mass and stiffness matrices are all positive. Since the system is undamped and therefore the responses have no delay or phase difference with respect to the applied force, one can write

$$\begin{pmatrix} x_1 \\ x_2 \end{pmatrix} = \begin{pmatrix} X_1 \\ X_2 \end{pmatrix} \sin \omega t.$$

In this equation, X_i $(i = 1, 2)$ are the amplitudes of instantaneous displacements x_i and symbolically not to be confused with the Laplace transform of x_i.

Substituting the above solution and its derivatives into Equation (4.17) and simplifying, one has

$$\begin{bmatrix} (k_{11} - m_1\omega^2) & k_{12} \\ k_{21} & (k_{22} - m_2\omega^2) \end{bmatrix} \begin{pmatrix} X_1 \\ X_2 \end{pmatrix} = \begin{pmatrix} F_1 \\ 0 \end{pmatrix} \tag{4.18}$$

or writing it in a more concise form

$$[Z(\omega)] \begin{pmatrix} X_1 \\ X_2 \end{pmatrix} = \begin{pmatrix} F_1 \\ 0 \end{pmatrix}$$

where the so-called *impedance matrix* is

$$[Z(\omega)] = \begin{bmatrix} (k_{11} - m_1\omega^2) & k_{12} \\ k_{21} & (k_{22} - m_2\omega^2) \end{bmatrix}.$$

Premultiplying on both sides of Equation (4.18) by the inverse of $[Z(\omega)]$, it becomes

$$\begin{pmatrix} X_1 \\ X_2 \end{pmatrix} = [Z(\omega)]^{-1} \begin{pmatrix} F_1 \\ 0 \end{pmatrix} = \frac{adj[Z(\omega)] \begin{pmatrix} F_1 \\ 0 \end{pmatrix}}{|Z(\omega)|} \quad \text{or}$$

$$\begin{pmatrix} X_1 \\ X_2 \end{pmatrix} = \frac{\begin{bmatrix} (k_{22} - m_2\omega^2) & -k_{12} \\ -k_{21} & (k_{11} - m_1\omega^2) \end{bmatrix} \begin{pmatrix} F_1 \\ 0 \end{pmatrix}}{|Z(\omega)|}. \tag{4.19a}$$

The determinant in the denominator on the rhs of Equation (4.19a) may be expressed

$$|Z(\omega)| = m_1 m_2 \left(\omega_1^2 - \omega^2\right)\left(\omega_2^2 - \omega^2\right) \tag{4.19b}$$

where ω_1 and ω_2 are the natural or normal mode frequencies of the 2-dof system.

- The proof of Equation (4.19b) and the expressions for the two natural frequencies are provided in Appendix 4A.

Applying Equation (4.19b), one can write the elements or entries of the amplitude displacement vector of Equation (4.19a), respectively, as

$$X_1 = \frac{(k_{22} - m_2\omega^2)F_1}{m_1 m_2 \left(\omega_1^2 - \omega^2\right)\left(\omega_2^2 - \omega^2\right)}, \tag{4.20a}$$

$$X_2 = \frac{-k_{21}F_1}{m_1 m_2 \left(\omega_1^2 - \omega^2\right)\left(\omega_2^2 - \omega^2\right)}. \tag{4.20b}$$

The above equations are very useful in practice for the elimination of vibration in machines and structural systems. To illustrate their use, the so-called *dynamic absorber* system is presented in the next subsection.

Meanwhile, attention is directed at the so-called *modal analysis* for mdof linear systems. Of course, there are more efficient modal analysis techniques [1], but they are outside the scope of this book.

One of the main reasons for performing such a modal analysis is that the system of coupled second-order differential equations of motion can be uncoupled such that every uncoupled second-order differential equation of motion may be simply analysed similarly to that for the single dof system. After all the uncoupled equations of motion are solved, they can be transformed back to the original coordinate system.

Returning to Equation (4.18), the *characteristic* or *frequency equation* of the system is given by setting the determinant of the impedance matrix to zero. That is,

$$\begin{vmatrix} (k_{11} - m_1\omega^2) & k_{12} \\ k_{21} & (k_{22} - m_2\omega^2) \end{vmatrix} = 0.$$

This equation can be simplified as

$$a\lambda^2 + b\lambda + c = 0$$

where $a = m_1 m_2$, $b = -(k_{11}m_2 + k_{22}m_1)$, and $c = k_{11}k_{22} - k_{12}k_{21}$.

Solving the quadratic equation one finds

$$\lambda_1 = \frac{-b + \sqrt{b^2 - 4ac}}{2a}, \quad \lambda_2 = \frac{-b - \sqrt{b^2 - 4ac}}{2a}.$$

These two roots of the quadratic equation are called the *eigenvalues* of the system. Note that $\lambda_1 = \omega_1^2$ and $\lambda_2 = \omega_2^2$, and ω_1^2 and ω_2^2 are defined by Equations (A2a) and (A2b) in Appendix 4A, respectively.

Substituting λ_1 into Equation (4.18) with the rhs applied forcing vector being set to zero, one obtains the amplitude ratio

$$\left(\frac{X_1}{X_2}\right)_{(1)} = \frac{k_{12}}{m_1\omega_1^2 - k_{11}} = \frac{m_2\omega_1^2 - k_{22}}{k_{21}}, \tag{4.21a}$$

where the subscript (1) on the lhs of Equation (4.21a) denotes the amplitude ratio associated with the first natural frequency.

Similarly, substituting λ_2 into Equation (4.18) with the rhs applied forcing vector being set to zero, one has

$$\left(\frac{X_1}{X_2}\right)_{(2)} = \frac{k_{12}}{m_1\omega_2^2 - k_{11}} = \frac{m_2\omega_2^2 - k_{22}}{k_{21}}. \tag{4.21b}$$

Note that these equations only enable one to obtain the amplitude ratios and not their absolute values.

In practice or numerical computation, one frequently selects unity or any other numerical value as one of the amplitudes so that it is referred to as the amplitude ratio being *normalized* to that value. This normalized amplitude ratio is called the *normal mode*. Mathematically, it is known as the *eigenvector*. Thus, the normal modes can be regarded as the steady-state vibration patterns or configurations of the system at their corresponding natural frequencies. In Section 4.5.3 examples are presented to illustrate the determination of the normal modes.

4.5.2 Vibration Analysis of a System with a Dynamic Absorber

- The objective in this subsection is to make use of Equation (4.20a) for the elimination of vibration of machine m_1 by attaching a *dynamic absorber* of mass m_2.

With reference to Figure 4.12, m_1 is the mass of the vibrating machine and m_2 is the mass of the dynamic absorber. By making use of Equation (4.19b) and recalling that the denominator on the rhs is the determinant of impedance matrix $[Z(\omega)]$, one has

$$X_1 = \frac{(k_{22} - m_2\omega^2)F_1}{\begin{vmatrix} (k_{11} - m_1\omega^2) & k_{12} \\ k_{21} & (k_{22} - m_2\omega^2) \end{vmatrix}} \quad \text{or}$$

$$X_1 = \frac{(k_{22} - m_2\omega^2)F_1}{(k_{11} - m_1\omega^2)(k_{22} - m_2\omega^2) - k_{12}k_{21}}. \tag{4.22a}$$

The numerator becomes zero when the following relation is satisfied

$$\omega^2 = \frac{k_{22}}{m_2}.$$

In other words, when the applied frequency is equal to the natural frequency of the dynamic absorber, the amplitude of machine vibration is zero.

In general, it is more convenient to express the amplitudes of the displacements in dimensionless forms. To this end, one can write Equation (4.21b) as

$$X_1 = \frac{k_{22}\left(1 - \frac{m_2}{k_{22}}\omega^2\right)F_1}{k_{11}k_{22}\left(1 - \frac{m_1}{k_{11}}\omega^2\right)\left(1 - \frac{m_2}{k_{22}}\omega^2\right) - k_{12}k_{21}}.$$

This equation can be non-dimensionalized by moving F_1 and k_{11} from the rhs to thc lhs such that

$$\frac{X_1 k_{11}}{F_1} = \frac{\left(1 - \frac{m_2}{k_{22}}\omega^2\right)}{\left(1 - \frac{m_1}{k_{11}}\omega^2\right)\left(1 - \frac{m_2}{k_{22}}\omega^2\right) - \frac{k_{12}k_{21}}{k_{11}k_{22}}}$$

This can be simplified by writing

$$\Omega_1^2 = \frac{k_{11}}{m_1}, \quad \Omega_2^2 = \frac{k_{22}}{m_2}$$

so that the dimensionless amplitude of the displacement of the vibrating machine becomes

$$\frac{X_1 k_{11}}{F_1} = \frac{1 - \left(\frac{\omega}{\Omega_2}\right)^2}{\left[1 - \left(\frac{\omega}{\Omega_1}\right)^2\right]\left[1 - \left(\frac{\omega}{\Omega_2}\right)^2\right] - \frac{k_{12}k_{21}}{k_{11}k_{22}}}. \tag{4.22b}$$

Similarly, the dimensionless amplitude of displacement of the dynamic absorber can be written as

$$\frac{X_2 k_{11}}{F_1} = \frac{-1}{\frac{k_{22}}{k_{12}}\left\{\left[1 - \left(\frac{\omega}{\Omega_1}\right)^2\right]\left[1 - \left(\frac{\omega}{\Omega_2}\right)^2\right] - \frac{k_{12}k_{21}}{k_{11}k_{22}}\right\}}.$$

Since in this 2-dof system $k_{22} = k_2 = -k_{12}$, it can be further simplified to give

$$\frac{X_2 k_{11}}{F_1} = \frac{1}{\left[1 - \left(\frac{\omega}{\Omega_1}\right)^2\right]\left[1 - \left(\frac{\omega}{\Omega_2}\right)^2\right] - \frac{k_2}{k_{11}}}. \tag{4.22c}$$

- Note that $\Omega_1 \neq \omega_1$ and $\Omega_2 \neq \omega_2$. The natural frequencies ω_1 and ω_2 of the 2-dof system are derived in Appendix 4A.

4.5.3 Normal Mode Analysis

At the end of Section 4.5.1 mode analysis and normal mode analysis were briefly introduced. To provide more illustration and the steps in obtaining the normal modes, the following two examples are presented.

Example 1

Consider a 2-dof system whose matrix equation of motion is given as

$$\begin{bmatrix} m_{11} & 0 \\ 0 & m_{22} \end{bmatrix}\begin{pmatrix} \ddot{x}_1 \\ \ddot{x}_2 \end{pmatrix} + \begin{bmatrix} k_{11} & k_{12} \\ k_{21} & k_{22} \end{bmatrix}\begin{pmatrix} x_1 \\ x_2 \end{pmatrix} = \begin{pmatrix} 0 \\ 0 \end{pmatrix}, \tag{i}$$

where $m_{11}=2m$, $m_{22}=\frac{2}{3}mL^2$, $k_{11}=2k$, $k_{12}=k_{21}=-\frac{kL}{2}$, $k_{22}=\frac{25kL^2}{8}$, $x_1=x$ is the translational displacement, and $x_2=\theta$ is the angular or rotational displacement.

a. Determine the natural frequencies and mode shapes of the 2-dof system.

b. For the given system with $2m = 2000$ kg, $2k = 40{,}000$ N/m, and $L = 1.0$ m, find the natural frequencies and normal modes of the 2-dof system.

Solution:

a. Since the system is undamped and therefore the responses have no delay or phase difference with respect to the applied force, one can write

$$\begin{pmatrix} x_1 \\ x_2 \end{pmatrix} = \begin{pmatrix} X_1 \\ X_2 \end{pmatrix}\sin\omega t = \begin{pmatrix} X \\ \Theta \end{pmatrix}\sin\omega t.$$

In this equation, X and Θ are the amplitudes of instantaneous displacements x_i and symbolically not to be confused with the Laplace transform of x_i.

Substituting the above solution and its derivatives into Equation (i) and simplifying, one has

$$\begin{bmatrix} (k_{11}-m_{11}\omega^2) & k_{12} \\ k_{21} & (k_{22}-m_{22}\omega^2) \end{bmatrix}\begin{pmatrix} X \\ \Theta \end{pmatrix} = \begin{pmatrix} 0 \\ 0 \end{pmatrix} \tag{ii}$$

or writing it in a more concise form

$$[Z(\omega)]\begin{pmatrix} X_1 \\ X_2 \end{pmatrix} = \begin{pmatrix} 0 \\ 0 \end{pmatrix}$$

where the so-called *impedance matrix* has been defined by Equation (4.18) as

$$[Z(\omega)] = \begin{bmatrix} (k_{11}-m_{11}\omega^2) & k_{12} \\ k_{21} & (k_{22}-m_{22}\omega^2) \end{bmatrix}.$$

Of course, one can determine the natural frequencies of the system by making use of Equations (A2a) and (A2b) in Appendix 4A. However, one can simply apply the frequency equation from Equation (ii); that is, equating the determinant of the impedance matrix $[Z(\omega)]$ to zero,

$$\begin{vmatrix} (k_{11}-m_{11}\omega^2) & k_{12} \\ k_{21} & (k_{22}-m_{22}\omega^2) \end{vmatrix} = 0.$$

Substituting the system parameters given above and writing $\lambda=\omega^2$,

$$\begin{vmatrix} (2k-2m\lambda) & -\dfrac{kL}{2} \\ -\dfrac{kL}{2} & \left(\dfrac{25kL^2}{8}-\dfrac{2}{3}mL^2\lambda\right) \end{vmatrix} = 0.$$

Operating on this equation, one has

$$(2k-2m\lambda)\left(\frac{25kL^2}{8}-\frac{2}{3}mL^2\lambda\right)-\left(\frac{kL}{2}\right)^2=0.$$

This quadratic equation in λ gives

$$\lambda_1=0.9498\,\frac{k}{m},\qquad \lambda_2=4.7377\,\frac{k}{m}$$

which gives the two natural frequencies as

$$\omega_1=\sqrt{0.9498\,\frac{k}{m}}=0.9746\sqrt{\frac{k}{m}}, \tag{iiia}$$

$$\omega_2=\sqrt{4.7377\,\frac{k}{m}}=2.1766\sqrt{\frac{k}{m}}. \tag{iiib}$$

In order to obtain the mode shapes, one substitutes ω_1 into Equation (ii), resulting in

$$\left(\frac{X}{\Theta}\right)_{(1)}=\frac{k_{12}}{m_{11}\omega_1^2-k_{11}}=\frac{-\frac{kL}{2}}{2m\left(0.9498\,\frac{k}{m}\right)-2k}=4.98L. \tag{iva}$$

where the subscript (1) on the lhs of Equation (iva) denotes the amplitude ratio associated with the first natural frequency. Similarly, substituting ω_2 into Equation (ii), one has

$$\left(\frac{X}{\Theta}\right)_{(2)}=\frac{k_{12}}{m_{11}\omega_1^2-k_{11}}=\frac{-\frac{kL}{2}}{2m\left(4.7377\,\frac{k}{m}\right)-2k}=-\,0.067L. \tag{ivb}$$

Equations (iva) and (ivb) give the two mode shapes of the 2-dof system.

b. For the given system with $2m = 2000$ kg, $2k = 40{,}000$ N/m, $L = 1.0$ m, Equations (iiia) and (iiib) give the two natural frequencies as

$$\omega_1=4.3584\ \text{rad/s}, \tag{va}$$

$$\omega_2=9.7342\ \text{rad/s}. \tag{vb}$$

If one assumes $\Theta L = 1$ so that Equations (iva) and (ivb) become

$$\left(\frac{X}{\Theta L}\right)_{(1)}=4.98 \tag{via}$$

$$\left(\frac{X}{\Theta L}\right)_{(2)}=-\,0.067. \tag{vib}$$

Thus, the required two normal modes can be written as

$$\Phi_1 = \begin{pmatrix} 4.980 \\ 1.000 \end{pmatrix} \tag{viia}$$

$$\Phi_2 = \begin{pmatrix} -0.067 \\ 1.000 \end{pmatrix} \tag{viib}$$

Example 2

Consider a 2-dof system whose matrix equation of motion is given as

$$\begin{bmatrix} m_{11} & 0 \\ 0 & m_{22} \end{bmatrix} \begin{pmatrix} \ddot{x}_1 \\ \ddot{x}_2 \end{pmatrix} + \begin{bmatrix} k_{11} & k_{12} \\ k_{21} & k_{22} \end{bmatrix} \begin{pmatrix} x_1 \\ x_2 \end{pmatrix} = \begin{pmatrix} 0 \\ 0 \end{pmatrix}, \tag{i}$$

where $m_{11} = m$, $m_{22} = mr^2$, $k_{11} = k_1 + k_2$, $k_{12} = k_{21} = k_2L_2 - k_1L_1$, $k_{22} = k_2L_2^2 + k_1L_1^2$, $x_1 = x$ is the translational displacement, and $x_2 = \theta$ is the angular or rotational displacement.

If $m = 4000$ kg, $k_1 = k_2 = 20{,}000$ N/m, $r = 0.8$ m, $m_{22} = mr^2 = 2560$ kg m^2, $L_1 = 1.4$ m, and $L_2 =$ 0.9 m, find the natural frequencies and normal modes of the 2-dof system.

Solution:

a. Since the system is undamped and therefore the responses have no delay or phase difference with respect to the applied force, one can write

$$\begin{pmatrix} x_1 \\ x_2 \end{pmatrix} = \begin{pmatrix} X_1 \\ X_2 \end{pmatrix} \cos\omega t = \begin{pmatrix} X \\ \Theta \end{pmatrix} \cos\omega t.$$

In this equation, X and Θ are the amplitudes of instantaneous displacements x_i.

Substituting the above solution and its derivatives into Equation (i) and upon simplification, one has

$$\begin{bmatrix} (k_{11} - m_{11}\omega^2) & k_{12} \\ k_{21} & (k_{22} - m_{22}\omega^2) \end{bmatrix} \begin{pmatrix} X \\ \Theta \end{pmatrix} = \begin{pmatrix} 0 \\ 0 \end{pmatrix} \tag{ii}$$

or writing it in a more concise form

$$[Z(\omega)] \begin{pmatrix} X_1 \\ X_2 \end{pmatrix} = \begin{pmatrix} 0 \\ 0 \end{pmatrix}$$

where the so-called *impedance matrix* has been defined by Equation (4.18) as

$$[Z(\omega)] = \begin{bmatrix} (k_{11} - m_{11}\omega^2) & k_{12} \\ k_{21} & (k_{22} - m_{22}\omega^2) \end{bmatrix}.$$

The frequency equation is

$$\begin{vmatrix} (k_{11} - m_{11}\omega^2) & k_{12} \\ k_{21} & (k_{22} - m_{22}\omega^2) \end{vmatrix} = 0.$$

Substituting the system parameters given above and writing $\lambda = \omega^2$,

$$\begin{vmatrix} 40,000-4000\lambda & 10,000 \\ 10,000 & 55,400-2560\lambda \end{vmatrix} = 0.$$

Operating on this equation, one has

$$(40,000-4000\lambda)(55,400-2560\lambda)-(10,000)^2=0.$$

This quadratic equation in λ gives $\lambda_1 = 9.2173$, $\lambda_2 = 59.8457$, so that the two natural frequencies are

$$\omega_1 = \sqrt{9.2173} = 3.036\,\text{rad/s}, \tag{iiia}$$

$$\omega_2 = \sqrt{59.8457} = 4.736\,\frac{\text{rad}}{\text{s}} \tag{iiib}$$

In order to obtain the mode shapes, one first substitutes ω_1 into Equation (ii), resulting in

$$\left(\frac{X}{\Theta}\right)_{(1)} = -3.18\,\text{m/rad} \tag{iva}$$

where the subscript (1) on the lhs of the equation denotes the amplitude ratio associated with the first natural frequency.

Similarly, substituting ω_2 into Equation (ii), one has

$$\left(\frac{X}{\Theta}\right)_{(2)} = 0.202\,\text{m/rad}. \tag{ivb}$$

Equations (iva) and (ivb) give the two mode shapes of the 2-dof system.

4.6 Vibration of Continuous Systems

- Continuous systems are those having continuously distributed mass and elasticity. These systems are known as distributed parameter models.
- To specify the position of every particle in the elastic body or continuous system, an infinite number of coordinates are necessary, and such a body therefore possesses an infinite number of dof.
- In practice, only a finite number of dof is considered.
- Basic assumptions are: the elastic body or system is homogeneous and isotropic, and it obeys Hooke's law within the elastic limit. In other words, linear dynamic system analysis is considered.

4.6.1 Vibrating Strings or Cables

A flexible string of mass per unit length μ is stretched under tension T.

- Assume that the lateral deflection y (strictly, this is not the coordinate y) of the string to be small such that the change in tension with deflection is negligible.

Consider the FBD of an elementary length dx of the string as shown in Figure 4.13.

The slope or angular displacement θ is a function of x, and therefore at point A it is $\theta(x)$, whereas at point B it is $\theta(x + dx)$.

By Taylor expansion,

$$\theta(x+dx)=\theta(x)+\frac{\partial\theta}{\partial x}dx+hot$$

where *hot* denotes the higher-order terms which will be disregarded for linear dynamic system analysis.

By Newton's law of motion, the equation of motion for the element shown in Figure 4.13 becomes

$$\mu dx\frac{\partial^2 y}{\partial t^2}=T\sin\left(\theta+\frac{\partial\theta}{\partial x}dx\right)-T\sin\theta.$$

Since θ is small

$$\sin\theta\approx\theta,\quad \sin\left(\theta+\frac{\partial\theta}{\partial x}dx\right)\approx\theta+\frac{\partial\theta}{\partial x}dx,$$

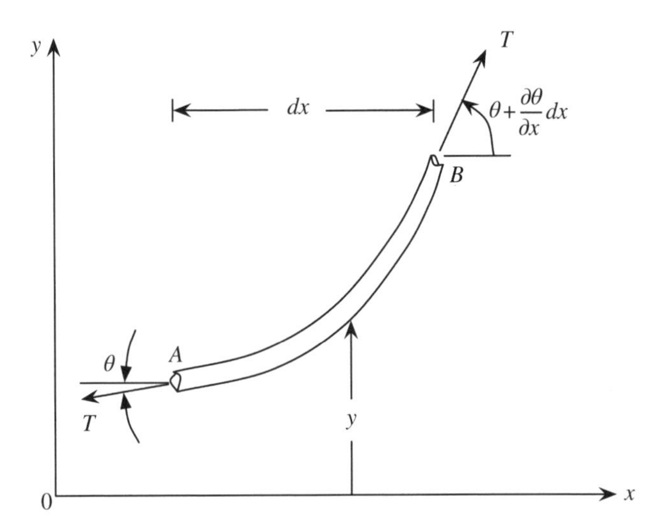

Figure 4.13 An elementary length of a vibrating string

so that the equation of motion becomes

$$\mu\, dx \frac{\partial^2 y}{\partial t^2} = T\left(\theta + \frac{\partial \theta}{\partial x} dx\right) - T\theta \quad \text{or} \quad \frac{\partial^2 y}{\partial t^2} = \left(\frac{T}{\mu}\right)\frac{\partial \theta}{\partial x}.$$

Since the slope of the string is $\theta = \frac{\partial y}{\partial x}$, the last equation can be written as

$$\frac{\partial^2 y}{\partial t^2} = c^2 \frac{\partial^2 y}{\partial x^2} \tag{4.23}$$

where $c = \sqrt{T/\mu}$ is known as the velocity of wave propagation of the string.

One approach to solving this linear second-order partial differential equation (p.d.e.) is that of the method of *separable variables*. In this method the solution is assumed to be a product of two single variable functions:

$$y(x,t) = Y(x)U(t). \tag{4.24}$$

By substituting into Equation (4.23) and writing $Y(t)$ and $U(t)$ simply as Y and U, and dividing both sides of the resulting equation by YU, one obtains

$$\frac{1}{U}\frac{d^2U}{dt^2} = \frac{c^2}{Y}\frac{d^2Y}{dx^2}. \tag{4.25}$$

Since the lhs of this equation is independent of x, whereas the rhs is independent of t, it follows that each side must be a constant. Letting this constant be $-\omega^2$, Equation (4.24) becomes

$$\frac{1}{U}\frac{d^2U}{dt^2} = \frac{c^2}{Y}\frac{d^2Y}{dx^2} = -\omega^2.$$

This gives two ordinary differential equations (o.d.e.) as $\frac{d^2U}{dt^2} + \omega^2 U = 0$, and $\frac{d^2Y}{dx^2} + \left(\frac{\omega}{c}\right)^2 Y = 0.$

The solution to the latter o.d.e. is given by

$$Y = A \sin\left(\frac{\omega}{c}x\right) + B \cos\left(\frac{\omega}{c}x\right), \tag{4.26a}$$

whereas the solution to the former o.d.e. is

$$U = C \sin \omega t + D \cos \omega t, \tag{4.26b}$$

where the arbitrary constants A, B, C, and D depend on the boundary and initial conditions of the system.

Example

If the string is stretched between two fixed points with distance L between them, the boundary conditions are $y(0, t) = y(L, t) = 0$. Determine the general solution $y(x,t)$ of the stretched string.

Solution:

With reference to Equation (4.26) and the condition that $y(0,t) = 0$, one requires that $B = 0$ so that the solution becomes

$$Y = A\sin\left(\frac{\omega}{c}x\right)(C\sin\omega t + D\cos\omega t) \tag{i}$$

The condition that $y(L,t) = 0$ leads to the equation $sin\left(\frac{\omega}{c}L\right) = 0$ which gives $\frac{\omega_n L}{c} = n\pi$, with $n = 1, 2, 3, \ldots$.

By definition, $c = f\lambda$ where f is the frequency of oscillation and λ is the wavelength. One can then write

$$\frac{\omega_n}{c} = \frac{n\pi}{L} \tag{ii}$$

Therefore, by making use of Equation (4.26a) and recalling that $B = 0$, the *mode shape* or *normal mode* or *eigenvector* of the vibrating string is given as

$$Y = A\sin\left(n\pi\frac{x}{L}\right). \tag{iii}$$

Note that this equation originates from the rhs of Equation (i).

In general, free vibration of a string contains many of the normal modes, and the equation for the displacement may be written as

$$y(x,t) = \sum_{n=1}^{\infty}\sin\left(n\pi\frac{x}{L}\right)(C_n\sin\omega_n t + D_n\cos\omega_n t). \tag{iv}$$

where the arbitrary constants C_n and D_n can be evaluated with the initial conditions of $y(x, 0)$ and $dy(x, 0)/dt$. The integer n is the number of mode shapes or eigenvectors.

4.6.2 Remarks

1. The equation of motion for longitudinal vibration of rods can be shown to be

$$\frac{\partial^2 u}{\partial t^2} = c^2\frac{\partial^2 u}{\partial x^2} \tag{4.27}$$

where u is the longitudinal deformation or displacement and $c = \sqrt{E/\rho}$ is the speed of wave propagation in the rod (E is the Young's modulus of elasticity and ρ is the mass density of the material).

2. Similarly, it can be shown that the equation of motion for torsional vibration of rods is

$$\frac{\partial^2\theta}{\partial t^2} = c_s^2\frac{\partial^2\theta}{\partial x^2}, \tag{4.28a}$$

$$c_s = \sqrt{\frac{G}{\rho}} \tag{4.28b}$$

where c_s is the velocity of propagation of shear waves in the material and G is the shear modulus of elasticity.

3. Note that Equations (4.23), (4.27), and (4.28a) are second-order p.d.e. and they have the same form. This means that the solution to Equations (4.27) and (4.28a) is similar to that for Equation (4.23).

4.7 Questions and Solutions

Eight questions are solved in this section. These include a single dof pendulum, a 2-dof airfoil section supported by translational and rotational springs, a 2-dof jig applied to sizing coal, an inverted pendulum hinged at a moving carriage, a 2-dof system modeling an unbalanced centrifugal pump on a foundation, the derivation of the equation of transverse motion of a vertical cable under gravity, the derivation of natural frequencies of a cord-spring-mass system, and a satellite consisting of two equal masses that are connected by a cable and rotating at constant angular velocity.

Example 1
A simple pendulum is pivoted at point O as shown in Figure 4E1a. Assuming the mass of the vertical rod is negligible and the oscillation is small:

a. derive the equation of motion for this simple pendulum;
b. determine the damped natural frequency of the pendulum; and

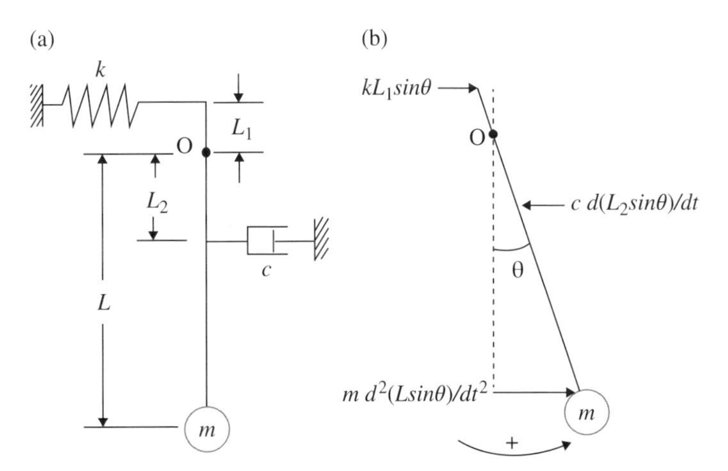

Figure 4E1 (a) A simple pendulum, and (b) FBD of (a)

c. with the result in (b), determine the value of damping coefficient c when the damped natural frequency of the pendulum becomes zero. Comment on this damping coefficient.

Solution:

a. With reference to the FBD in Figure 4E1b and summing moments about an axis through O (assume that moments are positive counter-clockwise and that the amplitude of oscillation is small such that $\sin\theta \approx \theta$),

$$\left(mL\ddot{\theta}\right)L = \sum M_o = -L_1(kL_1\theta) - c\left(L_2\dot{\theta}\right)L_2 - mg(L\theta).$$

Re-arranging terms, the equation of motion for the pendulum becomes

$$\left(mL^2\right)\ddot{\theta} + \left(cL_2^2\right)\dot{\theta} + \left(kL_1^2 + mgL\right)\theta = 0.$$

b. By definition, the undamped natural frequency of the pendulum is given by

$$\omega_n = \sqrt{\frac{kL_1^2 + mgL}{mL^2}}. \tag{i}$$

By analogy to the translational oscillation, one has

$$2\zeta\omega_n = \frac{cL_2^2}{mL^2}, \quad \text{or} \quad \zeta = \frac{cL_2^2}{2mL^2}\sqrt{\frac{mL^2}{kL_1^2 + mgL}}. \tag{ii}$$

If $\zeta < 1$, the damped natural frequency is defined by Equation (4.4) as

$$\omega_d = \omega_n\sqrt{1-\zeta^2} \quad \text{or}$$

$$\omega_d = \sqrt{\frac{kL_1^2 + mgL}{mL^2}\left[1 - \left(\frac{cL_2^2}{2mL^2}\right)^2\left(\frac{mL^2}{kL_1^2 + mgL}\right)\right]}.$$

Simplifying, it becomes

$$\omega_d = \sqrt{\left(\frac{kL_1^2 + mgL}{mL^2}\right) - \left(\frac{cL_2^2}{2mL^2}\right)^2}. \tag{iii}$$

This is the damped natural frequency of the given pendulum. Note that the second term inside the square root in Equation (iii) was incorrectly given as $\left(\frac{cL_1L_2^2}{2mL^2}\right)^2$ in the answer to

Supplementary Problem 63 in page 29 of [2]. The term $\left(\frac{cL_1 L_2^2}{2mL^2}\right)^2$ is dimensionally incorrect and therefore is inconsistent with the first term inside the square root.

c. When $\omega_d = 0$, from Equation (iii) one obtains

$$\left(\frac{cL_2^2}{2mL^2}\right)^2 = \left(\frac{kL_1^2 + mgL}{mL^2}\right).$$

Therefore:

$$c = \frac{2mL^2}{L_2^2}\sqrt{\frac{kL_1^2 + mgL}{mL^2}}. \tag{iv}$$

This is the required damping coefficient.
Equation (iv) can be written as

$$\frac{cL_2^2}{mL^2} = 2\omega_n.$$

But from Equation (ii), the rhs is equal to $2\zeta\omega_n$ or $\zeta = 1$. In other words, when the damping is critical, the damped natural frequency of the pendulum becomes zero.

Example 2

The support of an airfoil section frequently studied in a wind tunnel is modeled as a 2-dof system with a linear spring k and a torsional spring k_T. This simple model may also be applied to study turbulence control over a wing of an airplane. A sketch of this 2-dof system is included in Figure 4E2. It is given that the mass of the airfoil is m and J_o is the moment of inertia about O.

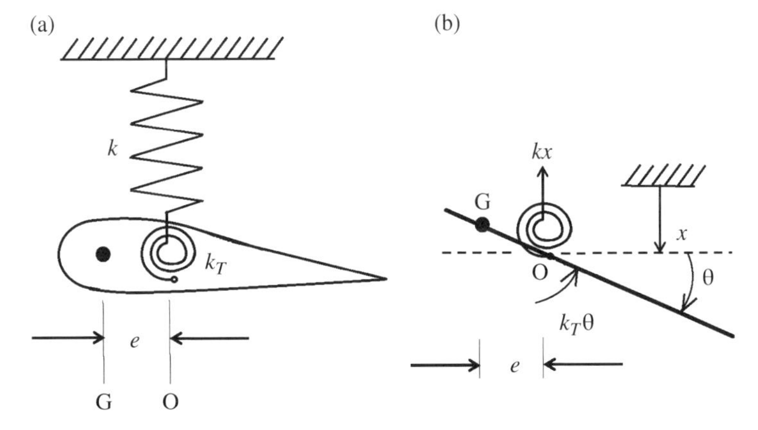

Figure 4E2 (a) An airfoil section with spring support and (b) its FBD

If the center of gravity G of the section is a distance e ahead of the point of support O, for small oscillations derive the differential equations of motion for the system.

Solution:
Applying Newton's law of motion along the y-direction and with the FBD shown in Figure 4E2b, and remembering that for small oscillation $\sin\theta \approx \theta$, one has

$$m\ddot{y} = \sum F_y = -ky + me\ddot{\theta} \quad \text{or} \quad m(\ddot{y} - e\ddot{\theta}) + ky = 0.$$

Summing moments about G, it gives

$$J_G\ddot{\theta} = -k_T\theta - e(ky) \quad \text{or} \quad J_G\ddot{\theta} + k_T\theta + e(ky) = 0,$$

where J_G is the moment of inertia about G, $J_o = J_G + me^2$ with J_o being the moment of inertia about O.

The above two required equations may be expressed in matrix form as

$$\begin{bmatrix} m & -me \\ 0 & J_o - me^2 \end{bmatrix}\begin{pmatrix} \ddot{y} \\ \ddot{\theta} \end{pmatrix} + \begin{bmatrix} k & 0 \\ ek & k_T \end{bmatrix}\begin{pmatrix} y \\ \theta \end{pmatrix} = \begin{pmatrix} 0 \\ 0 \end{pmatrix}.$$

With reference to this matrix equation it is clear that if the center of gravity G coincides with the point of support O such that $e = 0$, the translational dof is uncoupled from the rotational dof. Additionally, if $e \neq 0$ and one sums all moments about O instead of G, one can show that the resulting equations of motion in matrix form are

$$\begin{bmatrix} m & -me \\ -me & J_o \end{bmatrix}\begin{pmatrix} \ddot{y} \\ \ddot{\theta} \end{pmatrix} + \begin{bmatrix} k & 0 \\ 0 & k_T \end{bmatrix}\begin{pmatrix} y \\ \theta \end{pmatrix} = \begin{pmatrix} 0 \\ 0 \end{pmatrix}.$$

In this equation the mass and stiffness matrices are symmetric. This is called a system with dynamic coupling because the mass matrix is not diagonal.

Example 3
A jig applied to sizing coal contains a screen that reciprocates with a frequency of 400 rpm. It is shown in Figure 4.10a. The jig has a mass of 200 kg and a fundamental frequency of 20 Hz. If an absorber with a mass of 30.5 kg is to be attached to eliminate the vibration of the jig frame,

a. determine the absorber spring stiffness, and
b. find the two natural frequencies of the resulting system.

Solution:
The conceptual model of this 2-dof system and its FBD have already been presented in Figure 4.12. Fundamental frequency of the jig is $\Omega_1 = \sqrt{\dfrac{k_1}{m_1}} = 20(2\pi)\text{rad/s}$, and mass of the jig $m_1 = 200\,\text{kg}$.

Therefore, the stiffness, $k_1 = [20(2\pi)]^2 \times 200\,\text{N/m} = 3.1583 \times 10^6\,\text{N/m}$.

With reference to Figure 4.12, the present 2-dof system has the amplitudes of the displacement responses defined by Equations (4.21a) and (4.21b). Thus,

$$X_1 = \frac{(k_{22} - m_2\omega^2)F_1}{m_1 m_2 \left(\omega_1^2 - \omega^2\right)\left(\omega_2^2 - \omega^2\right)}, \tag{4.21a}$$

$$X_2 = \frac{-k_{21}F_1}{m_1 m_2 \left(\omega_1^2 - \omega^2\right)\left(\omega_2^2 - \omega^2\right)}. \tag{4.21b}$$

a. For the jig frame to have no steady-state vibration when the absorber is attached, $X_1 = 0$. From Equation (4.21a), one requires that $k_{22} = m_2\omega^2$.
   The equations of motion for this system are given by

$$\begin{bmatrix} m_1 & 0 \\ 0 & m_2 \end{bmatrix}\begin{pmatrix} \ddot{x}_1 \\ \ddot{x}_2 \end{pmatrix} + \begin{bmatrix} (k_1 + k_2) & -k_2 \\ -k_2 & k_2 \end{bmatrix}\begin{pmatrix} x_1 \\ x_2 \end{pmatrix} = \begin{pmatrix} 0 \\ 0 \end{pmatrix}.$$

   Hence, $k_{22} = k_2 = m_2\omega^2$. But the absorber mass $m_2 = 30.5$ kg and the applied frequency $\omega =$ 400 rpm.
   Therefore, $\omega = 400(2\pi)/(60)$ rad/s = 41.8879 rad/s..
   This gives the absorber stiffness,

$$k_2 = m_2\omega^2 = 30.5 \times 41.8879^2\,\text{N/m}$$
$$= 30.5 \times 41.8879^2\,\text{N/m} = 53515.1831\,\text{N/m}.$$

b. Substituting the above parameters, $m_1 = 200$ kg, $m_2 = 30.5$ kg, $k_1 = 3.1583\times10^6$ N/m, $k_2 =$ 53515.1831 N/m, into Equations (A2a) and (A2b) of Appendix 4A, one obtains

$$\omega_1 = 41.54\,\text{rad/s}, \quad \omega_2 = 126.84\,\text{rad/s}.$$

   These are the two required natural frequencies of the system.

Example 4

The inverted pendulum hinged at O of a carriage shown in Figure 4E4 is a classical system applied to the studies of stability in control engineering. The inverted pendulum consisted of a uniform rigid rod of length 2 and mass m. The mass of the carriage is M, which is excited by a force $f(t)$ along the horizontal direction. The friction coefficients of rotary motion and linear motion are, respectively, c_r and c. Derive the equations of motion for the inverted pendulum and carriage system.

Solution:

Let x and θ be the horizontal displacement of the pivot O on the carriage and angular displacement or rotation of the pendulum about O, respectively.

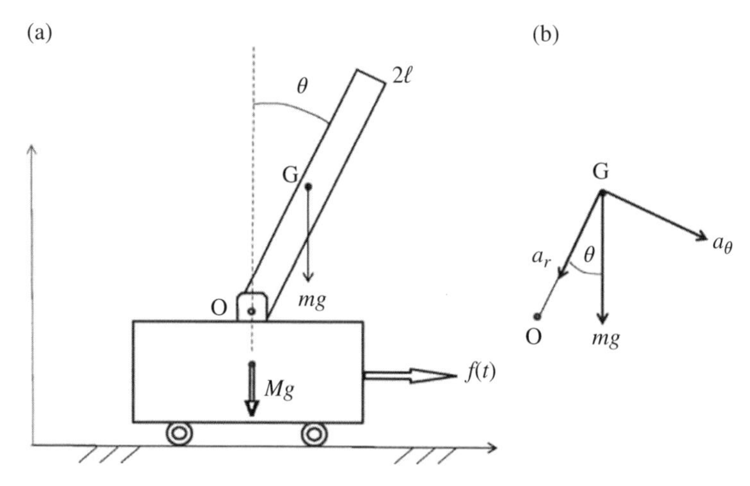

Figure 4E4 (a) The inverted pendulum and (b) FBD at G of (a)

At the center of gravity G the radial acceleration towards O is $a_r = \ddot{\ell} - \ell\dot{\theta}^2 = -\ell\dot{\theta}^2$, since the length of the pendulum is constant.

The transversal acceleration at G as indicated in the FBD is $a_\theta = \ell\ddot{\theta} - 2\dot{\ell}\dot{\theta} = \ell\ddot{\theta}$, since the length of the pendulum is constant.

With reference to the FBD in Figure 4E4b and applying Newton's law of motion in the horizontal direction, one has

$$(M+m)\ddot{x} + m\left(\ell\ddot{\theta}\cos\theta\right) - m\left(\ell\dot{\theta}^2\sin\theta\right) = \sum F_x = f(t) - c\dot{x}.$$

Re-arranging, one obtains

$$(M+m)\ddot{x} + m(\ell\cos\theta)\ddot{\theta} - m(\ell\sin\theta)\dot{\theta}^2 + c\dot{x} = f(t). \tag{i}$$

Taking moments about O, with clockwise moments being positive,

$$\left(\frac{m\ell^2}{3} + m\ell^2\right)\ddot{\theta} + m\ddot{x}(\ell\cos\theta) = -c_r\dot{\theta} + mg(\ell\sin\theta).$$

Re-arranging, one may expressed this in a more familiar form as

$$\left(\frac{4m\ell^2}{3}\right)\ddot{\theta} + m(\ell\cos\theta)\ddot{x} + c_r\dot{\theta} = (mg\ell)\sin\theta. \tag{ii}$$

Equations (i) and (ii) are highly nonlinear. However, if a small oscillation is assumed such that $sin\theta \approx \theta$, $cos\theta \approx 1$, and $\dot{\theta}^2$ can be disregarded, then the two equations of motion reduce to

$$(M+m)\ddot{x}+m\ell\ddot{\theta}+c\dot{x}=f(t). \tag{iii}$$

$$m\ell\ddot{x}+\left(\frac{4m\ell^2}{3}\right)\ddot{\theta}+c_r\dot{\theta}-(mg\ell)\theta=0. \tag{iv}$$

Writing in matrix form, one has

$$\begin{bmatrix} M+m & m\ell \\ m\ell & \dfrac{4m\ell^2}{3} \end{bmatrix}\begin{pmatrix}\ddot{x}\\ \ddot{\theta}\end{pmatrix}+\begin{pmatrix} c & 0\\ 0 & c_r\end{pmatrix}\begin{pmatrix}\dot{x}\\ \dot{\theta}\end{pmatrix}+\begin{bmatrix}0 & 0\\ 0 & -mg\ell\end{bmatrix}\begin{pmatrix}x\\ \theta\end{pmatrix}=\begin{pmatrix}f(t)\\ 0\end{pmatrix}. \tag{v}$$

This system is said to possess *dynamic coupling* because the off-diagonal terms in the inertia matrix are not zero. However, it is not a 2-dof oscillatory system in the usual sense. Owing to the fact that the diagonal term associated with θ is negative, this system is unstable. Furthermore, it may be appropriate to point out that the dimensions of x and θ are different. Therefore, if numerical integration techniques are to be applied to the solution of Equation (v), it may be more convenient to rewrite it as

$$\begin{bmatrix} M+m & m \\ m & \dfrac{4m}{3} \end{bmatrix}\begin{pmatrix}\ddot{x}\\ \ddot{\theta}\ell\end{pmatrix}+\begin{pmatrix} c & 0\\ 0 & c_r/\ell^2\end{pmatrix}\begin{pmatrix}\dot{x}\\ \dot{\theta}\ell\end{pmatrix}+\begin{bmatrix}0 & 0\\ 0 & -\dfrac{mg}{\ell}\end{bmatrix}\begin{pmatrix}x\\ \theta\ell\end{pmatrix}=\begin{pmatrix}f(t)\\ 0\end{pmatrix}. \tag{vi}$$

Now, the elements in the displacement, velocity, and acceleration vectors have identical dimensions.

Example 5

A centrifugal pump is supported on a rigid foundation of mass m_2 through isolator springs of stiffness k_1. This centrifugal pump has an unbalance of *me*. The 2-dof system is shown in Figure 4E5.

If the soil stiffness and damping are, respectively, k_2 and c_2, find the displacements of the pump and foundation for the system in which $m = 0.25$ kg, $e = 0.1525$ m (or 15.25 cm), $k_1 = 230$ MN/m, $k_2 = 115$ MN/m, $m_1 = 364$ kg, $m_2 = 909$ kg, $c_2 = 0.25$ MNs/m, and the speed of the pump is 1200 rpm.

Solution:

a. ***Given parameters***

Given that the speed of the pump is 1200 rpm, the applied frequency $\omega = 2\pi(1200)/60$ rad/s = 125.66 rad/s.

The applied force due to unbalance,

$$\begin{aligned} f_1 &= me\omega^2\sin\omega t \\ &= 0.25(0.1525)(125.66)^2\sin 125.66t\ \text{N} \\ &= 602\sin 125.66t\ \text{N} \end{aligned}$$

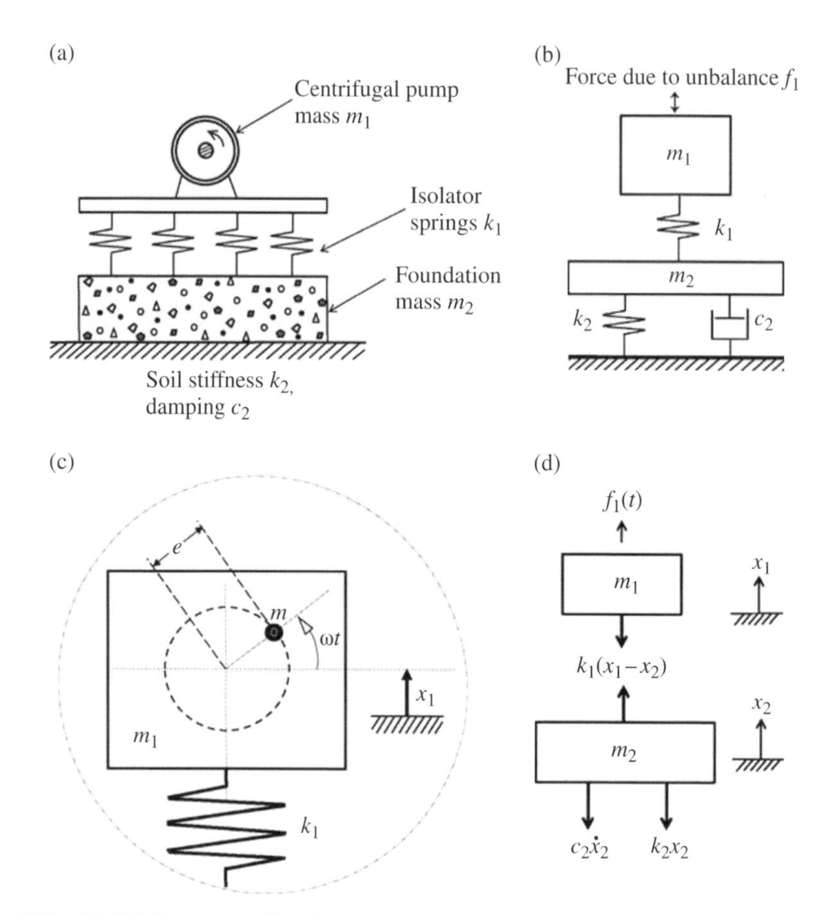

Figure 4E5 (a) Unbalanced centrifugal pump; (b) conceptual model; (c) kinematics of unbalance mass; and (d) FBD of system

The stiffness of the isolator springs, k_1 = 230 MN/m = 2.3×10^8 N/m.
The stiffness of the soil, k_2 = 115 MN/m = 1.15×10^8 N/m.
The mass of centrifugal pump, m_1 = 364 kg, and the mass of the foundation, m_2 = 909 kg.
The damping coefficient of soil, c_2 = 0.25 MNs/m = 0.25×10^6 Ns/m.

b. ***Equations of motion***

Let x_1 and x_2 be the absolute equilibrium vertical displacements of the centrifugal pump (not including the unbalance mass m) and foundation, respectively.

With reference to Figure 4E5c, the vertical displacement of m is $x_1 + e \sin \omega t$. With reference to this equation and the FBD in Figure 4E5d, by applying Newton's second law of motion for the centrifugal pump of mass m_1 and the unbalance mass m, one has

$$m_1\ddot{x}_1 + m\frac{d^2(x_1 + e\sin\omega t)}{dt^2} = \sum F = -k_1(x_1 - x_2).$$

Re-arranging terms, it becomes

$$(m_1+m)\ddot{x}_1+k_1x_1-k_1x_2=f_1=F_1\sin\omega t,\quad F_1=me\omega^2.$$

As m is much smaller than m_1, one can simply write

$$m_1\ddot{x}_1+k_1x_1-k_1x_2=f_1=F_1\sin\omega t.$$

Similarly, applying Newton's second law of motion for the foundation m_2, the equation of motion is given by

$$m_2\ddot{x}_2=-c_2\dot{x}_2-k_2x_2+k_1(x_1-x_2).$$

After re-arranging terms, it leads to

$$m_2\ddot{x}_2+c_2\dot{x}_2-k_1x_1+x_2(k_1+k_2)=0.$$

For subsequent analysis, one can express the two equations of motion in matrix form. Thus, the above two equations of motion can be written as

$$\begin{bmatrix} m_1 & 0 \\ 0 & m_2 \end{bmatrix}\begin{pmatrix} \ddot{x}_1 \\ \ddot{x}_2 \end{pmatrix}+\begin{pmatrix} 0 & 0 \\ 0 & c_2 \end{pmatrix}\begin{pmatrix} \dot{x}_1 \\ \dot{x}_2 \end{pmatrix}+\begin{bmatrix} k_1 & -k_1 \\ -k_1 & k_{22} \end{bmatrix}\begin{pmatrix} x_1 \\ x_2 \end{pmatrix}=\begin{pmatrix} F_1 \\ 0 \end{pmatrix}\sin\omega t \tag{i}$$

in which $k_{22}=k_1+k_2$.

c. ***Solution for responses***

For convenience, one can write $\sin\omega t$ as $Im\{e^{i\omega t}\}$ such that $x_i=X_i\sin\omega t$, $i=1, 2$, can be written as

$$x_i=Im\{X_ie^{i\omega t}\},$$

where $Im\{.\}$ denotes the imaginary part of the enclosing complex variable.

Without loss of generality, one can disregard $Im\{.\}$ for the time being so that the displacements can be written as $x_i=X_ie^{i\omega t}$. Then, the velocity and acceleration are given by $\dot{x}_i=i\omega X_ie^{i\omega t}$, $\ddot{x}_i=-\omega^2X_ie^{i\omega t}$.

Substituting the above expressions into Equation (i) and canceling $e^{i\omega t}$ on both sides of the resulting matrix equation of motion, one has

$$\begin{bmatrix} -\omega^2m_1 & 0 \\ 0 & -\omega^2m_2 \end{bmatrix}\begin{pmatrix} X_1 \\ X_2 \end{pmatrix}+\begin{pmatrix} 0 & 0 \\ 0 & i\omega c_2 \end{pmatrix}\begin{pmatrix} X_1 \\ X_2 \end{pmatrix}+\begin{bmatrix} k_1 & -k_1 \\ -k_1 & k_{22} \end{bmatrix}\begin{pmatrix} X_1 \\ X_2 \end{pmatrix}=\begin{pmatrix} F_1 \\ 0 \end{pmatrix}.$$

Upon simplification, one arrives at

$$\begin{bmatrix} k_1-\omega^2m_1 & -k_1 \\ -k_1 & k_1+k_2-\omega^2m_2+i\omega c_2 \end{bmatrix}\begin{pmatrix} X_1 \\ X_2 \end{pmatrix}=\begin{pmatrix} F_1 \\ 0 \end{pmatrix}. \tag{ii}$$

This is identical in form to Equation (4.18). Thus, one has

$$[Z(\omega)] = \begin{bmatrix} k_1-\omega^2 m_1 & -k_1 \\ -k_1 & k_1+k_2-\omega^2 m_2+i\omega c_2 \end{bmatrix}. \tag{iii}$$

The form of the impedance or coefficient matrix $[Z(\omega)]$ is identical to that in Equation (4.18) for a 2-dof system. That is, the elements of the coefficient matrix are

$$z_{11}=k_1-\omega^2 m_1, \quad z_{12}=z_{21}=-k_1, \quad z_{22}=k_1+k_2-\omega^2 m_2+i\omega c_2.$$

From Equation (4.19), the amplitudes of the displacements are

$$X_1=\frac{z_{22}F_1}{|Z(\omega)|}=\frac{z_{22}F_1}{z_{11}z_{22}-z_{12}z_{21}};$$

therefore

$$X_1=\frac{k_1F_1}{(k_1-\omega^2 m_1)\,(k_1+k_2-\omega^2 m_2+i\omega c_2)-k_1^2}, \tag{iv}$$

and

$$X_2=\frac{-z_{12}F_1}{|Z(\omega)|}=\frac{(k_1+k_2-\omega^2 m_2+i\omega c_2)F_1}{(k_1-\omega^2 m_1)\,(k_1+k_2-\omega^2 m_2+i\omega c_2)-k_1^2}. \tag{v}$$

Now, substituting the given data into the numerator and denominator terms in Equation (iv), one has

$$\begin{aligned}
&k_1+k_2-\omega^2 m_2+i\omega c_2\\
&\quad=\left[(2.3+1.15)\times10^8-125.66^2(909)+i(125.66\times0.25)\times10^6\right]602\\
&\quad=19.904919\times10^{10}+i(1.891183)\times10^{10}\ \text{N/m}\\
&(k_1+k_2-\omega^2 m_2+i\omega c_2)\,(k_1-\omega^2 m_1)-k_1^2\\
&\quad=7.4148238\times10^{16}-5.29\times10^{16}+i(0.7044886)\times10^{16}\ (\text{N/m})^2\\
&\quad=(2.1248238+i0.7044886)\times10^{16}\ (\text{N/m})^2.
\end{aligned}$$

Therefore,

$$X_1=\frac{19.904919\times10^{10}+i(1.891183)\times10^{10}}{(2.1248238+i0.7044886)\times10^{16}}\text{m}.$$

Rationalizing, one has

$$X_1=\frac{(19.904919+i1.891183)\,(2.1248238-i0.7044886)\times10^{-6}}{(2.1248238+i0.7044886)\,(2.1248238-i0.7044886)}\text{m}.$$

$$\begin{aligned}
X_1&=\frac{(43.5801-i\,10.0044)\times10^{-6}}{2.1248238^2+0.7044886^2}\ \text{m}\\
&=(8.6870-i\,1.9942)\times10^{-6}\ \text{m}.
\end{aligned}$$

From the theory of complex variables,

$$X_1 = (8.6870 - i\,1.9942) \times 10^{-6}\,\text{m} \;\; = X_{10} e^{i\varphi_1}\,\text{m},$$

where

$$X_{10} = \sqrt{8.6870^2 + 1.9942^2} \times 10^{-6}\,\text{m}, \quad \varphi_1 = \tan^{-1}\left(\frac{-1.9942}{8.6870}\right).$$

Therefore,

$$X_{10} = 8.9129 \times 10^{-6}\,\text{m}, \quad \varphi_1 = -12.9^\circ.$$

Thus,

$$\begin{aligned} x_1 &= Im\{X_1 e^{i\omega t}\} = Im\{X_{10} e^{i(\omega t + \varphi_1)}\} = X_{10}\sin(\omega t + \varphi_1). \\ x_1 &= 8.9129 \times 10^{-6} \sin(\omega t - 12.9^\circ)\,\text{m}. \end{aligned} \tag{vi}$$

Similarly, applying Equation (v), one obtains

$$X_2 = \frac{13.8460 \times 10^{-6}}{(2.1248238 + i0.7044886)}\text{m}.$$

Rationalizing, one has

$$\begin{aligned} X_2 &= \frac{13.8460\,(2.1248238 - i0.7044886) \times 10^{-6}}{(2.1248238 + i0.7044886)\,(2.1248238 - i0.7044886)}\text{m}. \\ &= (5.8660 - i\,1.9449) \times 10^{-6}\text{m}. \end{aligned}$$

Again, from the theory of complex variables,

$$X_2 = (5.8660 - i\,1.9449) \times 10^{-6}\,\text{m} = X_{20} e^{i\varphi_2}\,\text{m},$$

where

$$X_{20} = \sqrt{5.8660^2 + 1.9942^2} \times 10^{-6}\,\text{m}, \quad \varphi_2 = \tan^{-1}\left(\frac{-1.9942}{5.8660}\right).$$

Therefore,

$$X_{20} = 6.18 \times 10^{-6}\,\text{m}, \quad \varphi_1 = -18.3^\circ.$$

Thus,

$$\begin{aligned} x_2 &= Im\{X_2 e^{i\omega t}\} = Im\{X_{20} e^{i(\omega t + \varphi_2)}\} = X_{20}\sin(\omega t + \varphi_2). \\ x_2 &= 6.1800 \times 10^{-6} \sin(\omega t - 18.3^\circ)\,\text{m}. \end{aligned} \tag{vii}$$

Example 6

The upper end of a flexible cable is supported by a stationary helicopter (not included in the figure) in air, as shown in Figure 4E6.

The cable is free to oscillate under the influence of gravity. Assuming small oscillation, show that the equation of lateral (in the horizontal direction with elastic displacement or deformation u) motion is

$$\frac{\partial^2 u}{\partial t^2} = g\left(y\frac{\partial^2 u}{\partial y^2} + \frac{\partial u}{\partial y}\right).$$

Solution:

With the FBD indicated in Figure 4E6b, considering the vertical forces for the elementary length dy of the cable, one obtains

$$(T + dT)\cos\left(\theta + \frac{\partial \theta}{\partial y}dy\right) = T\cos\theta + \mu g dy,$$

where μ is the length density in mass per unit length, and θ is the angular displacement as indicated in Figure 4E6b.

For small oscillations, the angular displacement θ is small such that the above equation reduces to

$$dT = \mu g dy \quad \text{or} \quad T = \mu g y. \tag{i}$$

Now, consider forces along the horizontal direction by using Newton's law of motion (that is, along x here):

$$(\mu dy)\frac{\partial^2 u}{\partial t^2} = (T + dT)\sin\left(\theta + \frac{\partial \theta}{\partial y}dy\right) - T\sin\theta.$$

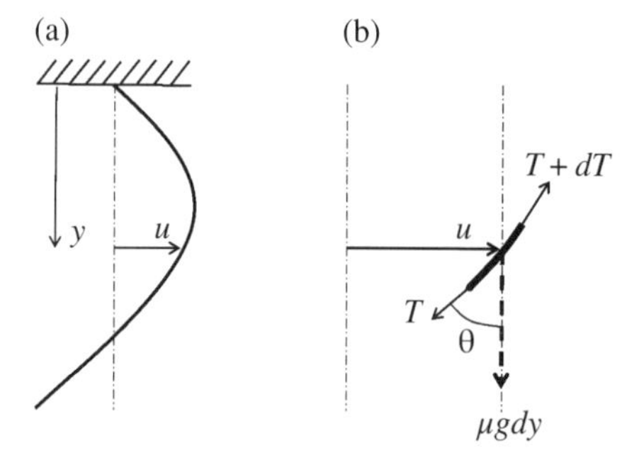

Figure 4E6 (a) A flexible cable fixed at the upper end and (b) FBD of elementary length dy of the cable

For small oscillations,

$$(\mu dy)\frac{\partial^2 u}{\partial t^2} = (T+dT)\left(\theta + \frac{\partial \theta}{\partial y}dy\right) - T\theta$$

$$(\mu dy)\frac{\partial^2 u}{\partial t^2} = T\theta + T\frac{\partial \theta}{\partial y}dy + dT\theta + dT\frac{\partial \theta}{\partial y}dy - T\theta$$

$$(\mu dy)\frac{\partial^2 u}{\partial t^2} = T\frac{\partial \theta}{\partial y}dy + \left(\theta + \frac{\partial \theta}{\partial y}dy\right)dT.$$

Substituting for Equation (i), the above equation becomes

$$(\mu dy)\frac{\partial^2 u}{\partial t^2} = T\frac{\partial \theta}{\partial y}dy + \left(\theta + \frac{\partial \theta}{\partial y}dy\right)\mu g dy.$$

It may be appropriate to note that at this stage one should not delete the common factor dy on both sides of the equation. If such action is taken one would arrive at an erroneous equation. In order to obtain the correct equation of motion, one should expand the rhs of the last equation and re-arrange terms to give

$$(\mu dy)\frac{\partial^2 u}{\partial t^2} = T\frac{\partial \theta}{\partial y}dy + (\theta \mu g)dy + \frac{\partial \theta}{\partial y}(dy)^2 \mu g.$$

Since only small oscillations are considered, the higher-order term associated with $(dy)^2$ can be disregarded such that the above equation reduces to

$$(\mu dy)\frac{\partial^2 u}{\partial t^2} = T\frac{\partial \theta}{\partial y}dy + (\theta \mu g)dy.$$

Now, the common factor dy may be canceled on both sides so that it becomes

$$\mu\frac{\partial^2 u}{\partial t^2} = T\frac{\partial \theta}{\partial y} + (\theta \mu g). \tag{ii}$$

From Equation (i), $T = \mu g y$ and recall that $\theta = \frac{\partial u}{\partial y}$; one therefore has

$$\mu\frac{\partial^2 u}{\partial t^2} = \mu g y\frac{\partial^2 u}{\partial y^2} + \left(\frac{\partial u}{\partial y}\mu g\right) \quad \text{or}$$

$$\frac{\partial^2 u}{\partial t^2} = g\left(y\frac{\partial^2 u}{\partial y^2} + \frac{\partial u}{\partial y}\right). \tag{iii}$$

This is the required equation of motion. Note that the rhs of Equation (iii) contains the second order-partial derivative with a time-dependent coefficient.

Example 7

A cord of length P and mass per unit length μ is under tension T, with the left end fixed and the right end attached to a spring-mass system as shown in Figure 4E7. Find the equation for the natural frequencies of this cord-spring-mass system.

Solution:

For a general solution of the transverse vibration of the cord, one can use

$$y(x,t) = \left[A\sin\left(\frac{\omega x}{c}\right) + B\cos\left(\frac{\omega x}{c}\right)\right]\sin\omega t,$$

where $c = \sqrt{T/\mu}$ is the velocity of wave propagation along the cord.

At $x = 0$, $y(0, t) = 0$, because it is fixed at this end. This leads to $B = 0$.

At $x = \ell$, $y(\ell, t) = Y$.

With reference to the FBD (Figure 4E7b) and by Newton's law of motion applied at $x = \ell$ to the spring-mass system, one has

$$m\frac{\partial^2 Y}{\partial t^2} = -kY - T\sin\theta|_{x=\ell}.$$

Since θ is small, $\sin\theta \approx \theta$. Also, $\theta = \frac{\partial y}{\partial x}$. Thus, the last equation can be written as

$$m\frac{\partial^2 Y}{\partial t^2} + kY = -T\frac{\partial y}{\partial x}\bigg|_{x=\ell}. \tag{i}$$

As small oscillation or linear vibration is assumed, the method of separable variables can be applied. Thus, one can write

$$y(x,t) = W(x)\sin\omega t, \quad \text{where } W(x) = A\sin\left(\frac{\omega x}{c}\right).$$

$$\frac{\partial y}{\partial x}\bigg|_{x=\ell} = A\left(\frac{\omega}{c}\right)\cos\left(\frac{\omega\ell}{c}\right)\sin\omega t.$$

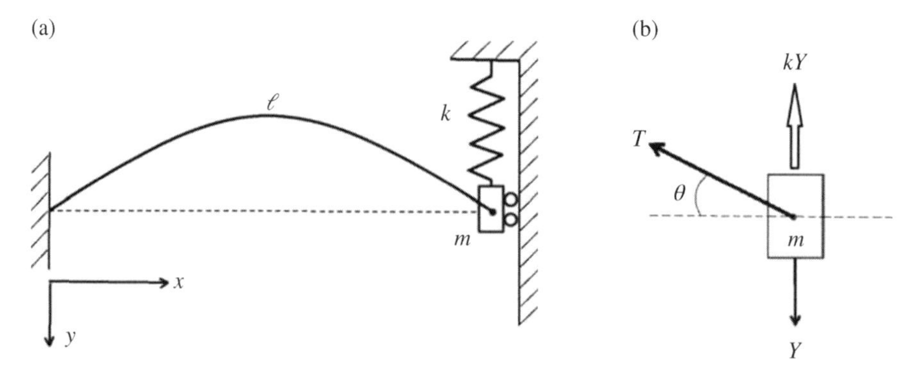

Figure 4E7 (a) A cord fixed at one end and attached to a mass-spring device at the other end, and (b) FBD of the mass-spring device

Substituting this equation into Equation (i),

$$m\frac{\partial^2 Y}{\partial t^2}+kY=-T\left(\frac{\omega}{c}\right)A\cos\left(\frac{\omega\ell}{c}\right)\sin\omega t \quad \text{or}$$

$$-m\omega^2 W(\ell)\sin\omega t+kW(\ell)\sin\omega t=-T\left(\frac{\omega}{c}\right)A\cos\left(\frac{\omega\ell}{c}\right)\sin\omega t.$$

Simplifying,

$$W(\ell)=\frac{-T\left(\frac{\omega}{c}\right)A\cos\left(\frac{\omega\ell}{c}\right)}{k-m\omega^2}. \tag{ii}$$

Equation (ii) is required for the determination of natural frequencies of the given cord-spring-mass system.

Example 8

A satellite is considered as a system consisting of two equal masses of m, connected by a cable of length $2L$. This simplified model is shown in Figure 4E8a. The system rotates in space with a constant angular velocity Ω. If the variation in the cable tension is disregarded, show that the differential equation of lateral motion of the cable is

$$\frac{\partial^2 y}{\partial x^2}=\frac{\rho}{mL\Omega^2}\left(\frac{\partial^2 y}{\partial t^2}-\Omega^2 y\right).$$

Solution:

Consider the kinematics of the satellite as shown in Figure 4E8b. Note that when the angular velocity is zero the system still has vibration, and therefore one still has motion, y and $\ddot{y}$. Thus, if one assumes the mode shape or motion pattern as that given in Figure 4E8b, the acceleration in the y-direction is $\ddot{y}-\Omega^2 y$ for $\Omega\neq 0$. The following is concerned with the derivation of this quantity.

The velocity at point B of the cable is

$$\boldsymbol{v}_B=\boldsymbol{v}_A+\boldsymbol{v}_{B/A}.$$

But the differentiation of the relative velocity with respect to time t is

$$\frac{d\boldsymbol{v}_{B/A}}{dt}=\frac{d\boldsymbol{\Omega}\times\mathbf{r}}{dt}=\frac{d\boldsymbol{\Omega}}{dt}\times\mathbf{r}+\boldsymbol{\Omega}\times\frac{d\mathbf{r}}{dt}.$$

Since $\boldsymbol{\Omega}$ is constant and the first term on the rhs becomes zero, then

$$\frac{d\boldsymbol{v}_{B/A}}{dt}=\boldsymbol{\Omega}\times\frac{d\mathbf{r}}{dt}$$

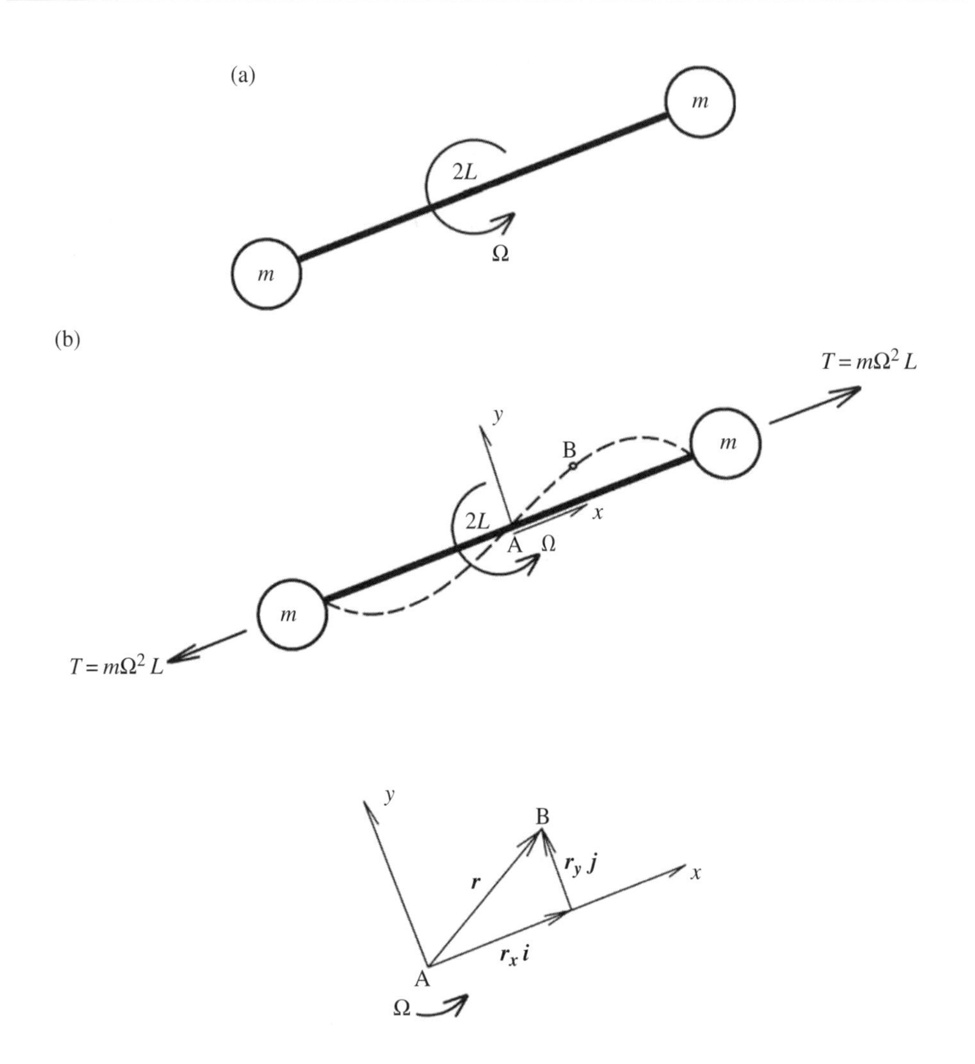

Figure 4E8 (a) A satellite model rotating with an angular velocity; (b) kinematics of the satellite model

where

$$\frac{d\mathbf{r}}{dt} = r_x \frac{d\boldsymbol{i}}{dt} + r_y \frac{d\boldsymbol{j}}{dt} = \boldsymbol{\Omega} \times \left(r_x \boldsymbol{i} + r_y \boldsymbol{j}\right) = \boldsymbol{\Omega} \times \mathbf{r}.$$

Note that Axyz is the rotating frame of reference in which the z-axis is perpendicular to the x-y plane and pointing outward from the x-y plane in accordance with the right-hand screw rule. Thus,

$$\frac{d\boldsymbol{v}_{B/A}}{dt} = \boldsymbol{\Omega} \times (\boldsymbol{\Omega} \times \mathbf{r}).$$

Operating on this equation, one has

$$\frac{d\boldsymbol{v}_{B/A}}{dt} = \boldsymbol{\Omega} \times (\boldsymbol{\Omega} \times \mathbf{r}) = \Omega \boldsymbol{k} \times \left[\Omega \boldsymbol{k} \times \left(r_x \boldsymbol{i} + r_y \boldsymbol{j}\right)\right]$$
$$= \Omega \boldsymbol{k} \times \left(\Omega r_x \boldsymbol{j} - \Omega r_y \boldsymbol{i}\right) = \Omega^2 \left(-r_x \boldsymbol{i} - r_y \boldsymbol{j}\right).$$

Note that according to the symbol given in the example, $r_x = x$ and $r_y = y$. Since in the present example only lateral motion is required, only the component in the y direction is of interest. This component is in addition to the acceleration term $\frac{\partial^2 y}{\partial t^2}$ or $\ddot{y}$ in Equation (4.23) in which $\Omega = 0$. In other words, the total acceleration in the y direction is now

$$\frac{\partial^2 y}{\partial t^2} - \Omega^2 y.$$

With reference to Equation (4.23), one has

$$T \frac{\partial^2 y}{\partial x^2} = \rho \left(\frac{\partial^2 y}{\partial t^2} - \Omega^2 y \right).$$

Since the tension in the cable is $T = mL\Omega^2$, this equation becomes

$$\frac{\partial^2 y}{\partial x^2} = \frac{\rho}{mL\Omega^2} \left(\frac{\partial^2 y}{\partial t^2} - \Omega^2 y \right).$$

This is the required equation.

Appendix 4A: Proof of Equation (4.19b)

Consider first the determinant of the impedance matrix $[Z(\omega)]$:

$$|Z(\omega)| = \begin{vmatrix} (k_{11} - m_1\omega^2) & k_{12} \\ k_{21} & (k_{22} - m_2\omega^2) \end{vmatrix},$$

$$|Z(\omega)| = (k_{11} - m_1\omega^2)(k_{22} - m_2\omega^2) - k_{12}k_{21}.$$

Therefore, the characteristic equation (c.e.) of the system is

$$k_{11}k_{22} - k_{11}m_2\omega^2 - m_1\omega^2 k_{22} + m_1 m_2 \omega^4 - k_{12}k_{21} = 0.$$

Re-arranging into a polynomial in ω, one has

$$(m_1 m_2)\omega^4 - (k_{11}m_2 + k_{22}m_1)\omega^2 + (k_{11}k_{22} - k_{12}k_{21}) = 0.$$

Writing as

$$a\omega^4 + b\omega^2 + c = 0 \tag{A1}$$

where $a = m_1m_2$, $b = -(k_{11}m_2 + k_{22}m_1)$, and $c = k_{11}k_{22} - k_{12}k_{21}$.
Solving for Equation (A1), one obtains

$$\omega_1^2 = \frac{-b + \sqrt{b^2 - 4ac}}{2a} \tag{A2a}$$

$$\omega_2^2 = \frac{-b - \sqrt{b^2 - 4ac}}{2a}. \tag{A2b}$$

The objective now is to show that Equation (4.20) is true, that is,

$$|Z(\omega)| = m_1m_2\left(\omega_1^2 - \omega^2\right)\left(\omega_2^2 - \omega^2\right).$$

To this end, one expands the rhs to give $(m_1m_2)\omega^4 + m_1m_2\left(-\omega_1^2 - \omega_2^2\right)\omega^2 + m_1m_2\,\omega_1^2\omega_2^2$. Since ω_1 and ω_2 are the natural frequencies of the 2-dof system, the last equation should agree with Equation (A1). This means that one has to show that

$$b = m_1m_2\left(-\omega_1^2 - \omega_2^2\right) \tag{A3a}$$

$$c = m_1m_2\omega_1^2\omega_2^2 \tag{A3b}$$

Now, substituting Equations (A2a) and (A2b) into the rhs of Equation (A3a) results in

$$-m_1m_2\left(\frac{-2b}{2a}\right) = m_1m_2\left(\frac{b}{a}\right).$$

However, from Equation (A1), $a = m_1m_2$ and therefore the rhs of Equation (A3a) = lhs of Equation (A3a).

Substituting Equations (A2a) and (A2b) into the rhs of Equation (A3b),

$$m_1m_2\omega_1^2\omega_2^2 = m_1m_2\left(\frac{-b + \sqrt{b^2 - 4ac}}{2a}\right)\left(\frac{-b - \sqrt{b^2 - 4ac}}{2a}\right)$$

$$= m_1m_2\left(\frac{1}{4a}\right)\left(b^2 - b^2 + 4ac\right) = m_1m_2\left(\frac{c}{a}\right) = c.$$

Thus, the rhs of Equation (A3b) = lhs of Equation (A3b). This completes the proof of Equation (4.19b).

Exercise Questions

Q1. A homogeneous sphere of radius r and mass m is free to roll without slipping on a spherical surface of radius R. If the motion of the sphere is restricted to a vertical plane and it is displaced from its equilibrium position as shown in Figure 4Q1, determine the natural frequency of small oscillation of the sphere.

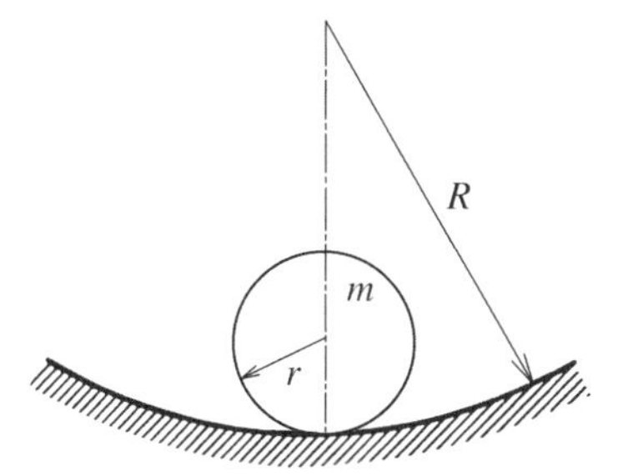

Figure 4Q1 Sphere rolls without slipping on a spherical surface

If now the sphere is replaced by a cylinder of same radius and same mass as the sphere and it is free to roll without slipping on a cylindrical surface whose radius remains R, such that the cross-sectional view in this case is still similar to that shown Figure 4Q1, determine the natural frequency of small oscillation of the cylinder.

In the above two cases which has a higher natural frequency?

Q2. A uniform rigid slender rod of mass m and length L is hinged at the middle. Both ends of the rod are attached by springs, each one of which has a spring coefficient k as shown in Figure 4Q2. Assuming the rod is free to rotate about an axis perpendicular to the plane of the figure and through the hinge, find the natural frequency of small oscillation of the system.

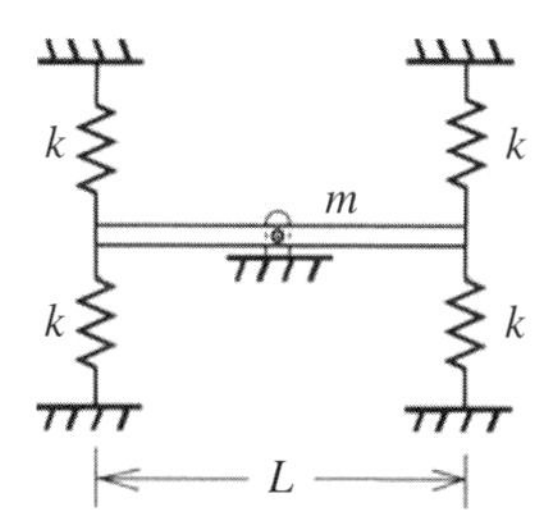

Figure 4Q2 Slender rod and springs arranged as an "H" plane frame

Q3. The single degree-of-freedom system shown in Figure 4.6 is excited by a force of periodic square wave as shown in Figure 4Q3. Determine the steady-state response x of the system.

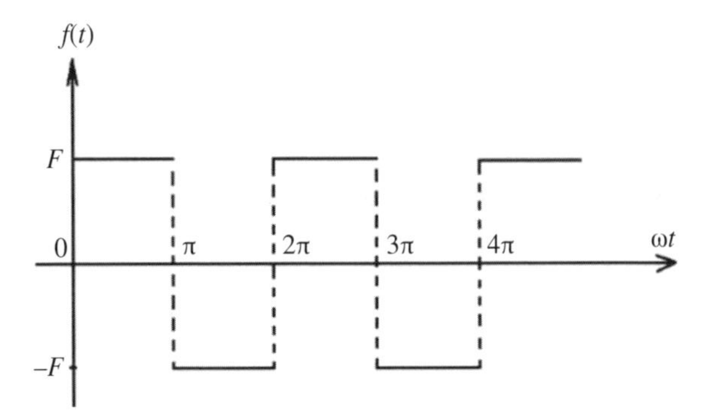

Figure 4Q3 Periodic square wave force

Q4. A helicopter of mass m and its landing gear system represented as a spring constant of k tries to land on a rigid ground. However, owing to the downdraft, it falls from a distance h before it touches the ground. The single degree-of-freedom model is shown in Figure 4Q4. Determine the time elapsed from the first contact of the landing gear system until it breaks contact again. Find the time when $h = \frac{mg}{2k}$ and its implication.

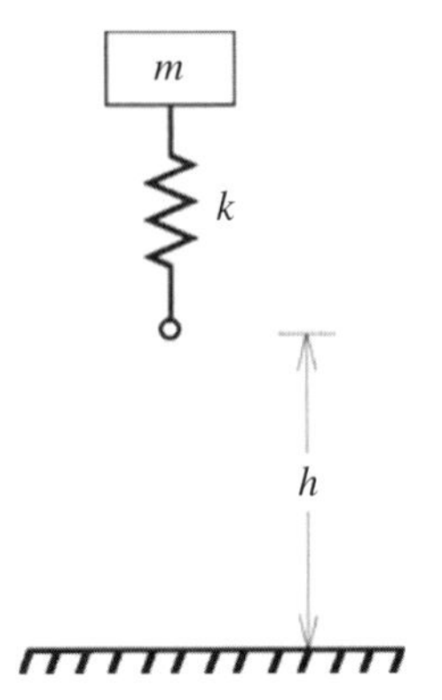

Figure 4Q4 A simplified landing helicopter model

Q5. Applying Newton's second law of motion, derive the equation of motion for the system shown in Figure 4Q5. Determine the undamped natural frequencies of the system in term of coefficient of spring k, mass m, and length L.

If the stiffness of the spring system attached to the right-hand end of the rigid bar approaches an infinitely large value, and the remaining spring constant (that is, the stiffness of the dynamic absorber) does not change, what is its implication for the two natural frequencies of the system?

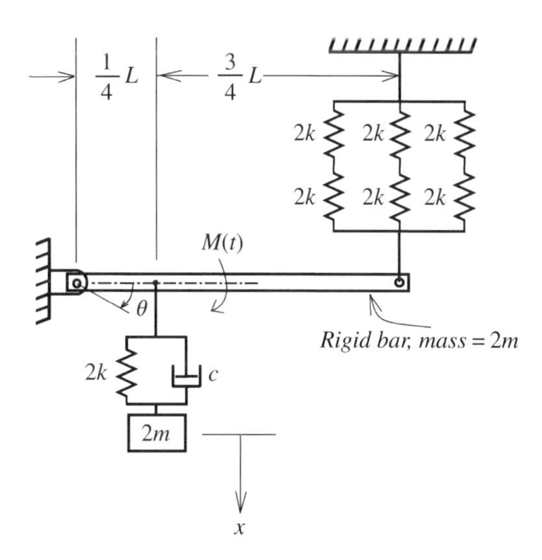

Figure 4Q5 A hinged rigid bar with a dynamic absorber

Q6. Applying Newton's second of law of motion and related principles, derive the equations of motion, in the plane of the paper (assume the rods, masses, and connected springs are on a frictionless table), of the system shown in Figure 4Q6. It is assumed that the horizontal rods are rigid and their masses are zero, and the amplitudes of oscillation are small. Each of the two horizontal rods is hinged at one end, and the other end is attached to a concentrated mass m.

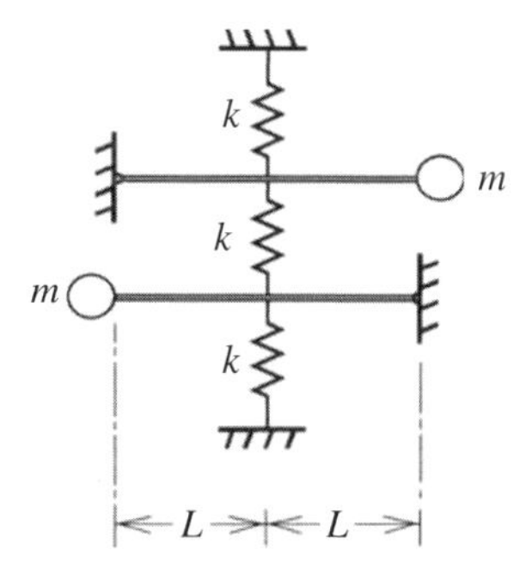

Figure 4Q6 A system having two hinged rods and connected to springs

Q7. A double pendulum of lengths L_1 and L_2 with masses m_1 and m_2 is shown in Figure 4Q7. Applying Newton's second law of motion, derive the equations of motion. Find the natural frequencies of the double pendulum.

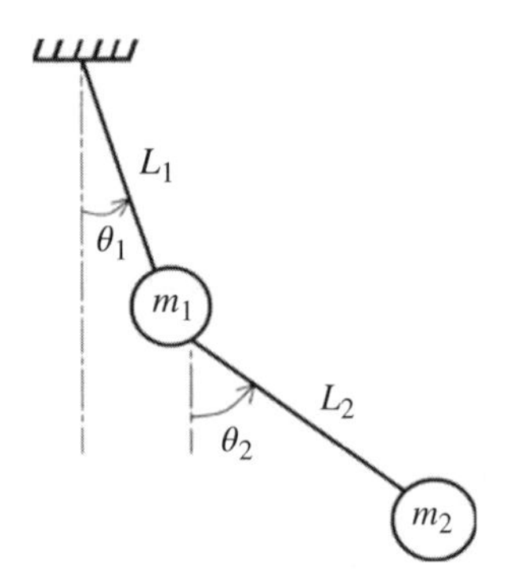

Figure 4Q7 A double pendulum

Q8. A single cylinder reciprocating engine of mass m_1 is mounted on a fixed-fixed beam of length L, width w, thickness h, and Young's modulus E, as shown in Figure 4Q8. A spring-mass system having mass m_2 and stiffness k_2 is attached to the beam at its center, as indicated in the figure.

Assuming the beam has no mass and the steady-state harmonic force from the reciprocating engine acts upon the mid-position of the beam:

a. Derive the equations of motion for the above system.
b. Find the relation between m_2 and k_2 that leads to no steady-state vibration of the reciprocating engine when a harmonic force from the engine is given by $f_1(t) = f_1 = F \cos \omega t$.

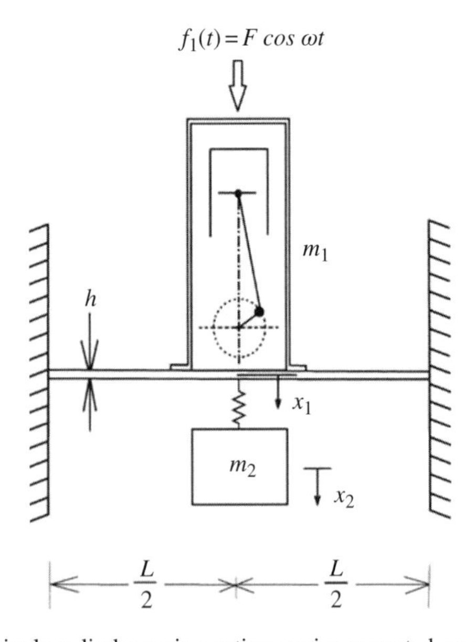

Figure 4Q8 A single cylinder reciprocating engine mounted on a fixed-fixed beam

c. Evaluate the steady-state displacement X_1 if $L = 10$ m, $w = 1$ m, $h - 2.10858$ cm, $E = 2 \times 10^{11}$ Pa, $k_2 = 10$ kN/m, $m_1 = 1000$ kg, $m_2 = 100$ kg, $F = 100$ N, and $\omega = 5$ rad/s on a fixed-fixed beam.

Q9. A uniform cable of length L and high initial tension is statically displaced h units from the center as shown in Figure 4Q9. The cable is released so that it undergoes oscillation. The governing equation of motion in the cable becomes

$$a^2 \frac{\partial^2 v(x,t)}{\partial x^2} = \frac{\partial^2 v(x,t)}{\partial t^2}$$

in which $v(x, t)$ or v is the displacement or deformation of the cable in the y-direction and the remaining symbols have their usual meaning.

For this problem the boundary conditions are: at $x = 0$ and $v(0, t) = 0$ at $x = L$, $v(L, t) = 0$. The initial conditions are: at $t = 0$,

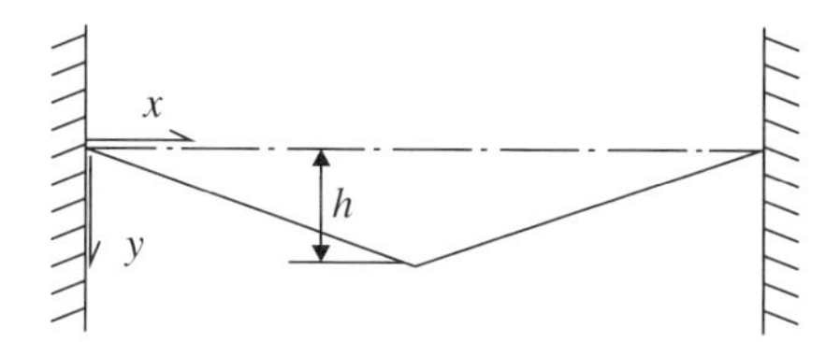

Figure 4Q9 A cable being given a triangular initial input

$$v(x,0) = \begin{cases} \dfrac{2hx}{L}, & 0 \le x \le \dfrac{L}{2} \\[2ex] \dfrac{2h}{L}(L-x), & \dfrac{L}{2} \le x \le L, \end{cases}$$

$$\left.\frac{\partial v(x,t)}{\partial t}\right|_{t=0} = 0.$$

Determine the expression for the displacement $v(x, t)$ of the cable.

Q10. A simplified model of a rocket may be considered as a uniform bar of length L, and its propulsion system may be represented as a force that generates a co-sinusoidal movement $V_o \cos \omega t$. In practice, the change of mass of the rocket is substantial due to fuel consumption in the process. However, for simplicity, the mass of the rocket is considered constant within a relatively short duration. This simple model is shown in Figure 4Q10.

Assuming small oscillation, find the steady-state vibration and determine the applied frequencies ω that cause resonance of the system.

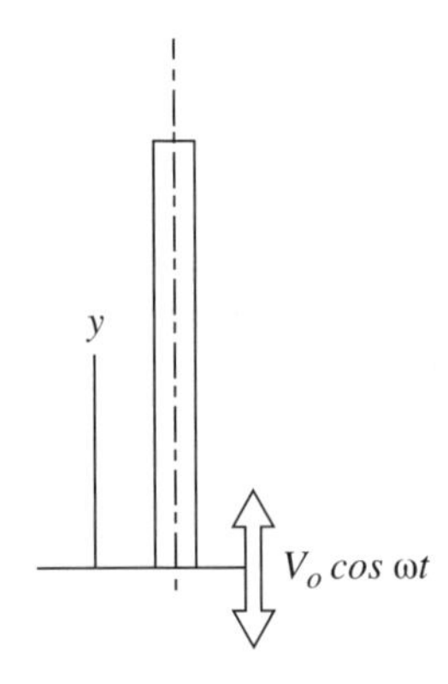

Figure 4Q10 A simple model of a rocket

References

[1] Thomson, W.T. (1993). *Theory of Vibration with Applications*, 4th edn, Prentice-Hall, Englewood Cliffs, NJ.
[2] Seto, W.W. (1964). *Theory and Problems of Mechanical Vibrations*, McGraw-Hill, New York.

5

Formulation and Dynamic Behavior of Thermal Systems

This chapter introduces the formulation and dynamic behavior of thermal systems. The elements of thermal systems are presented in Section 5.1, while basic yet useful thermal systems are studied in Section 5.2. Problem examples are included in Section 5.3.

5.1 Elements of Thermal Systems

In general, a thermal system can consist of a combination of resistance, capacitance, and radiation elements. These basic elements are shown in Figure 5.1 and are introduced in the following subsections.

5.1.1 Thermal Resistance

For heat conduction through a wall with surface temperatures T_1 and T_2 and thickness d, the heat flow is proportional to the temperature gradient $(T_1 - T_2)/d$. Thus, one has the following relation

$$q = \frac{kA(T_1 - T_2)}{d} \tag{5.1}$$

where q is the heat flow rate in units of heat per unit time, and k is the coefficient of thermal conductivity.

Introduction to Dynamics and Control in Mechanical Engineering Systems, First Edition. Cho W. S. To.

Companion website: www.wiley.com/go/to/dynamics

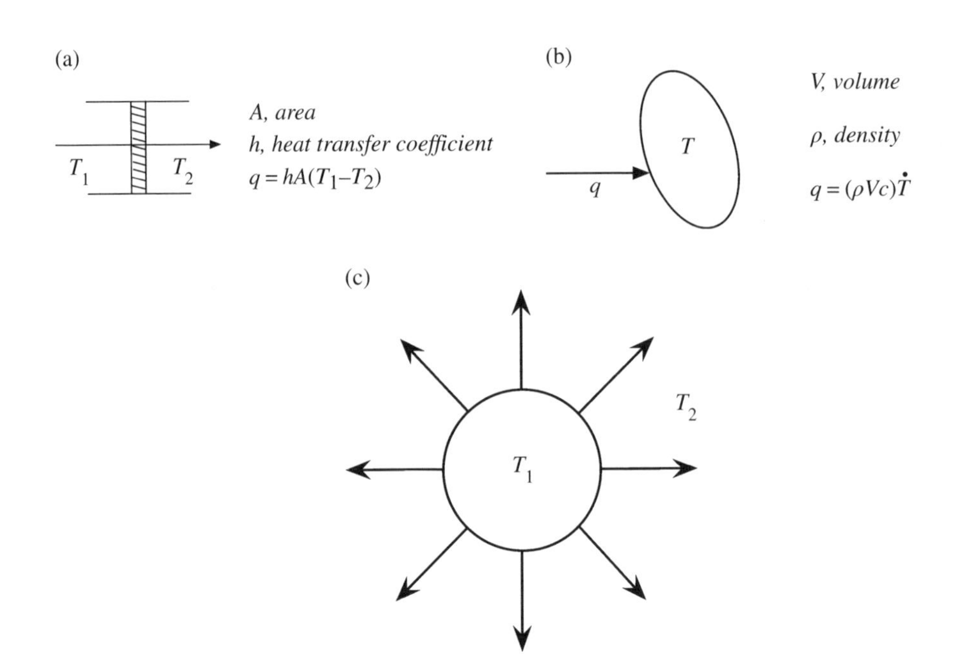

Figure 5.1 Elements of thermal systems: (a) resistance, (b) capacitance, and (c) radiation

In Equation (5.1) the term, $h = k/d$ is the equivalent heat transfer coefficient. Therefore, Equation (5.1) may be written in terms of the heat transfer coefficient and resistance, respectively:

$$q = hA(T_1 - T_2) \tag{5.2a}$$

$$q = \frac{1}{R_t}(T_1 - T_2) \tag{5.2b}$$

where the thermal resistance, or simply resistance, $R_t = 1/(hA)$. The latter is analogous to the resistance in the hydraulic tank system considered in Chapter 3.

5.1.2 Thermal Capacitance

Let q be the net heat flow rate into a volume V of a material with density ρ (in kg/m^3) and specific heat c (which is defined as the amount of heat required to raise the temperature of a unit mass by 1°). This net inflow q of heat per unit time must equal to the change per unit time (that is, the rate of change) of heat stored in V. Since the mass is ρV, the heat required for a 1° rise in temperature is $c\rho V$, and hence the heat stored at a temperature T is $c\rho V\, T$.

Assuming ρ, V, and c are constant, the rate of change is $c\rho V$. In other words, one can write

$$q = C_t \frac{dT}{dt} = C_t \dot{T}, \tag{5.3a}$$

$$C_t = c\rho V \tag{5.3b}$$

where C_t is known as the thermal capacitance.

5.1.3 Thermal Radiation

For thermal radiation, the heat flow rate is given by

$$q=\beta\left(T_1^4-T_2^4\right), \tag{5.4}$$

where β is a constant. This equation is called the Stefan-Boltzmann law.

5.2 Thermal Systems

In order to illustrate the important use of the elements described in Sections 5.1.1–5.1.3, three different thermal systems are considered in this section.

5.2.1 Process Control

One application of the elements is in process control. Figure 5.2 shows a tank or compartment of volume V filled with an incompressible fluid of density ρ and specific heat c.

Let f_i be the volume flow rate entering the tank, and f_o out of the tank. Note that in Chapter 3 the same quantities are represented by q_i and q_o. The corresponding inflow temperature is T_i.

It is assumed that the tank is well ventilated or stirred so that the outlet temperature T_o equals the tank temperature T. Since the tank is filled with incompressible fluid, $f_i=f_o$, the mass flow rate entering and leaving the tank is ρf_i. Therefore, the heat inflow rate is $f_i\, c\rho T_i$ and the heat outflow rate is $f_i\, c\rho T$, such that the net heat inflow rate is

$$f_i\rho c(T_i-T). \tag{5.5}$$

This is the net inflow of heat per unit time, and so it must be equal to the change per unit time (that is, the rate of change) of heat stored in the tank.

This stored heat is $c\rho VT$, and its rate of change is $c\rho V\dot{T}$. Therefore, the equation governing the above process becomes

$$\rho cV\frac{dT}{dt}=\rho cV\dot{T}=f_i\rho c(T_i-T).$$

This equation may be rewritten as

$$\left(\frac{V}{f_i}\right)\frac{dT}{dt}+T=T_i. \tag{5.6}$$

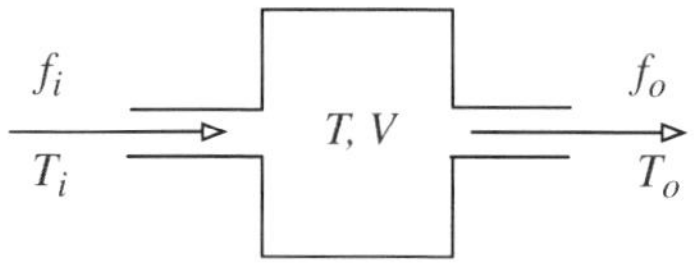

f_i T_i T, V f_o T_o

Figure 5.2 Process control of a compartment

Taking Laplace transforms and re-arranging, one can show that the transfer function of this process is

$$\frac{\mathcal{L}\{T\}}{\mathcal{L}\{T_i\}} = \frac{1}{\left(\frac{V}{f_i}\right)s + 1}. \tag{5.7}$$

The system governed by this transfer function is known as the *simple lag* system. The form of this expression is the same as that for the hydraulic tank.

5.2.2 Space Heating

Another application of the elements considered in Sections 5.1–5.3 is in space heating, as shown schematically in Figure 5.3.

By the principle of heat balance, that is, the difference between the heat inflow rate and outflow rate must be equal to the rate of heat accumulated in the space that is being heated, one has (note that the form of the equation is similar to that of the hydraulic tank problem that was considered earlier in Chapter 3):

$$q_i - q_o = \rho c V \frac{dT}{dt}, \tag{5.8}$$

where T is now the difference of temperature between the temperature in the space and the ambient temperature, q_o is the rate of heat loss to the ambient, and q_i is the heat inflow rate from an electrical heater or heating element.

In the above equation, the rate of heat loss to the ambient and thermal capacitance are, respectively

$$q_o = \frac{T}{R_t} \tag{5.9a}$$

$$C_t = \rho c V \tag{5.9b}$$

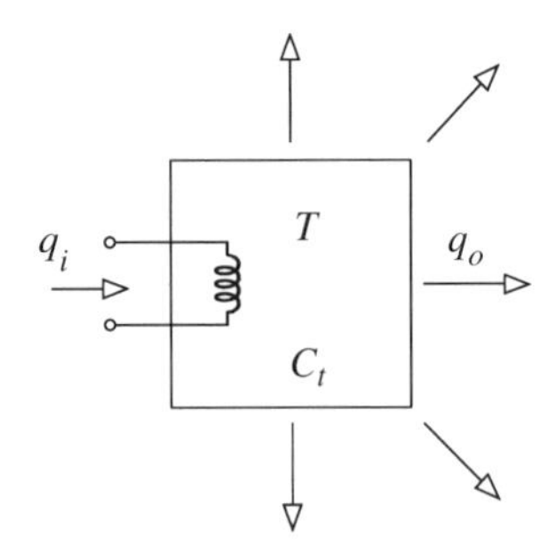

Figure 5.3 Space heating

Taking the Laplace transform of Equation (5.8) and re-arranging, one can obtain the system transfer function as

$$\frac{\mathcal{L}\{T\}}{\mathcal{L}\{q_i\}} = \frac{R_t}{(R_tC_t)s+1}, \tag{5.10}$$

in which R_tC_t is the time constant of the system, and therefore the above equation is similar in form to the transfer function of the hydraulic tank.

5.2.3 Three-Capacitance Oven

One important application of the thermal elements is in the three-capacitance oven, shown schematically in Figure 5.4. The underlining principle and assumptions are listed below.

- *Principle*: The net heat flow rate is balanced or equal to the rate of change of heat for every capacitance.
- *Assumptions*:
  1. The fluid in the oven is well mixed or stirred and incompressible.
  2. The volume does not change.

Notation

f	=	volumetric flow rate (note that in the hydraulic tank in Chapter 3 it is denoted as q),
C_o	=	$c_o \rho_o V$, the capacitance of the oven, Equation (5.3b),
h_o	=	coefficient of heat transfer,
A_o	=	surface area of oven,
V	=	volume of oven,
T_o	=	oven or output temperature.

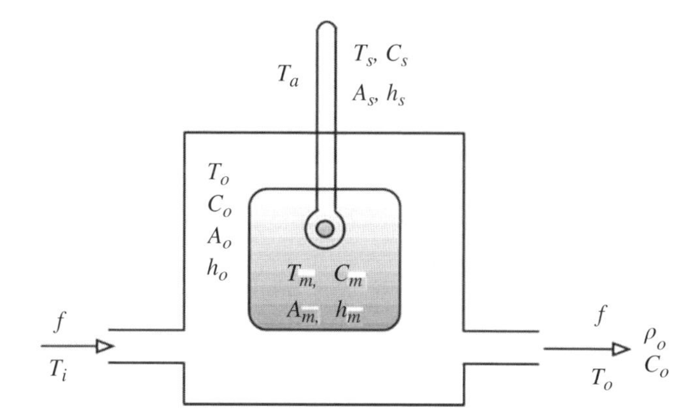

Figure 5.4 Three-capacitance oven

Subscripts		
o	=	oven/output (since it is well stirred),
a	=	ambient,
s	=	sensor/thermometer/thermocouple,
m	=	material.

Applying the above principle to the oven, material, and sensor, respectively, one has

$$C_o\frac{dT_o}{dt}=f\rho_o c_o(T_i-T_o)-A_o h_o(T_o-T_a)-A_m h_m(T_o-T_m), \tag{5.11}$$

$$C_m\frac{dT_m}{dt}=A_m h_m(T_o-T_m) \tag{5.12}$$

$$C_s\frac{dT_s}{dt}=A_s h_s(T_m-T_s) \tag{5.13}$$

- Note that in Equation (5.12) the contribution due to heat loss to the sensor is disregarded because it is relatively negligible.

Explanation

In the following, Equations (5.11)–(5.13) will be explained.

- Consider Equation (5.11) first. The mass flow rate entering and leaving the oven is $f\rho$, in which the density ρ is constant.

Therefore, the heat inflow rate is $fc\rho T_i$ or with more specific symbols $fc_o\, \rho_o\, T_i$.
The heat outflow rate is $fc_o\, \rho_o\, T_o$.
Thus, the net heat inflow rate is $f\, \rho_o\, c_o\, (T_i - T_o)$.

This must be equal to the change per unit time of heat stored in the oven of volume V, plus the heat loss of the oven, plus the heat required to heat up the material in the oven. Now, the change per unit of heat stored in the oven, by Equation (5.3a), is

$$q=(\rho_o c_o V)\frac{dT_o}{dt}.$$

The heat loss of the oven, by Equations (5.2a) or (5.2b), is

$$T_o-T_a=q_o R_t \quad \text{or} \quad q_o=h_o A_o(T_o-T_a),$$

where now

$$R_t=\frac{1}{h_o A_o}.$$

The heat required to heat up the material, by Equation (5.2a) or (5.2b), is

$$T_o - T_m = q_m R_t \quad \text{or} \quad q_m = h_m A_m (T_o - T_m),$$

where now

$$R_t = \frac{1}{h_m A_m}.$$

Adding all the above terms, one has Equation (5.11).

- Equations (5.12) and (5.13) are obtained by applying Equations (5.2a) and (5.3a).

For the material in oven, by Equation (5.2b), one has

$$T_o - T_m = q_m R_t = \frac{q_m}{h_m A_m}.$$

But q_m is given by Equation (5.3a) as $q_m = C_m \frac{dT_m}{dt}$

Therefore

$$C_m \frac{dT_m}{dt} = A_m h_m (T_o - T_m).$$

Similarly, one obtains the equation for the sensor as

$$C_s \frac{dT_s}{dt} = A_s h_s (T_m - T_s).$$

- Note that Equations (5.11) and (5.12) are coupled. Equation (5.13) requires T_m to solve for T_s.

Thus Equations (5.12) and (5.13) constitute a non-interacting system, while Equations (5.11) and (5.12) make up an interacting system. That is,

a. sensor and material are non-interacting;
b. material and oven are interacting.

Returning to Equation (5.13), after taking the Laplace transform and re-arranging, one obtains

$$\frac{\mathcal{L}\{T_s\}}{\mathcal{L}\{T_m\}} = \frac{1}{\tau_s s + 1}, \quad \tau_s = \frac{C_s}{A_s h_s}. \tag{5.14}$$

Similarly, for Equation (5.12), one can show that

$$\frac{\mathcal{L}\{T_m\}}{\mathcal{L}\{T_o\}} = \frac{1}{\tau_m s + 1}, \quad \tau_m = \frac{C_m}{A_m h_m}. \tag{5.15}$$

Substituting this for T_m in the Laplace transformed Equation (5.11), and bringing all terms for the Laplace transformed oven temperature to one side, gives

$$\beta \mathcal{L}\{T_o\} = f\rho_o c_o \mathcal{L}\{T_i\} + A_o h_o \mathcal{L}\{T_a\},$$

where

$$\beta = C_s s + (f\rho_o c_o + A_o h_o) + A_m h_m \left(1 - \frac{1}{\tau_m + 1}\right),$$

and after some algebraic manipulation one has

$$\mathcal{L}\{T_o\} = \frac{(\tau_m s + 1)(a_1 \mathcal{L}\{T_i\} + a_2 \mathcal{L}\{T_a\})}{b_2 s^2 + b_1 s + a_1 + a_2}, \tag{5.16}$$

in which

$$a_1 = f\rho_o c_o, \quad a_2 = A_o h_o, \quad b_2 = C_o \tau_m, \quad b_1 = (a_1 + a_2)\tau_m + C_o + C_m.$$

Note that on the rhs of Equation (5.16) there are two input temperatures. In addition to the temperature of the heat inflow T_i, the variations of ambient temperature T_a act as a disturbance input to the system.

5.3 Questions and Solutions

The following examples are presented in this section to illustrate applications of the thermal elements and systems introduced in the foregoing sections.

Example 1
A two-layered wall is constructed to reduce the amount of heat loss from one side of the wall to the other side, as sketched conceptually in Figure 5E1. If the thermal resistances of the two layers of the wall are R_{t1} and R_{t2}. Starting from first principles, find the equivalent thermal resistance of the wall.

Solution:
Let the heat flow rate through the wall be q, and the interface temperature between the two layers T_{12} such that, by using Equation (5.2b) for the first layer, one has

$$q = \frac{T_1 - T_{12}}{R_{t1}}. \tag{i}$$

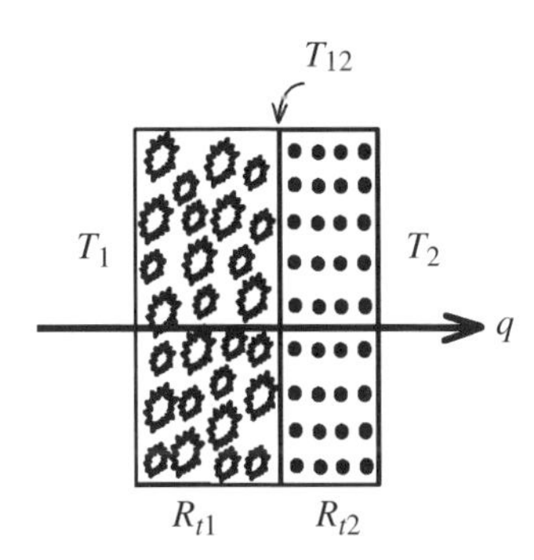

Figure 5E1 Thermal resistances of a two-layered wall

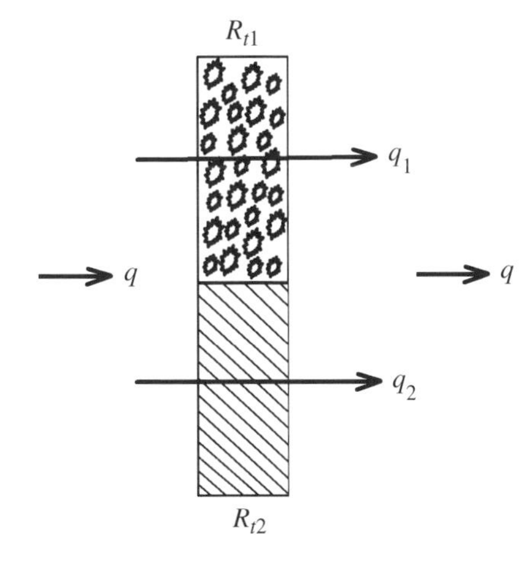

Figure 5E2 Thermal resistances of two walls in parallel

Similarly, the heat flow rate through the second layer becomes

$$q = \frac{T_{12} - T_2}{R_{t2}}. \tag{ii}$$

Eliminating the interface temperature from Equations (i) and (ii), one obtains

$$q = \frac{T_1 - T_2}{R_{t1} + R_{t2}} = \frac{T_1 - T_2}{R_{te}}, \tag{iii}$$

where $R_{te} = R_{t1} + R_{t2}$ is the equivalent thermal resistance of the wall. This means that the resistances are in series.

Example 2

A wall is constructed of two components; one is near the foundation of the building (not shown) and the other above the first component as indicated in Figure 5E2. The thermal resistances of the two components are, respectively, R_{t1} and R_{t2}. Find the equivalent thermal resistance of the wall.

Solution:
With reference to Figure 5E2, the heat flow rate q is given by

$$q = q_1 + q_2 \tag{i}$$

and by applying Equation (5.2b):

$$q_1 = \frac{T_1 - T_2}{R_{t1}} \tag{ii}$$

$$q_2 = \frac{T_1 - T_2}{R_{t2}} \tag{iii}$$

Substituting Equations (ii) and (iii) into Equation (i), one has

$$q = \frac{T_1 - T_2}{R_{t1}} + \frac{T_1 - T_2}{R_{t2}} = \left(\frac{1}{R_{t1}} + \frac{1}{R_{t2}}\right)(T_1 - T_2) = \frac{1}{R_{te}}(T_1 - T_2), \tag{iv}$$

where $\frac{1}{R_{te}} = \frac{1}{R_{t1}} + \frac{1}{R_{t2}}$ and R_{te} is the equivalent thermal resistance of the wall. In other words, the resistances in this case are in parallel.

Example 3
Two identical rooms are perfectly insulated at the ground and ceiling levels (not shown in the figure). These rooms are constructed in a cold environment with ambient temperature T_o. The cross-sectional view of the walls is sketched in Figure 5E3, in which Room 2 has an electric heater producing a heat flow rate of q_h. The walls can be considered as thermal resistances. Derive the dynamic equations for the two rooms and provide comments on the dynamic behaviors and resistances of the walls.

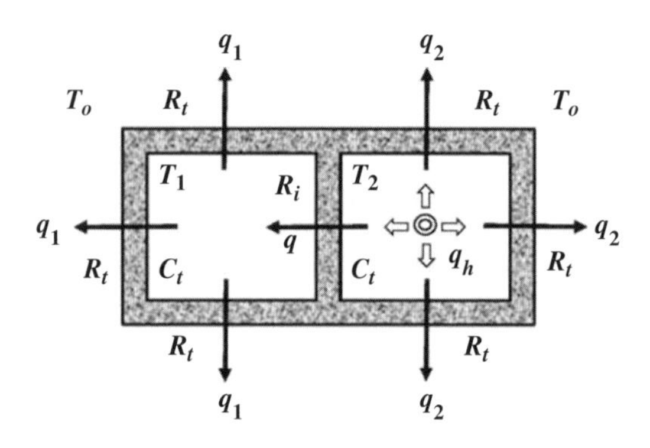

Figure 5E3 Two identical rooms with an electric heater

Solution:

With reference to Figure 5E3 and applying Equations (5.3a) or (5.8) to Room 1, one has

$$C_t \frac{dT_1}{dt} = q - q_1. \tag{i}$$

Similarly, for Room 2, one obtains

$$C_t \frac{dT_2}{dt} = q_h - q_2 - q. \tag{ii}$$

Applying Equation (5.2b), the thermal resistance relation for Room 1 becomes

$$R_t = \frac{1}{q_1}(T_1 - T_o). \tag{iii}$$

Similarly, for Room 2, the thermal resistance relation is

$$R_t = \frac{1}{q_2}(T_2 - T_o). \tag{iv}$$

For the partition or interfacing wall between Rooms 1 and 2:

$$R_i = \frac{1}{q}(T_2 - T_1). \tag{v}$$

Substituting Equations (iii)–(v) into Equations (i) and (ii), one can show that

$$C_t \frac{dT_1}{dt} + \left(\frac{1}{R_i} + \frac{1}{R_t}\right) T_1 - \frac{T_2}{R_i} = \frac{T_o}{R_t}, \tag{vi}$$

and

$$C_t \frac{dT_2}{dt} - \frac{T_1}{R_i} + \left(\frac{1}{R_i} + \frac{1}{R_t}\right) T_2 = \frac{T_o}{R_t} + q_h. \tag{vii}$$

Equations (vi) and (vii) are the dynamic equations required. They can be written in matrix form as

$$\begin{bmatrix} C_t & 0 \\ 0 & C_t \end{bmatrix} \begin{pmatrix} \dfrac{dT_1}{dt} \\ \dfrac{dT_2}{dt} \end{pmatrix} + \begin{bmatrix} \dfrac{1}{R_e} & \dfrac{-1}{R_i} \\ \dfrac{-1}{R_i} & \dfrac{1}{R_e} \end{bmatrix} \begin{pmatrix} T_1 \\ T_2 \end{pmatrix} = \begin{pmatrix} \dfrac{T_o}{R_t} \\ \dfrac{T_o}{R_t} + q_h \end{pmatrix}, \tag{viii}$$

where $\dfrac{1}{R_e} = \dfrac{1}{R_i} + \dfrac{1}{R_t}$ and R_e are the equivalent thermal resistances of the external and partition walls.

With reference to Equation (viii), one can conclude that the thermal resistances of the external and partition walls act as if they are connected in parallel. The two rooms function as interacting systems since the two first-order differential equations are coupled. The coefficient matrix associated with the temperature vector is symmetric.

Example 4

A long metal pipe of length L has internal radius r_i and external radius r_o as shown in Figure 5E4. The coefficient of thermal conductivity of the metal is k. Determine the heat flow rate conducted through the cylindrical wall of the pipe.

Solution:

Since the pipe is long, the end effects are negligible. Consider an elementary pipe of thickness dr at a distance r from the axis of the pipe. Let dT be the drop in temperature across the element of the metal. The amount of heat q that is conducted across the pipe wall per unit time is equal to that conducted through the elementary pipe. Thus, according to Equation (5.1):

$$q = k(2\pi rL)\left(-\frac{dT}{dr}\right)$$

in which the negative sign is applied to indicate that the temperature decreases as the radius of the pipe increases.

Rearranging gives

$$\frac{dr}{r} = -\frac{k(2\pi L)}{q}\,dT.$$

Integrating both sides,

$$\int_{r_i}^{r_o} \frac{dr}{r} = -\frac{k(2\pi L)}{q}\int_{T_i}^{T_o} dT$$

where T_o is the temperature at $r = r_o$ and T_i is the temperature at $r = r_i$. Upon performing the integration on both sides of the last equation, it gives

$$\log_e\left(\frac{r_o}{r_i}\right) = \frac{k(2\pi L)}{q}(T_i - T_o)$$

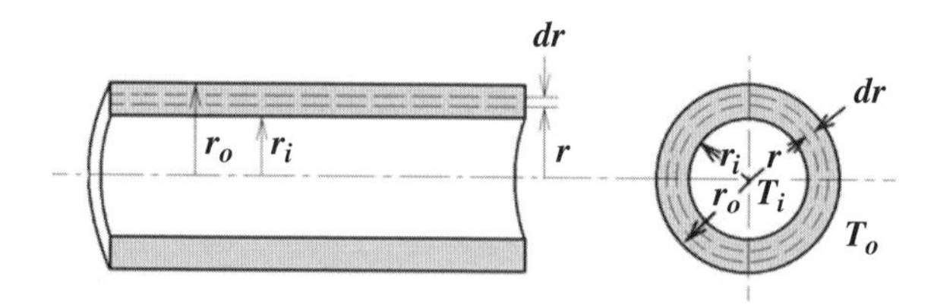

Figure 5E4 Heat conduction through the cylindrical wall of a pipe

Therefore, the heat flow rate through the pipe wall is

$$q = \frac{k(2\pi L)}{log_e\left(\frac{r_o}{r_i}\right)}(T_i - T_o).$$

Exercise Questions

Q1. A turkey of mass m and specific heat c is suddenly placed inside an oven at $t = 0$. The temperature of the turkey at $t = 0$ is T_i and the constant oven temperature T_o is considered as a step input to the turkey. Derive the equation of the temperature $T(t)$ or T of m and solve for the temperature T. It is assumed that the constant surface area of the turkey is A and the constant coefficient of heat transfer is h.

Q2. A simplified model of a concert chamber is shown in Figure 5Q2. In winter, heat is provided inside the chamber through a resistance heater at a heat flow rate of q_c. In order to maintain a comfortable environment inside the chamber, a constant mass flow rate ρf_i of air with specific heat c is delivered to the chamber, which contains a mass m of the air. As shown in the figure, the inflow temperature is T_i, while the outflow and the chamber temperature is T_o. Derive the equation for the chamber temperature T_o and find the transfer function $\mathcal{L}\{T_o\}/\ \mathcal{L}\{q_c\}$.

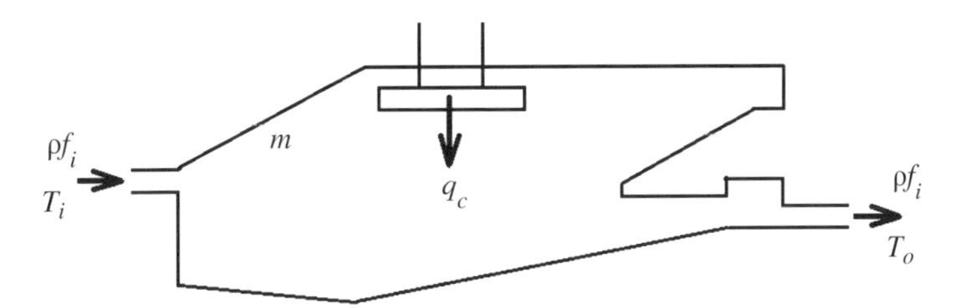

Figure 5Q2 A heated concert chamber

Q3. For a heat exchanger, shown in Figure 5Q3, in a nuclear power plant, owing to the high flow rate of the fluid through the chamber, the temperature T_c of the outer chamber is

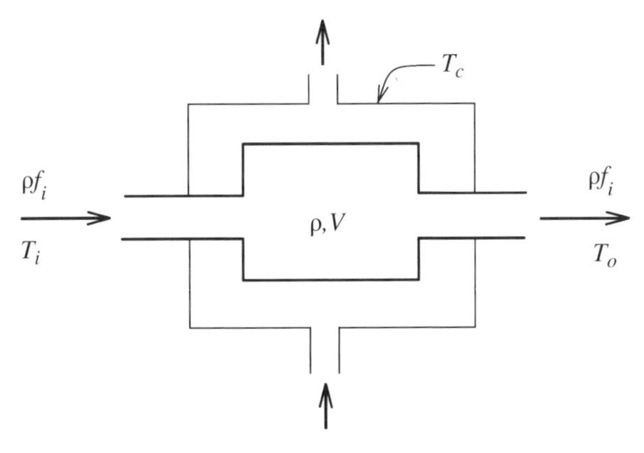

Figure 5Q3 Heat exchanger in a nuclear power plant

considered constant. The mass flow rate ρf through the inner chamber is also considered constant. If the volume of the chamber is V, its surface area A, inflow and outflow temperatures T_i and T_o, respectively, derive the equation for T_o which is also the temperature of the inner chamber. The specific heat of the fluid is c and the coefficient of heat transfer h. Find the transfer function $\mathcal{L}\{T_o\}/\mathcal{L}\{T_i\}$.

Q4. A simplified model of an automobile brake block is shown in Figure 5Q4. When a force N is applied against the rotor of the wheel, friction $F = \mu N$ is generated, where μ is the coefficient of friction. The surface velocity between the block and the rotor is U. The power due to friction is converted into heat. The conversion factor that changes mechanical power into thermal power is H. The mass of the brake block is m and its specific heat c. In addition, the brake block loses heat to the ambient T_a through the surface area A, whose coefficient of heat transfer is h. Find the transfer function for the temperature T of the block and the applied force N, $\mathcal{L}\{T\}/\mathcal{L}\{N\}$, assuming all thermal power enters the block.

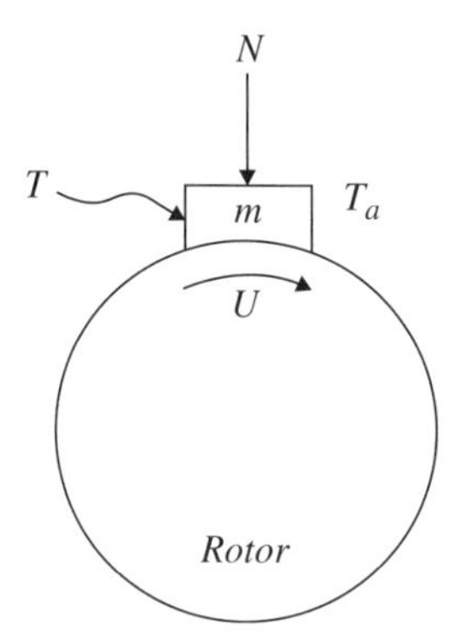

Figure 5Q4 An automobile brake block with rotor in plane view

Q5. Consider a composite wall of three layers of different materials as shown in Figure 5Q5. Their thicknesses are d_i and constant thermal conductivities k_i, with $i = 1,2,3$. Every layer has a constant surface area A. Determine the heat flow rate across the wall, q. Assume the side effects of the wall can be disregarded.

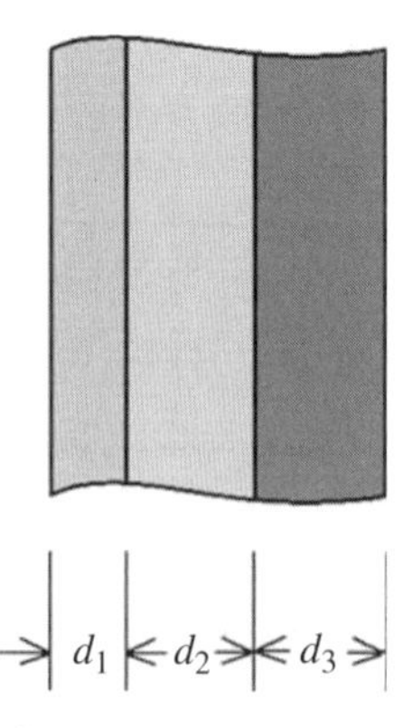

Figure 5Q5 Heat flow through a three-layer composite wall

Q6. A long, uniform composite pipe of length L is made of three layers of different materials whose radii are shown in Figure 5Q6, and coefficients of thermal conductivities are k_i, with $i = 1, 2, 3$. Assuming T_4 is the temperature at r_4 and T_1 is the temperature at r_1, find the heat flow rate q across the composite wall of the pipe.

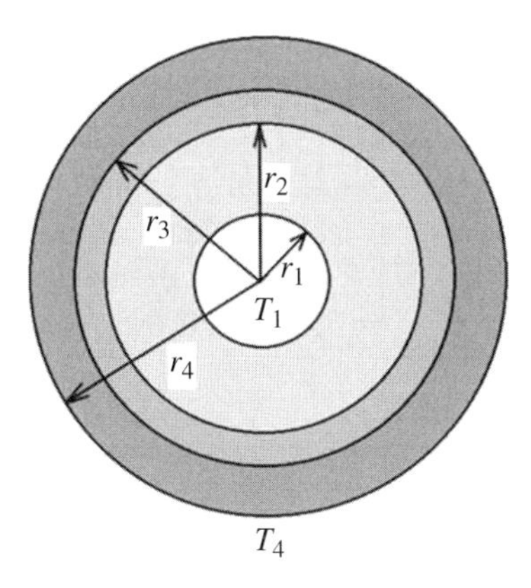

Figure 5Q6 Cross-section of the composite pipe

Q7. A satellite in space is subjected to solar heating. It has a constant mass m and constant surface area A. The heat flow rate input to the satellite from the sun is given by $q_s = \alpha Aq(t)$ where α is the constant coefficient of thermal absorptivity, and $q(t)$ or q the instantaneous incident energy of the sun. The satellite is simultaneously losing heat in accordance with the Stefan-Boltzmann law at a rate of $q_L = \varepsilon A\beta T^4$ in which ε is the emissivity of the body and is assumed constant, β is the Stefan-Boltzmann constant, and T the temperature of the satellite. By assuming $T = T_o + \widetilde{T}$ and $q = q_o + \widetilde{q}$, where T_o and q_o are constant, and $\widetilde{T}$ and $\widetilde{q}$ are deviation or fluctuation temperature and energy, respectively, derive the equation governing the deviation temperature of the satellite. Obtain the linearized equation of deviation temperature of the satellite and find the transfer function of the system, $\mathcal{L}\{\widetilde{T}\}/\mathcal{L}\{\widetilde{q}\}$.

6

Formulation and Dynamic Behavior of Electrical Systems

Basic electrical elements, fundamentals of electrical circuits, simple electrical circuits and networks, and electromechanical systems are introduced in Sections 6.1–6.4, respectively. The focus in this chapter is on the simple electrical circuits and networks, and the derivations of transfer functions for various representative electromechanical systems, such as the armature-controlled DC motor, field-controlled DC motor, and DC generator.

6.1 Basic Electrical Elements

An electrical system, in general, can have a combination of the three basic elements which are the resistor, capacitor, and inductor, as shown in Figure 6.1.

The resistance R of a resistor is defined as

$$R = \frac{v}{i} \quad \text{or} \quad v = iR \tag{6.1}$$

where $v(t)$ or simply v is the voltage across the resistor and $i(t)$ or simply i is the current flowing through the resistor (not to be confused with the imaginary number $i = \sqrt{-1}$). The unit of resistance is the ohm, represented by the symbol Ω.

The function of a capacitor is to store electrical charge, and therefore the capacitance C is a measure of the quantity of charge q that can be stored for a given voltage v in a capacitor. Thus,

$$C = \frac{q}{v} \quad \text{or} \quad q = vC. \tag{6.2}$$

Introduction to Dynamics and Control in Mechanical Engineering Systems, First Edition. Cho W. S. To.

Companion website: www.wiley.com/go/to/dynamics

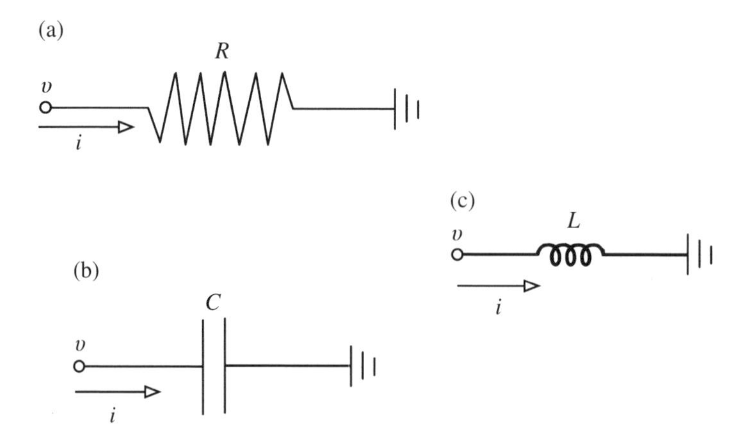

Figure 6.1 Representations of basic electrical elements: (a) resistance, (b) capacitance, and (c) inductance

The unit for capacitance is the farad, denoted as F. Note that the charge q is related to the current i by the following equation.

$$i = \frac{dq}{dt} \quad \text{or} \quad q = \int_0^t i\,dt. \tag{6.3}$$

The third basic electrical element is the inductor. The function of the latter is to store magnetic energy. The property of an inductor is the inductance L. Generally, the induced voltage v in an inductor is proportional to the rate of change of current i, and the proportionality constant is the inductance L. Thus, symbolically,

$$v = \left(\frac{di}{dt}\right)L. \tag{6.4}$$

The unit for inductance is the henry or H.

6.2 Fundamentals of Electrical Circuits

In this section, resistors connected in series and parallel, and the current and voltage laws of Kirchhoff are introduced.

6.2.1 *Resistors Connected in Series*

Resistors can be connected in series as indicated in Figure 6.2a. In this configuration, the voltage between the two end-points is

$$v = v_1 + v_2 + v_3, \tag{6.5}$$

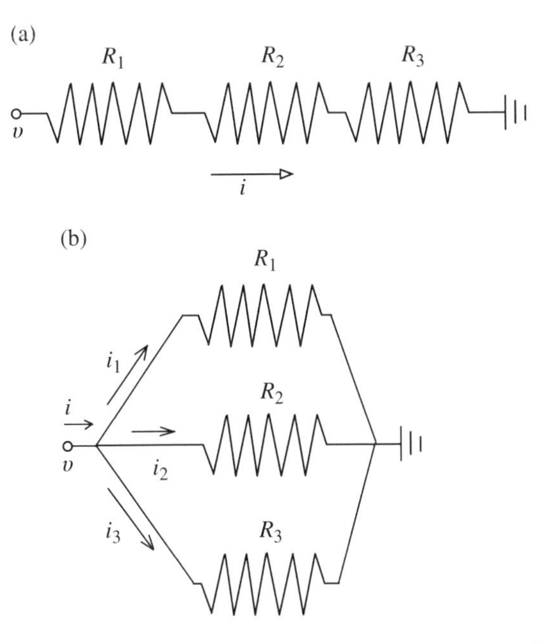

Figure 6.2 Resistors connected in (a) series and (b) parallel

and by applying Equation (6.1) the voltages across the resistors are

$$v_j = iR_j \quad \text{with} \quad j = 1, 2, 3. \tag{6.6}$$

Substituting Equation (6.6) into Equation (6.5), one has

$$\frac{v}{i} = R_1 + R_2 + R_3 = R_e. \tag{6.7}$$

Thus, the equivalent or combined resistance R_e of resistors connected in series is the sum of the individual resistances.

6.2.2 *Resistors Connected in Parallel*

Resistors can also be connected in parallel as shown in Figure 6.2b. In this case the current i through the resistors is

$$i = i_1 + i_2 + i_3, \tag{6.8}$$

where the currents by Equation (6.1) become

$$i = \frac{v}{R_1} + \frac{v}{R_2} + \frac{v}{R_3} \quad \text{or} \quad i = v\left(\frac{1}{R_1} + \frac{1}{R_2} + \frac{1}{R_3}\right) = \frac{v}{R_e}.$$

That is, the equivalent resistance R_e of the resistors connected in parallel is defined by the following relation:

$$\frac{1}{R_e} = \frac{1}{R_1} + \frac{1}{R_2} + \frac{1}{R_3}. \tag{6.9}$$

6.2.3 Kirchhoff's Laws

In the analysis of many circuits that deal with resistances, capacitances, inductances, and electromotive forces, two basic laws are often applied. They are Kirchhoff's voltage law and current law.

Kirchhoff's voltage law is known as the loop method, which states that *the sum of voltage drops around a closed loop is equal to zero.*

Kirchhoff's current law is known as the node method, which states that *the sum of currents at a node or junction of a circuit is equal to zero.*

6.3 Simple Electrical Circuits and Networks

In electrical circuits and networks, Kirchhoff's voltage and current laws are frequently applied to derive the system equations so that, upon using Laplace transformation and rearranging, the transfer function of the system can be obtained. In this section, equations and transfer functions of simple electrical circuits and networks shown in Figure 6.3 are introduced.

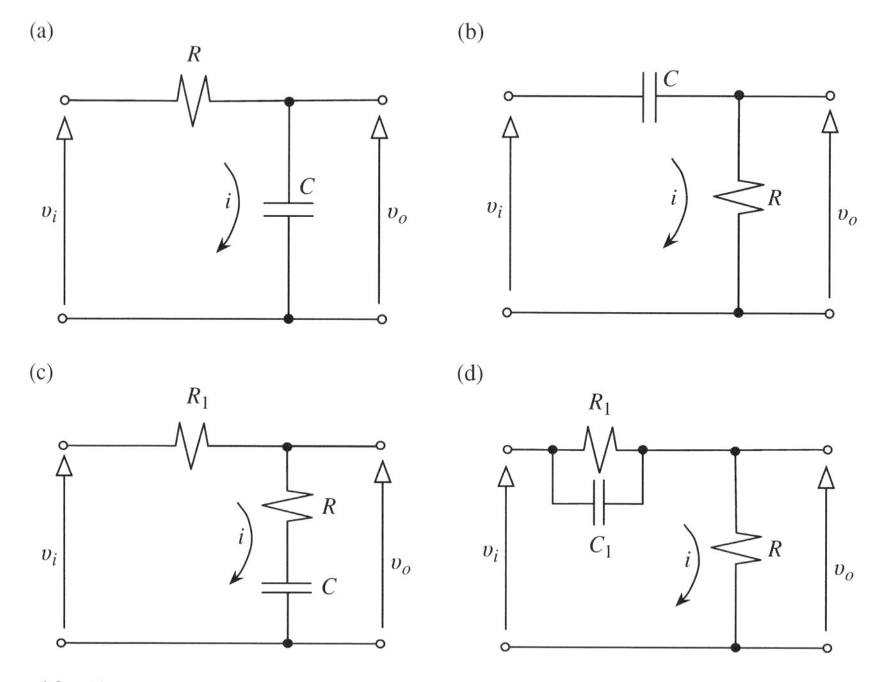

Figure 6.3 Simple electrical circuits: (a) simple lag, (b) transient lead, (c) phase lag, and (d) phase lead

Consider the circuit shown in Figure 6.3a in which the input and output voltages are v_i and v_o, respectively. By applying the loop method, one has

$$iR + \frac{\int_0^t i\,dt}{C} - v_i = 0 \quad \text{or} \quad v_i = iR + \frac{\int_0^t i\,dt}{C} \tag{6.10}$$

and the output voltage is given by

$$v_o = \frac{\int_0^t i\,dt}{C}. \tag{6.11}$$

Applying Laplace transformation to Equations (6.10) and (6.11) results in:

$$V_i = IR + \frac{I}{Cs} \tag{6.12}$$

$$V_o = \frac{I}{Cs} \tag{6.13}$$

Applying Equations (6.12) and (6.13), the transfer function of the system becomes

$$\frac{V_o}{V_i} = \frac{1}{RCs + 1}. \tag{6.14}$$

This is the transfer function of the *simple lag* circuit.

Similarly, for the circuit in Figure 6.3b the loop method gives

$$\frac{\int_0^t i\,dt}{C} + iR - v_i = 0 \quad \text{or} \quad v_i = \frac{\int_0^t i\,dt}{C} + iR \tag{6.15}$$

and the output voltage becomes

$$v_o = iR. \tag{6.16}$$

Taking the Laplace transformation of Equations (6.15) and (6.16), one obtains the transfer function of the circuit as

$$\frac{V_o}{V_i} = \frac{RCs}{RCs + 1}. \tag{6.17}$$

This is known as the transfer function of the *transient lead* circuit.

Again, by applying Kirchhoff's voltage law or loop method, the input voltage can be shown to be

$$v_i = iR_1 + iR + \frac{\int_0^t i\,dt}{C} \tag{6.18}$$

and the output voltage

$$v_o = iR + \frac{\int_0^t i\,dt}{C}. \tag{6.19}$$

By taking Laplace transform and rearranging, the transfer function becomes

$$\frac{V_o}{V_i} = \frac{RCs+1}{(R_1+R)Cs+1}. \tag{6.20}$$

This is the so-called *phase lag* transfer function of the circuit.

Now, consider the circuit in Figure 6.3d. For the parallel resistor R_1 and capacitor C_1 elements, the equivalent resistance R_e is

$$\frac{1}{R_e} = \frac{1}{R_1} + \frac{1}{R_c} \tag{6.21}$$

where $R_c = \dfrac{1}{C_1 s}$. The Laplace transforms of the input and output voltages can be shown as

$$V_i = I\left(\frac{R_1 + R + RR_1C_1s}{1+R_1C_1s}\right) \tag{6.22}$$

$$V_o = IR \tag{6.23}$$

By making use of Equations (6.22) and (6.23), one can find the transfer function of the circuit in Figure 6.3d as

$$\frac{V_o}{V_i} = \frac{(RR_1C_1)s + R}{(RR_1C_1)s + R + R_1}. \tag{6.24}$$

This is the transfer function of a *phase lead* circuit.

6.4 Electromechanical Systems

In this section, three frequently applied electromechanical systems are introduced. These are the armature-controlled DC motor, field-controlled DC motor, and the DC generator. Their equations and transfer functions are derived in the following subsections. Unless stated

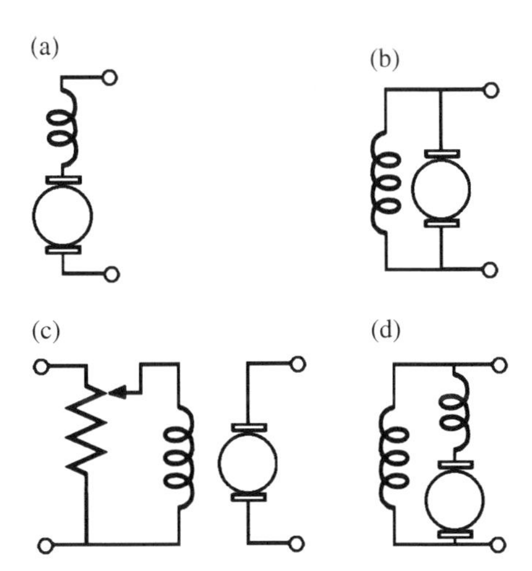

Figure 6.4 Classes of DC machines having electromagnetic excitation current: (a) series, (b) shunt, (c) separately excited, and (d) compound

otherwise, all systems are assumed to start from rest. Solved questions are included in Section 6.5.

The DC machines considered in this section are those applied in control systems and therefore, in general, they are very small, typically below 5 h.p. (1 h.p. = 745.7 W). Those employing permanent magnets have demagnetization and stabilization problems. Compared with machines possessing electromagnetic excitation, the permanent magnet machines have higher efficiencies, and lower weights, and no field windings. Naturally, they are more expensive [1, 2]. Most DC machines have electromagnetic excitation and are classified according to the source of excitation current [1]. The four classes, as shown in Figure 6.4, are: self-excited series, self-excited shunt, separately excited, and compound.

6.4.1 Armature-Controlled DC Motor

The schematic diagram of an armature-controlled DC motor is shown in Figure 6.5. The equation for the armature loop is given by

$$v_a = R_a i_a + L_a \frac{di_a}{dt} + v_m. \tag{6.25a}$$

The electromotive force (emf) voltage v_m is assumed to be proportional to the shaft speed $\frac{d\theta}{dt}$. Thus,

$$v_m = K_e \frac{d\theta}{dt}, \tag{6.25b}$$

where K_e is the proportionality constant.

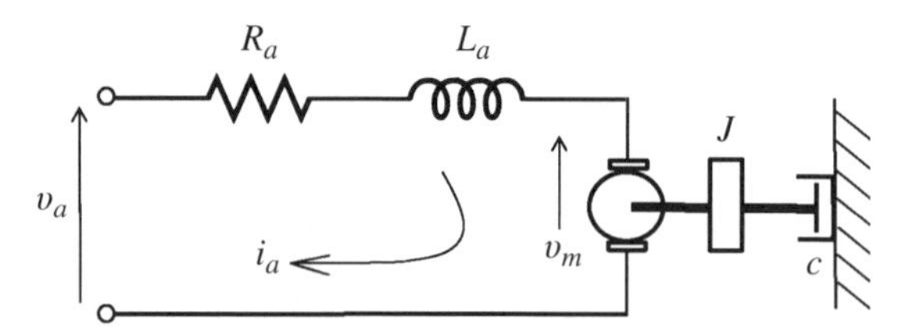

Figure 6.5 An armature-controlled DC motor

The motor torque t_m and motor shaft position θ are related by the following equation

$$t_m = J\frac{d^2\theta}{dt^2} + c\frac{d\theta}{dt} \tag{6.26a}$$

where J and c are the inertia and damping constant of a damper, respectively. Laplace transformation of Equation (6.26a) gives

$$T_m = s(Js + c)\Theta \tag{6.26b}$$

in which $T_m = T_m(s) = \mathcal{L}\{t_m\}$ and $\Theta = \Theta(s) = \mathcal{L}\{\theta\}$.

The Laplace transforms of Equations (6.25a) and (6.25b) are

$$V_a = (R_a + L_a s)I_a + V_m \tag{6.27a}$$

$$V_m = K_e\, s\Theta \tag{6.27b}$$

The developed torque t_m is assumed to be proportional to the armature current i_a. That is,

$$t_m = K\, i_a \tag{6.28a}$$

with K being the proportionality constant, and the Laplace transform of this equation becomes

$$T_m = KI_a. \tag{6.28b}$$

In order to derive the transfer function, one substitutes I_a from the above equation and V_m from Equation (6.27b) into Equation (6.27a) to give

$$V_a = (R_a + L_a s)\frac{T_m}{K} + K_e\, s\Theta.$$

Substituting T_m from Equation (6.26b) into the above equation, it results in the transfer function

$$\frac{\Theta}{V_a} = \frac{1}{\dfrac{(R_a + L_a s)s(Js + c)}{K} + K_e\, s}.$$

Upon simplification,

$$\frac{\Theta}{V_a} = \left(\frac{K}{s}\right) \left[\frac{1}{(L_a J)s^2 + (R_a J + L_a c)s + R_a c + K\ K_e}\right].$$

Upon further simplification, the transfer function becomes

$$\frac{\Theta}{V_a} = \left(\frac{1}{K_e s}\right) \left[\frac{1}{(\tau_a \tau_m)s^2 + (\tau_m + \tau_a \zeta)s + \zeta + 1}\right], \tag{6.29}$$

with an armature time constant $\tau_a = \frac{L_a}{R_a}$, motor time constant $\tau_m = \frac{JR_a}{K\ K_e}$, and damping factor $\zeta = \frac{R_a c}{K\ K_e}$.

6.4.2 Field-Controlled DC Motor

The schematic diagram of a field-controlled DC motor is shown in Figure 6.6. The equation for the field loop is given by

$$v_f = R_f i_f + L_f \frac{di_f}{dt}. \tag{6.30a}$$

The Laplace transformed equation becomes

$$V_f = \left(R_f + L_f s\right) I_f. \tag{6.30b}$$

With constant armature voltage, the developed torque t_m in the motor is assumed to be proportional to the field current such that

$$t_m = K_1 i_f \tag{6.31a}$$

where K_1 is the motor torque constant and its Laplace transformed equation

$$T_m = K_1 I_f. \tag{6.31b}$$

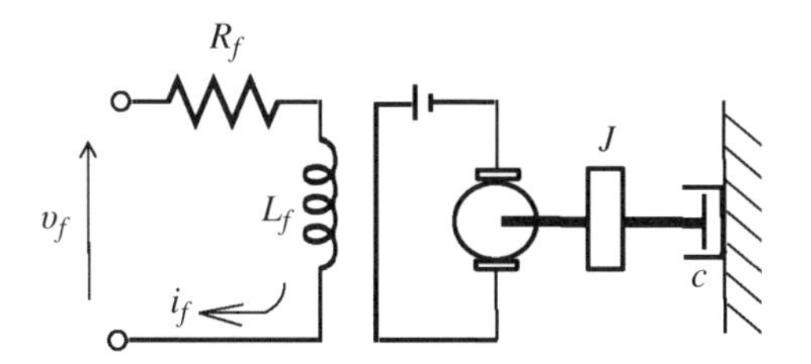

Figure 6.6 Field-controlled DC motor

Substituting I_f from Equation (6.31b) into Equation (6.30b), and, in turn, substituting for T_m from Equation (6.26b), one has

$$\frac{\Theta}{V_f} = \left(\frac{K_1}{s}\right)\left[\frac{1}{(L_f J)s^2 + (R_f J + L_f c)s + R_f c}\right].$$

Simplifying, it becomes

$$\frac{\Theta}{V_f} = \left(\frac{K_1}{R_f c\ s}\right)\left[\frac{1}{\left(\frac{L_f J}{R_f c}\right)s^2 + \left(\frac{J}{c} + \frac{L_f}{R_f}\right)s + 1}\right].$$

Further simplification gives

$$\frac{\Theta}{V_f} = \left(\frac{K_1}{R_f c\ s}\right)\left[\frac{1}{(\tau_f \tau_m)s^2 + (\tau_m + \tau_f)s + 1}\right] \tag{6.32}$$

in which the motor time constant $\tau_m = \frac{J}{c}$, and field time constant $\tau_f = \frac{L_f}{R_f}$.

6.4.3 DC Generator

The schematic DC generator is shown in Figure 6.7 and its field loop equation is given by Equation (6.30a). Naturally, the Laplace transformed field loop equation becomes Equation (6.30b).

The developed generator voltage v_g is assumed to be proportional to field current. Thus, one can write

$$v_g = K_g i_f \tag{6.33a}$$

where K_g is the proportionality constant. The Laplace-transformed Equation (6.33a) is

$$V_g = K_g I_f. \tag{6.33b}$$

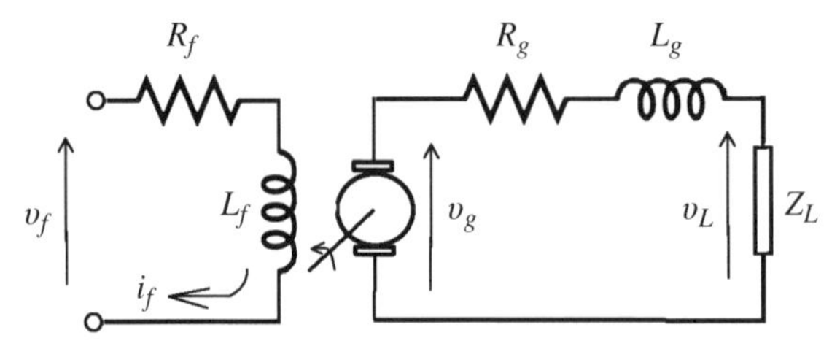

Figure 6.7 DC generator

The voltage across the load is given by

$$v_L = Z_L i_g \tag{6.34a}$$

where Z_L is the load impedance. Therefore, the Laplace transformed Equation (6.34a) becomes

$$V_L = Z_L I_g. \tag{6.34b}$$

With reference to Figure 6.6, the loop equation of the generator and its Laplace transformed version are, respectively:

$$v_g = R_g i_g + L_g \frac{di_g}{dt} + v_L \tag{6.35a}$$

and

$$V_g = R_g I_g + L_g s I_g + V_L = \left(L_g s + R_g + Z_L\right) I_g. \tag{6.35b}$$

Therefore, applying Equations (6.33b) and (6.30b), one has

$$\frac{V_g}{V_f} = \frac{K_g}{L_f s + R_f}. \tag{6.36}$$

On the other hand, applying Equations (6.34b) and (6.35b), it gives

$$\frac{V_L}{V_g} = \frac{Z_L}{L_g s + R_g + Z_L}. \tag{6.37}$$

By applying Equations (6.36) and (6.37), the transfer function of the DC motor is

$$\frac{V_L}{V_f} = \left(\frac{V_L}{V_g}\right)\left(\frac{V_g}{V_f}\right) = \left(\frac{Z_L}{L_g s + R_g + Z_L}\right)\left(\frac{K_g}{L_f s + R_f}\right).$$

Simplifying this equation, the transfer function of the DC motor becomes

$$\frac{V_L}{V_f} = \frac{K_g Z_L}{\left(L_f L_g\right) s^2 + \left(L_f R_g + L_f Z_L + L_g R_f\right) s + R_f R_g + R_f Z_L}. \tag{6.38}$$

Note that this system corresponds to a second-order dynamic system and therefore its dynamic behavior can be studied with the results obtained in Chapters 3 and 4.

6.5 Questions and Solutions

In this section, two questions and their solutions are presented. One is concerned with a shunt DC machine and the other deals with a series or self-excited series machine.

Example 1

A 3 kW, 110 V DC generator has an efficiency of 75.0% on full-load. The armature and shunt-field resistances are $R_a = 0.5\ \Omega$ and $R_f = 50\ \Omega$, respectively. Determine (a) the power taken by the machine when run light as a motor from a 110 V supply and (b) the shaft output when run as a motor operating on 110 V mains and with an armature current i_a of full-load value.

Solution:

As a generator, the power input becomes

$$p_i = \frac{3000}{0.75}\,\text{W} = 4000\ \text{W}.$$

The currents are $i_f = 110/50$ A = 2.2 A, and $i = 3000/110$ A = 27.27 A.
Thus, the rated full-load armature current, $i_a = i + i_f = 29.47$ A.
The output power is

$$\begin{aligned} p_o &= i_a^2 R_a + i_f^2 R_f + 3000\ \text{W} \\ &= 29.47^2 \times 0.5 + 2.2^2 \times 50 + 3000\ \text{W} = 3676.32\ \text{W}. \end{aligned}$$

Therefore, the loss of power due to rotation is $p_L = p_i - p_o = 4000 - 3676.32\ \text{W} = 323.68$ W.

a. *As a motor, running light*

$$\text{Armature power input,}\quad p_a = p_L + i_a^2 R_a.$$
$$110\,i_a = 323.68 + i_a^2(0.5) \quad \text{or} \quad i_a^2 - 220 i_a + 647.36 = 0.$$

This gives i_a = 3 A or 227 A. However, only i_a = 3.0 A is admissible.
Therefore, power input running light, p_r = (2.2 + 3.0)110 W = 572 W.

b. *As a motor on full-load*

$$i_a = i + i_f = 29.47\text{A},\quad i_f = 2.2\text{A},\quad \text{and}$$
$$\text{therefore,}\quad i_L = i_a + i_f = 31.67\text{A}.$$

Input power, $p_i = v_g i_L$ = 110 × 31.67 W = 3483.7 W.
Output power now,

$$p_o = 3483.7 - \left(i_a^2 \times R_a + i_f^2 R_f + p_L\right)$$

or

$$p_o = 3483.7 - \left(29.47^2 \times 0.5 + 2.2^2 \times 50 + 323.68\right)\text{W}.$$

Therefore, output power, p_o = 2483.78 W = 3.3308 h.p.

Example 2

A DC series motor of 230 V rotates at 1000 rpm when taking 110 A. If the armature resistance $R_a = 0.10\ \Omega$ and the field resistance $R_f = 0.05\ \Omega$, evaluate (a) the torque developed and (b) the speed at reduced load when the machine takes 50 A. According to the magnetization curve (not presented in this chapter), it gives 2.18×10^{-4} Wb/A at 110 A and 2.3×10^{-4} Wb/A at 50 A.

Solution:

a. The generated voltage is given by

$$v_g = v - i_a\left(R_a + R_f\right)$$

where v is the terminal voltage. Therefore, from the given data, one has

$$v_g = 230 - 110(0.10 + 0.05)V = 213.5\ \text{V}.$$

Thus, the torque t_m developed is given by the following power relation,

$$\frac{2\pi(1000)t_m}{60} = i_a v_g = 110(213.5).$$

This gives the torque developed, t_m = 224.2652 Nm.

b. At reduced load and when the machine takes 50 A, the new voltage generated is given by

$$v_g = 230 - 50(0.10 + 0.05)V = 222.5\ \text{V}.$$

In a DC machine,

$$v_g \propto \omega\varphi$$

where ω is the rotating speed of the machine, while φ is related to the given magnetization coefficient. Accordingly, one can write

$$\frac{\omega_1}{\omega_2} = \frac{v_{g1}\varphi_2}{v_{g2}\varphi_1}.$$

Therefore, the new speed ω_2 is given by

$$\frac{\omega_1 v_{g2}\varphi_1}{v_{g1}\varphi_2} = \omega_2 = \frac{1000(222.5)\left(100 \times 2.18 \times 10^{-4}\right)}{213.5\left(50 \times 2.3 \times 10^{-4}\right)}\ \text{rpm}.$$

Simplifying, this gives ω_2 = 1975.56 rpm.

This means the new speed is almost twice the original one of 1000 rpm.

Exercise Questions

Q1. The main component of a sound system is the amplifier and speaker. The latter is represented by the load, which draws current from the amplifier in order to produce sound. The circuit representation of this component of the sound system is shown in Figure 6Q1. The resistance of the load is R_L, and the current and voltage from the amplifier are, respectively, i_a and v_a. The associated (internal) resistance of the amplifier is R_a. In general, one of the goals for the designer is to optimize efficiency, which in turn means to maximize the power supplied to the speaker, for a given set of v_a and R_a. Determine the value of the load resistance R_L such that the power transfer to the speaker is maximized.

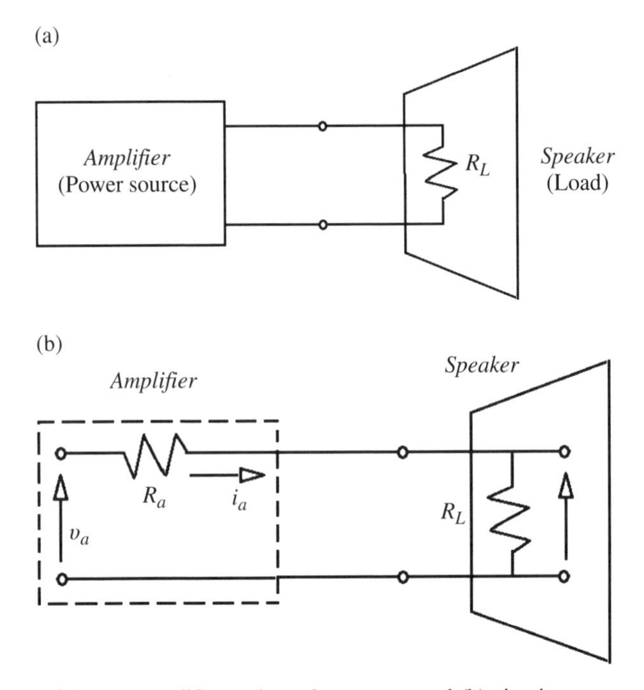

Figure 6Q1 (a) Amplifier and speaker system and (b) circuit representation

Q2. A switch S in an electrical circuit, as shown in Figure 6Q2, is open for $t < 0$. It is assumed that the system is in steady state. When S is closed at $t = 0$, determine the current $i(t)$ for $t \geq 0$.

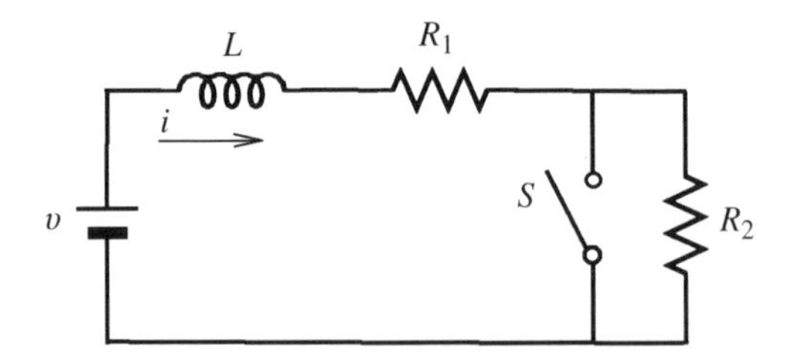

Figure 6Q2 Switch in an electrical circuit

Q3. A DC motor system is shown in Figure 6Q3, in which Gear 2 that drives the rotor of mass moment of inertia J_1 is supported by appropriate bearings (not shown in the figure). Determine the transfer function of the system, $\frac{\Theta_2(s)}{V_a(s)}$ in which $\Theta_2(s)$ is the Laplace transform of the angular displacement output from the rotor $\Theta_2(t)$, and $V_a(s)$ is the Laplace transform of the applied armature voltage $v_a(t)$, or simply v_a. The torque T is developed by the motor, while i_f the field current is assumed to be constant. The back emf is $v_b(t)$. The angular displacement of the motor shaft is θ_1. The numbers of teeth in Gears 1 and 2 are, respectively denoted by N_1 and N_2.

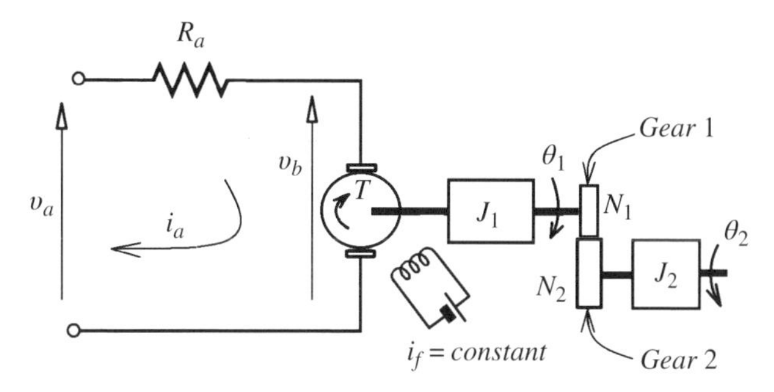

Figure 6Q3 A DC motor system

Q4. An amplifier of gain K and output resistance R is employed to supply the field voltage of a separately-excited DC motor. The armature is, however, supplied from a constant-current source as indicated in Figure 6Q4. Assume that the combined inertia of armature and load (rotor) is J, which is constant. The frictional damping is c. Derive the transfer functions $T_s(s)/V_f(s)$, and $\Omega(s)/T_s(s)$, where $T_s(s)$ is the Laplace transform of the shaft torque $t_s(t)$, $V_f(s)$ the Laplace transform of the field voltage $v_f(t) = Kv_a(t)$, with $v_a(t)$ being the input voltage to the amplifier, and $\Omega(s)$ the Laplace transform of the angular velocity ω or $\omega(t)$.

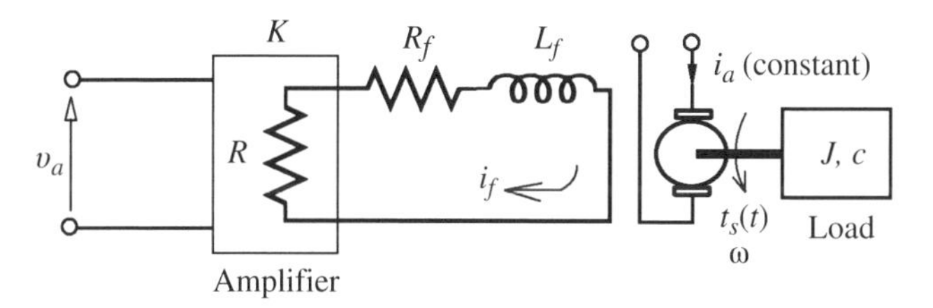

Figure 6Q4 Separately-excited DC motor

References

[1] Steven, R.E. (1970). *Electromechanics and Machines*. Chapman & Hall, London.

[2] Puchstein, A.F. (1961). *The Design of Small Direct Current Motors*. John Wiley & Sons, Inc., New York.

7

Dynamic Characteristics of Transducers

The word *transducer* as used in this chapter has a variety of names in the literature. These latter names are: *sensor, probe, gauge,* and *pick-up.* In general terms and in the present context, a transducer is defined as *a device or system that measures one form of energy and converts it into another form of energy, typically electrical energy so that it can be recorded, stored, transmitted, and analysed.* The literature of transducers or sensors is vast [1], and therefore in the present chapter only a selected numbers of transducers that are frequently applied to dynamic measurement and feedback control systems are introduced. The term *control* refers to the process of *modifying* the *dynamic behavior* of a system in order to achieve some *desired outputs.* A *system* is a combination of components or elements so constructed to achieve an objective or multiple objectives. These definitions and applications of control and system are to appear again from Chapter 8 onwards.

It should be mentioned that modern machines and dynamic systems employ many micro-electromechanical systems (MEMSs) as well as nano-electromechanical systems (NEMSs) for many applications. These systems are often designed to serve as sensors or transducers.

The organization of this chapter is as follows: the basic theory of the tachometer is presented in Section 7.1, while Section 7.2 is concerned with the principles and applications of oscillatory motion transducers. Principles and applications of microphones are introduced in Section 7.3. The microphones considered are the moving-coil microphone and the condenser microphone. Principles and applications of the piezoelectric hydrophone are included in Section 7.4. Three problem examples are presented in Section 7.5.

7.1 Basic Theory of the Tachometer

The simplest and most common transducer for rotational speed measurement is the tachometer, as sketched in Figure 7.1, whose equation of motion may be derived similarly to an electric

Introduction to Dynamics and Control in Mechanical Engineering Systems, First Edition. Cho W. S. To.

Companion website: www.wiley.com/go/to/dynamics

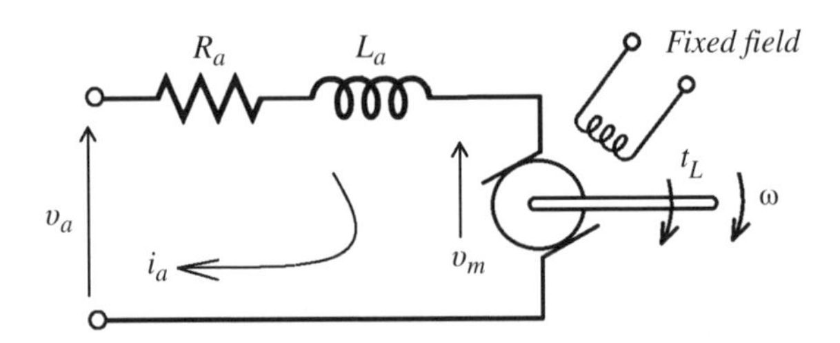

Figure 7.1 A tachometer as a DC generator with fixed field

motor. The difference is that instead of input voltage in a motor, here the load torque t_L is considered as the input. Thus, by reference to Equation (6.25a), which is repeated here for direct application

$$v_a = R_a i_a + L_a \frac{di_a}{dt} + v_m, \tag{7.1}$$

where the electromotive force (emf) voltage $v_m = K_e \omega$ in which ω is the rotational speed which is being measured and K_e is a constant.

Applying Equation (7.1) and recalling that there is no applied voltage v_a in the tachometer, so $v_a = 0$, and at steady-state condition $\frac{di_a}{dt} = 0$ one has

$$0 = R_a i_a + K_e \omega. \tag{7.2}$$

Note that $v_t = -R_a i_a$ is simply the voltage across the resistor R_a. Therefore, by measuring the voltage v_t one is able to determine the angular velocity or rotational speed ω. In other words,

$$v_t = -R_a i_a = K_e \omega.$$

Taking the Laplace transform for this equation and rearranging, one can find the transfer function for the tachometer as

$$\frac{V_t(s)}{\Omega(s)} = K_e, \tag{7.3}$$

in which $V_t(s) = \ell\{v_t\}$ and $\Omega(s) = \ell\{\omega\}$. In Equation (7.3) K_e is called the *tachometer constant* and is usually given as a catalog parameter (from the manufacturer) in volts per 1000 rpm.

7.2 Principles and Applications of Oscillatory Motion Transducers

To limit the scope, and since it is an introduction to the principles and applications of oscillatory motion transducers, the following is confined to two types of oscillatory motion transducers, the accelerometer and the seismometer. A typical oscillatory motion transducer such as the accelerometer is sketched in Figure 7.2 and its model and corresponding free-body diagram

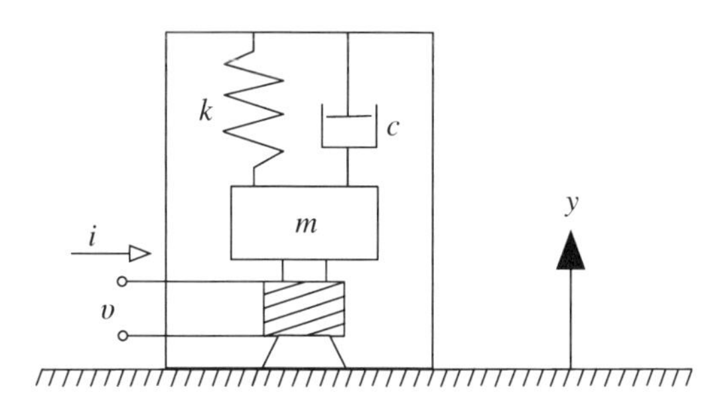

Figure 7.2 A typical transducer

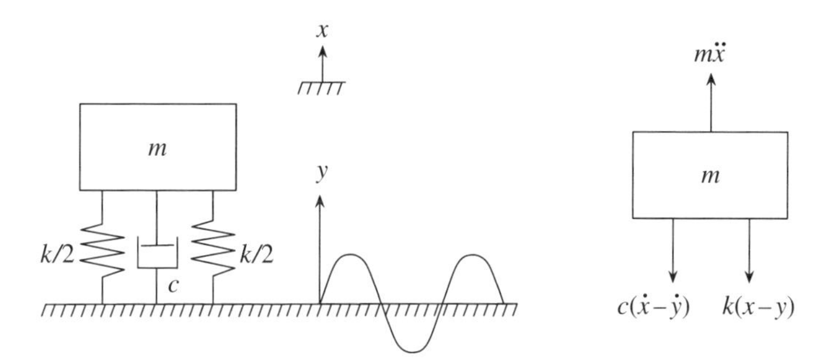

Figure 7.3 Transducer model and its FBD

(FBD) is included in Figure 7.3. In the following two subsections the equation of motion and design considerations of the two types of oscillatory motion transducers are presented.

7.2.1 Equation of Motion

The equation of motion for the system in Figure 7.3 is

$$m\ddot{x} + c(\dot{x} - \dot{y}) + k(x - y) = 0. \tag{7.4}$$

Making the substitution $z = x - y$, Equation (7.4) becomes

$$m\ddot{z} + c\dot{z} + kz = -m\ddot{y} = m\omega^2 Y \sin \omega t \tag{7.5}$$

where y is the motion at the base and it has been assumed that $y = Y \sin \omega t$, with Y being the amplitude of the base motion y, and ω the angular frequency of the base motion, and z is the relative displacement. Note that the base motion is that to be measured. For example, if the oscillatory motion y of a wing of an aeroplane is to be recorded, the transducer or accelerometer is to be attached to the vibrating wing.

- The form of this equation is similar to the single degree-of-freedom (dof) system for oscillatory motion in Chapter 4. In the latter system x there is replaced by z here, and F_o there is

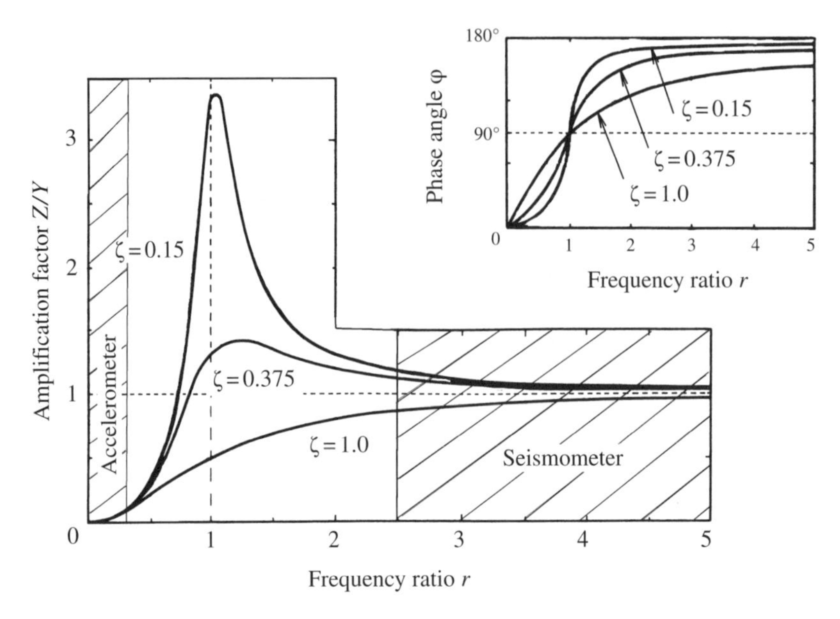

Figure 7.4 Magnitude and phase characteristics of a transducer

replaced by $m\omega^2 Y$. Thus, the solution to Equation (7.5) can be readily written, with reference to Equations (4.10) and (4.11), as

$$z = Z\sin(\omega t - \varphi),$$

$$Z = \frac{m\omega^2 Y}{\sqrt{(k - m\omega^2)^2 + (c\omega)^2}}, \quad \text{and} \quad \tan\varphi = \frac{c\omega}{k - m\omega^2}.$$

As in Chapter 4, these two quantities can be expressed as

$$\frac{Z}{Y} = \frac{r^2}{\sqrt{(1 - r^2)^2 + (2\zeta r)^2}} \tag{7.6}$$

$$\tan\varphi = \frac{2\zeta r}{1 - r^2} \tag{7.7}$$

where the frequency ratio $r = \dfrac{\omega}{\omega_n}$ and $\dfrac{Z}{Y}$ is the magnitude ratio.

Results of these expressions are plotted in Figure 7.4.

7.2.2 Design Considerations of Two Types of Transducer

In this subsection, design considerations of two types of oscillatory motion transducer are introduced.

For optimal dynamic system analysis the damping ratio $\zeta = \frac{1}{\sqrt{2}} = 0.707$, and when the natural frequency ω_n of the transducer is low compared with the vibrating frequency ω, the frequency ratio r approaches a large value, and the amplitude ratio or amplification factor $\frac{Z}{Y}$ is shown in Figure 7.4. This case corresponds to the *seismometer*, which is a transducer for measuring earthquake signals.

- One feature of a seismometer is its large mass (therefore lower natural frequency). As the relative amplitude Z approaches Y in the range that characterizes seismic activity, the relative motion of the seismometer must be of the same order of magnitude as that of the vibration to be measured.
- The relative magnitude Z is usually converted to an electrical voltage by making the mass m a magnet relative to the coil fixed in the case as shown in Figure 7.2, such that the voltage generated is proportional to the rate of cutting the magnetic field. This, in turn, means that the output of the transducer is proportional to the velocity of the vibrating body.
- Typical natural frequencies of seismometers are from 1 to 5 Hz, with a useful range of 10–2000 Hz.

When the natural frequency ω_n of the transducer is high compared with that of the frequency ω to be measured, the transducer is an accelerometer whose feature is indicated in Figure 7.4. In an accelerometer, as r approaches 0, since ω_n is high, the relative magnitude Z is approximately equal to $r^2 Y$.

- For acceleration measurement with transducer having high natural frequency, usually implies small mass, the operating range of r is from 0 to 0.06 f_n with $f_n = \frac{\omega_n}{2\pi}$, without distortion.
- Typical accelerometers of the piezoelectric type have natural frequencies in the 100 kHz range, such that the useful frequency range is between 0 and 3 kHz.

7.3 Principles and Applications of Microphones

Microphones are transducers that convert dynamic air pressure or acoustic signals into electrical energy. Generally, they may be classified into two categories: one is the *constant-velocity* and the other *constant-amplitude* types. The former type includes the moving-coil, velocity-ribbon, and magnetostriction microphones, while the latter category consists of, for example, the carbon, condenser, and crystal microphones. Microphones are either phase-shift operated, pressure-operated, or pressure-gradient operated. Each of the three types of operation will either accept or discriminate against acoustic signals from a particular direction.

A simplified moving-coil microphone is considered in Section 7.3.1, while the condenser microphone is included in Section 7.3.2. More detailed discussions on microphones may be found in references [2–5].

7.3.1 Moving-Coil Microphone

The schematic cross-section view of a simplified moving-coil microphone is presented in Figure 7.5a, whose oscillatory model representation and associated FBD is included in

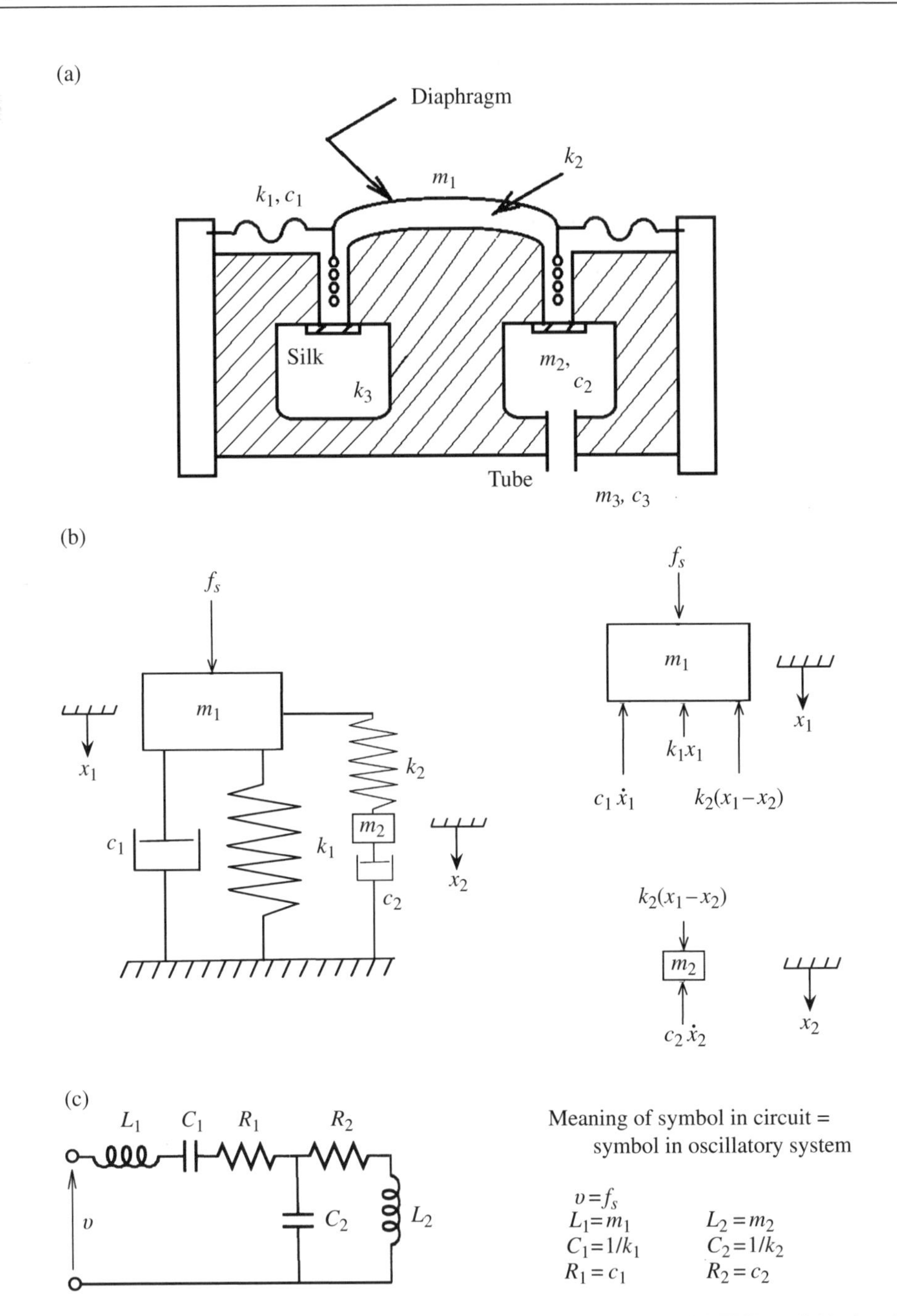

Figure 7.5 (a) Sketch of a moving-coil microphone; (b) oscillatory model with FBD; and (c) electrical circuit analog

Figure 7.5b. In this model, f_s is the force due to the dynamic acoustic pressure impinging on the diaphragm, m_1 is the mass of the light metal diaphragm, k_1 and c_1 are the stiffness and resistance or damping contributed by the corrugated annulus supporting the diaphragm, k_2 is the stiffness due to compression of the air trapped in the chamber beneath the diaphragm, m_2 and c_2 are the mass and resistance or damping due to viscous forces opposing the flow of air through the capillaries of the silk cloth, and k_3 is the stiffness due to the air chamber below the silk cloth. The latter stiffness is relatively small such that it is disregarded in the present analysis. Similarly, its associated mass and damping, m_3 and c_3 are relatively small such that they are disregarded in the analysis. The electrical circuit analog for this microphone is given in Figure 7.5c.

With reference to the FBD in Figure 7.5b and for the mass m_1, the equation of motion by applying the Newton's second law of motion becomes

$$m_1\ddot{x}_1 = -c_1\dot{x}_1 - k_2(x_1 - x_2) - k_1x_1 + f_s.$$

Re-arranging terms, one has

$$\begin{aligned} m_1\ddot{x}_1 + c_1\dot{x}_1 + k_2(x_1 - x_2) - k_1x_1 &= f_s \quad \text{or} \\ m_1\ddot{x}_1 + c_1\dot{x}_1 + (k_1 + k_2)x_1 - k_2x_2 &= f_s. \end{aligned} \tag{7.8}$$

Similarly, applying Newton's second law of motion for mass m_2, the equation of motion is given by

$$m_2\ddot{x}_2 = -c_2\dot{x}_2 + k_2(x_1 - x_2).$$

After re-arranging terms, it leads to

$$m_2\ddot{x}_2 + c_2\dot{x}_2 - k_2x_1 + k_2x_2 = 0. \tag{7.9}$$

Expressing Equations (7.8) and (7.9) in matrix form:

$$\begin{bmatrix} m_1 & 0 \\ 0 & m_2 \end{bmatrix}\begin{pmatrix} \ddot{x}_1 \\ \ddot{x}_2 \end{pmatrix} + \begin{bmatrix} c_1 & 0 \\ 0 & c_2 \end{bmatrix}\begin{pmatrix} \dot{x}_1 \\ \dot{x}_2 \end{pmatrix} + \begin{bmatrix} k_1 + k_2 & -k_2 \\ -k_2 & k_2 \end{bmatrix}\begin{pmatrix} x_1 \\ x_2 \end{pmatrix} = \begin{pmatrix} f_s \\ 0 \end{pmatrix}. \tag{7.10}$$

This is the matrix equation of motion for the simplified moving-coil microphone model. To obtain the transfer function $\dfrac{\mathcal{L}\{x_1\}}{\mathcal{L}\{f_s\}} = \dfrac{X_1(s)}{F_s(s)}$ of the microphone system, one may apply the method of Laplace transforms as in the following.

Assuming the system starts from rest and taking the Laplace transform for the second part of Equation (7.10), one obtains

$$X_2 = \frac{k_2X_1}{m_2s^2 + c_2s + k_2}. \tag{7.11}$$

Similarly, taking the Laplace transform of the first equation of Equation (7.10):

$$X_1\left(m_1s^2 + c_1s + k_1 + k_2\right) - k_2X_2 = F_s.$$

Substituting Equation (7.11) into the last equation and after rearranging, one has the transfer function of the moving-coil microphone as

$$\frac{X_1}{F_s} = \frac{X_1(s)}{F_s(s)} = \frac{m_2 s^2 + c_2 s + k_2}{(m_1 s^2 + c_1 s + k_1 + k_2)(m_2 s^2 + c_2 s + k_2) - k_2^2}.$$

After simplifying, one obtains

$$\frac{X_1}{F_s} = \frac{X_1(s)}{F_s(s)} = \frac{m_2 s^2 + c_2 s + k_2}{a_4 s^4 + a_3 s^3 + a_2 s^2 + a_1 s + a_0}, \tag{7.12}$$

in which

$$a_4 = m_1 m_2, \quad a_3 = m_1 c_2 + m_2 c_1, \quad a_2 = m_2 k_1 + k_2(m_1 + m_2) + c_1 c_2,$$
$$a_1 = c_1 k_2 + c_2(k_1 + k_2), \quad a_0 = k_1 k_2.$$

In practice, the masses m_1 and m_2 are small to the extent that their product is much smaller and the term associated with a_4 in Equation (7.12) can be disregarded, and Equation (7.12) may be approximated as

$$\frac{X_1}{F_s} = \frac{c_2 s + k_2}{a_3 s^3 + a_2 s^2 + a_1 s + a_0}. \tag{7.13}$$

7.3.2 Condenser Microphone

The operation of a condenser microphone depends on the variation in capacitance between a tightly stretched metal diaphragm and a fixed plate. While condenser microphones have some important disadvantages, such as very high internal impedance and polarizing voltage, they are frequently used in acoustic measurements and noise control research. Thus, in this subsection a simple presentation is included.

Consider the typical condenser microphone shown in Figure 7.6a. The cross-section sketch of a simple condenser microphone is included in Figure 7.6b. It consists of a rigid backing plate separated by a small distance $\boldsymbol{d}$ from a parallel thin stretched metal diaphragm, usually of aluminum or steel. This metal diaphragm has a radius $\boldsymbol{a}$. The backing plate is insulated from the remainder of the microphone. A polarizing voltage is supplied by a battery between the plate and the diaphragm. The corresponding circuit is shown in Figure 7.6c with the polarizing voltage v_o, load resistance R_L, and capacitance C. When a sound wave reaches the diaphragm and displaces it, resulting in a change of capacitance of the microphone, a signal voltage v_L is induced across the resistor R_L. The instantaneous capacitance can be written as

$$C = C_0 + C_1 \sin \omega t \tag{7.14}$$

where C_0 is the capacitance when there is no acoustic pressure on the metal diaphragm and C_1 is the amplitude of changing capacitance as a result of impinging sinusoidal sound pressure.

(a)

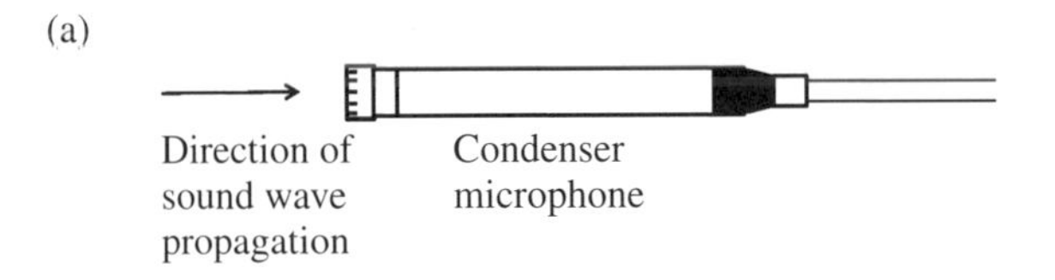

(b)

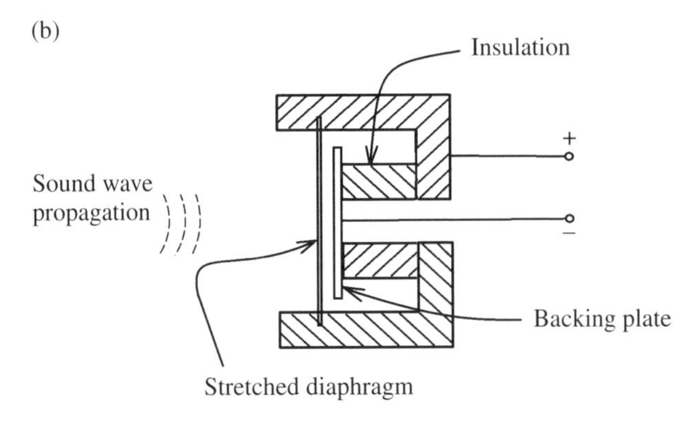

(c)

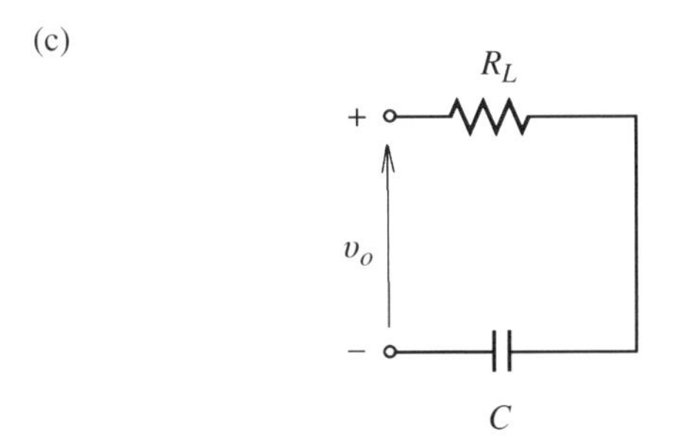

Figure 7.6 (a) A typical condenser microphone, (b) sketch of a simple condenser microphone, and (c) equivalent electrical circuit

With reference to the circuit in Figure 7.6c, one has the voltage equation in terms of current i as

$$v_o = iR_L + \left(\frac{1}{C}\right)\int i\,dt. \tag{7.15}$$

Substituting Equation (7.14) into Equation (7.15) and taking the time derivatives, it becomes

$$(C_0 + C_1 \sin\omega t)R_L \frac{di}{dt} + (1 + R_L C_1 \omega \cos\omega t)i = v_o C_1 \omega \cos\omega t. \tag{7.16a}$$

A series solution to this first-order linear differential equation with time-dependent coefficients can be obtained by assuming the current

$$i = \sum_{r=1}^{\infty} A_r \sin(r\omega t + \varphi_r), \tag{7.16b}$$

not to be confused with the imaginary number i.

For practical condenser microphones, even for cases with very intense acoustical pressure, $C_1 = C_0$. Consequently, the amplitudes $A_2, A_3, A_4, \ldots$ of higher harmonics are negligible compared with A_1. The solution to Equation (7.16a) may therefore be approximated as [3]:

$$i \approx \left(\frac{v_o C_1}{C_0}\right) \frac{\sin(\omega t + \varphi_1)}{\sqrt{\left(\frac{1}{\omega C_0}\right)^2 + R_L^2}} \tag{7.17a}$$

$$\tan \varphi_1 = \frac{1}{\omega C_0 R_L} \tag{7.17b}$$

The proof of Equation (7.17a) is presented in Appendix 7A.

The voltage across the resistor is therefore

$$v_L = i\,R_L = \left(\frac{v_o C_1 R_L}{C_0}\right) \frac{\sin(\omega t + \varphi_1)}{\sqrt{\left(\frac{1}{\omega C_0}\right)^2 + R_L^2}}. \tag{7.18}$$

For a circular diaphragm of radius a and low-frequency acoustic signals, Equation (7.18) can be applied to derive the open-circuit voltage response [3]:

$$W_o = \frac{v_o a^2}{8Td}, \tag{7.19}$$

where T is the tension in the stretched diaphragm.

As a numerical example, consider a typical condenser microphone having the following properties: $d = 40$ μm, $a = 12.7$ mm, $T = 20$ kN/m, and the polarizing voltage $v_o = 240$ V.

Applying Equation (7.19), one has the open-circuit voltage response

$$W_o = \frac{240 \times 0.0127^2}{8 \times 40 \times 10^{-6}\ 20 \times 10^3} \frac{\mathrm{V}}{\mathrm{N}\,m^{-2}} = 6.0484 \times 10^{-3} \frac{\mathrm{V}}{\mathrm{Pa}}.$$

7.4 Principles and Applications of the Piezoelectric Hydrophone

A hydrophone is similar to a microphone, but the former is a transducer for converting pressure variations associated with acoustic wave propagation in liquids into electrical voltage. The derivation of the relationship between the input voltage v_i generated by the acoustic wave and the output voltage v_o of the piezoelectric hydrophone may be represented by the equivalent electrical circuit shown in Figure 7.7.

With reference to Figure 7.7 and applying the loop method introduced in Chapter 6, one has

$$iR_1 + iR_2 - v_i = 0 \quad \text{or} \quad v_i = iR_1 + iR_2 \tag{7.20a}$$

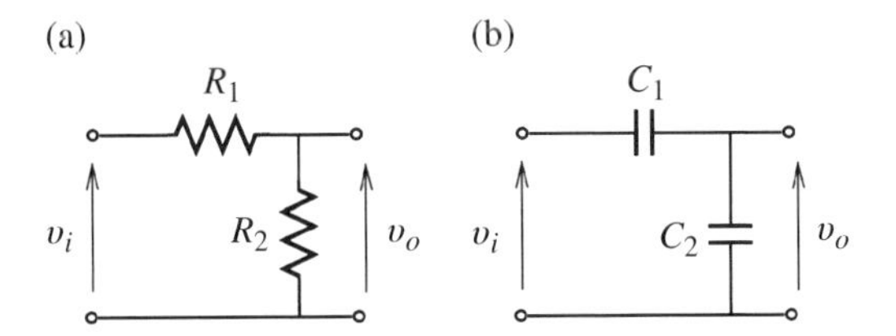

Figure 7.7 (a) General circuit and (b) low-frequency circuit of a piezoelectric hydrophone

and the output voltage is given by

$$v_o = iR_2. \tag{7.20b}$$

Applying Laplace transformation to Equations (7.20a) and (7.20b) gives

$$V_i(s) = I(s)R_1 + I(s)R_2 \tag{7.21a}$$

$$V_o(s) = I(s)R_2 \tag{7.21b}$$

From Equation (7.21b), one has $I(s) = V_o(s)/R_2$. Upon substituting the latter into Equation (7.21a) and disregarding the argument in the equations for conciseness, one obtains

$$V_i = \frac{R_1 V_o}{R_2} + V_o \quad \text{or} \quad \frac{V_o}{V_i} = \frac{R_2}{R_1 + R_2}. \tag{7.22}$$

As an application of Equation (7.22) to the special case in which the frequency of the input acoustic signal is well below the fundamental natural frequency of the piezoelectric material, the electrical circuit in this case is shown in Figure 7.7b. Thus, applying the loop method in Chapter 6 results in

$$\frac{\int_0^t i\,dt}{C_1} + \frac{\int_0^t i\,dt}{C_2} - v_i = 0 \quad \text{or} \quad v_i = \frac{\int_0^t i\,dt}{C_1} + \frac{\int_0^t i\,dt}{C_2} \tag{7.23a}$$

and the output voltage is given by

$$v_o = \frac{\int_0^t i\,dt}{C_2}. \tag{7.23b}$$

Taking the Laplace transformation of Equations (7.23a) and (7.23b) and disregarding the argument in the equations, one obtains

$$V_i = \frac{I}{C_1 s} + \frac{I}{C_2 s} \tag{7.24a}$$

$$V_o = \frac{I}{C_2 s} \tag{7.24b}$$

Applying these equations and rearranging, the transfer function of the low-frequency piezoelectric hydrophone becomes

$$\frac{V_o}{V_i} = \frac{C_1}{C_1 + C_2}. \tag{7.25}$$

For a more detailed discussion, reference [3] may be consulted.

7.5 Questions and Solutions

In this section, three questions and their solutions are included. The first question is concerned with the use of an accelerometer, while the second question deals with the transfer function of a moving-coil microphone. The third question is an illustration of the use of a hydrophone in the computation of the speed of a submarine.

Example 1

An accelerometer is applied to measure the signal of a vibrating machine. The motion of the accelerometer may be described by a linear second-order differential equation. Since the accelerometer is considered as a linear single degree-of-freedom (dof) system, its static displacement is proportional to the applied force. If the undamped natural frequency of the system is 3600 Hz and the total viscous damping is 75% of critical (that is, $\zeta = 0.75$), find the frequency range(s) over which the ratio of dynamic amplitude to static amplitude (the so-called inherent error) deviates from unity by an amount no greater than 6%.

Solution:

From Equation (4.12), the amplification ratio or magnification factor is defined as

$$\frac{kX}{F} = \frac{1}{\sqrt{(1-r^2)^2 + (2\zeta r)^2}}$$

which is also the ratio of dynamic amplitude to static amplitude.

Therefore, the ratio of dynamic amplitude to static amplitude deviates from unity by an amount no greater than 6%, which implies that

$$\frac{kX}{F} = 1.06 \quad \text{or} \quad \frac{kX}{F} = 0.94.$$

For $\frac{kX}{F} = 1.06$, one has

$$1.06 = \frac{1}{\sqrt{(1-r^2)^2 + (2 \times 0.75\, r)^2}}$$

which gives

$$\frac{1}{1.06^2} = 0.8899 = \left(1 - r^2\right)^2 + (2 \times 0.75\ r)^2.$$

Writing $u = r^2$, this equation becomes $0.8899 = (1 - u)^2 + (2 \times 0.75)^2 u$.
Solving this quadratic equation it gives

$$u = \left(\frac{-0.25}{2}\right) \pm \frac{1}{2}\sqrt{0.25^2 - 4(0.11)}.$$

This leads to imaginary roots.
Thus, one should try the second condition,

$$\frac{kX}{F} = 0.94 = \frac{1}{\sqrt{(1 - r^2)^2 + (2 \times 0.75\ r)^2}}.$$

Following a similar procedure to the foregoing, one finds that $u = 0.2588$, so that

$$u = r^2 = \left(\frac{\omega}{\omega_n}\right)^2$$

which leads to $\omega = \omega_n \sqrt{0.2588}$ where ω_n is given as 3600 Hz.
Therefore, $\omega = 11507.06$ rad/s or $f = 1831.41$ Hz. That is, the required frequency range is $0 \leq f \leq 1831.41$ Hz.

Example 2
A moving-coil microphone or so-called dynamic microphone as shown in Figure 7.5 has the following system parameters:

$$m_1 = 6 \times 10^{-4}\text{kg}, \quad m_2 = 3 \times 10^{-4}\text{kg},$$
$$c_1 = 1.0\frac{\text{Ns}}{\text{m}}, \quad c_2 = 24\frac{\text{Ns}}{\text{m}}, \quad k_1 = 10^4\frac{\text{N}}{\text{m}}, \quad \text{and} \quad k_2 = 10^6\frac{\text{N}}{\text{m}}.$$

Using these parameters, determine the transfer function of the dynamic microphone.

Solution:
With the given parameters and applying Equation (7.12), one finds

$$a_4 = m_1 m_2 = 1.8 \times 10^{-7}(\text{kg})^2, \quad a_3 = m_1 c_2 + m_2 c_1 = 1.47 \times 10^{-2}\left(\frac{\text{kg}^2}{\text{s}}\right),$$
$$a_2 = m_2 k_1 + k_2(m_1 + m_2) + c_1 c_2 = 927\left(\frac{\text{kg}}{\text{s}}\right)^2,$$
$$a_1 = c_1 k_2 + c_2(k_1 + k_2) = 25.24 \times 10^6\left(\frac{\text{kg}^2}{\text{s}^3}\right),$$
$$a_0 = k_1 k_2 = 10^{10}\left(\frac{\text{N}}{\text{m}}\right)^2.$$

Therefore, the transfer function given by Equation (7.12) becomes

$$\frac{X_1(s)}{F_s(s)} = \frac{(3\times10^{-10})s^2 + 2.4\times10^{-5}s + 1}{(1.8\times10^{-13})s^4 + (1.47\times10^{-8})s^3 + 9.27\times10^{-4}s^2 + 25.24\ s + 10^4}.$$

The equation can be approximated to

$$\frac{X_1(s)}{F_s(s)} = \frac{1}{25.24\ s + 10^4}.$$

The approximated transfer function indicates that this moving-coil microphone may be considered as a first-order system.

Example 3

The frequency of a return echo signal from a submarine detected by a hydrophone is 40,800 Hz, whereas the driving frequency supplied to the sonar transducer aboard a frigate is 40,500 Hz. If the frigate is speeding at 35 knots (17.9167 m/s), determine the speed of the submarine. Note that the speed of sound c in seawater is 1500 m/s.

Solution:

Owing to the Doppler effect, the detected frequency is given by

$$f_d = f\left(1 + 2\frac{v_{s/h}}{c}\right)$$

where f is the true frequency generated at the source (frigate) in Hz, and $v_{s/h}$ is the relative speed between the source (frigate) and receiver (submarine) in m/s.

Thus,

$$40\ 800 = 40\ 500\left(1 + 2\frac{v_{s/h}}{1500}\right)$$

which leads to

$$v_{s/h} = 5.5556\ \mathrm{m/s}$$

and the speed of the submarine is

$$v_h = v_s - v_{s/h} = (17.9167 - 5.5556)\mathrm{m/s} = 12.3611\ \mathrm{m/s}\ \text{or 24.2375 knots.}$$

Appendix 7A: Proof of Approximated Current Solution

For $r = 1$, Equation (7.16b) reduces to

$$i = A_1\ \sin(\omega t + \varphi_1). \tag{A1}$$

Substituting this equation into Equation (7.16a) gives

$$(C_0 + C_1 \sin\omega t)R_L \frac{dA_1 \sin(\omega t + \varphi_1)}{dt} + (1 + R_L C_1 \omega \cos \omega t)i = v_o C_1 \omega \cos \omega t.$$

Dividing both sides by C_0,

$$\left(1 + \frac{C_1}{C_0}\sin \omega t\right)R_L \frac{dA_1 \sin(\omega t + \varphi_1)}{dt} + \left(\frac{1}{C_0} + R_L \frac{C_1}{C_0}\omega \cos \omega t\right)i = \left(\frac{v_o C_1}{C_0}\right)\omega \cos \omega t.$$

Since $C_1 << C_0$, the above equation can be approximated as

$$R_L \frac{dA_1 \sin(\omega t + \varphi_1)}{dt} + \left(\frac{1}{C_0}\right)i = \left(\frac{v_o C_1}{C_0}\right)\omega \cos \omega t.$$

Performing differentiation of the first term on the lhs, then

$$R_L A_1 \omega \cos(\omega t + \varphi_1) + \left(\frac{1}{C_0}\right)i = \left(\frac{v_o C_1}{C_0}\right)\omega \cos \omega t.$$

Substituting A_1 from Equation (A1), it becomes

$$R_L \frac{i}{\sin(\omega t + \varphi_1)}\omega \cos(\omega t + \varphi_1) + \left(\frac{1}{C_0}\right)i = \left(\frac{v_o C_1}{C_0}\right)\omega \cos \omega t.$$

This gives the current i as

$$i = \left(\frac{v_o C_1}{C_0}\right)\left[\frac{(\omega \cos\omega t)\ \sin(\omega t + \varphi_1)}{R_L \omega\ \cos(\omega t + \varphi_1) + \frac{\sin(\omega t + \varphi_1)}{C_0}}\right]. \tag{A2}$$

But

$$\cos(\omega t + \varphi_1) = (\cos \omega t)\cos \varphi_1 - (\sin \omega t)\sin \varphi_1, \tag{A3a}$$

$$\sin(\omega t + \varphi_1) = (\sin \omega t)\cos\varphi_1 + (\cos \omega t)\sin \varphi_1. \tag{A3b}$$

Recall from Equation (7.17b) that

$$\tan \varphi_1 = \frac{1}{\omega C_0 R_L}$$

which, in turn, gives

$$\sin\varphi_1 = \frac{1}{\sqrt{1+(\omega C_0 R_L)^2}}, \quad \cos\varphi_1 = \frac{\omega C_0 R_L}{\sqrt{1+(\omega C_0 R_L)^2}}.$$

Substituting these into Equations (A3a) and (A3b) respectively gives

$$\cos(\omega t+\varphi_1) = (\cos\omega t)\frac{\omega C_0 R_L}{D} - (\sin\omega t)\frac{1}{D}, \tag{A4a}$$

$$\sin(\omega t+\varphi_1) = (\sin\omega t)\frac{\omega C_0 R_L}{D} + (\cos\omega t)\frac{1}{D}, \tag{A4b}$$

where $D=\sqrt{1+(\omega C_0 R_L)^2}$.

Substituting Equations (A4a) and (A4b) into the denominator term within the square brackets on the rhs of Equation (A2), it becomes

$$R_L\omega\cos(\omega t+\varphi_1) + \frac{\sin(\omega t+\varphi_1)}{C_0} = (\cos\omega t)\frac{(\omega R_L)^2 C_0}{D} - (\sin\omega t)\frac{\omega R_L}{D} + (\sin\omega t)\frac{\omega R_L}{D} + (\cos\omega t)\frac{1}{DC_0}.$$

$$R_L\omega\cos(\omega t+\varphi_1) + \frac{\sin(\omega t+\varphi_1)}{C_0} = (\cos\omega t)\frac{(\omega R_L)^2 C_0}{D} + (\cos\omega t)\frac{1}{DC_0}.$$

Substituting this into Equation (A2) and simplifying, one has

$$i = \left(\frac{v_o C_1}{C_0}\right)\left[\frac{\omega\sin(\omega t+\varphi_1)}{\frac{(\omega R_L)^2 C_0}{D}+\frac{1}{DC_0}}\right] = \left(\frac{v_o C_1}{C_0}\right)\left[\frac{\omega\sin(\omega t+\varphi_1)}{\frac{(\omega R_L C_0)^2+1}{DC_0}}\right].$$

But $(\omega R_L C_0)^2 + 1 = \mathrm{D}^2$. Therefore,

$$i = \left(\frac{v_o C_1}{C_0}\right)\left[\frac{\omega\sin(\omega t+\varphi_1)}{\frac{D}{C_0}}\right] = \left(\frac{v_o C_1}{C_0}\right)\left[\frac{\omega\sin(\omega t+\varphi_1)}{\frac{\omega C_0}{C_0}\sqrt{\left(\frac{1}{\omega C_0}\right)^2+R_L^2}}\right].$$

Upon further simplifying, one arrives at

$$i = \left(\frac{v_o C_1}{C_0}\right)\frac{\sin(\omega t+\varphi_1)}{\sqrt{\left(\frac{1}{\omega C_0}\right)^2+R_L^2}}. \tag{A5}$$

This is Equation (7.17a).

Exercise Questions

Q1. An accelerometer applied to measure the signal of a vibrating machine has its undamped natural frequency given as 3600 Hz. During the course of the measurement the accelerometer was accidently dropped and damaged. Its viscous damping ratio has changed to an unknown value. If the accelerometer is now subjected to a harmonic input of 2400 Hz, the phase angle between the output and input is measured as 45°. Determine the inherent error of the transducer when it is used to measure a harmonic motion of 1800 Hz. What is the phase angle between the output and input at this frequency?

Q2. Vibration preamplifiers are required in vibration signal measurements. Vibration preamplifiers fulfill the essential role of converting the high impedance output of a piezoelectric accelerometer into a low impedance signal suitable for direct transmission to measuring and analyzing instrumentation. There are two basic types of preamplifiers which may be used with piezoelectric accelerometers. These are the charge preamplifiers and the voltage preamplifiers. The former produce an output voltage proportional to the input charge, and the latter produce an output voltage proportional to the input voltage. Charge preamplifiers are frequently applied due to the fact that both very short and very long cables can be employed without changing the overall sensitivity of the system. The simplified equivalent circuit of an accelerometer with a charge preamplifier is shown in Figure 7Q2 (in which the time constant $\tau = RC$ may be used). Derive an equation relating the output voltage $v(t)$ or simply v to the input charge $q(t)$ or q for this accelerometer with preamplifier system. By applying the Laplace transformation, find the transfer function $\frac{V(s)}{Q(s)}$, where $V(s)$ is the Laplace transform of the output voltage v, and $Q(s)$ is the Laplace transform of the charge q that is generated in the accelerometer by the vibration being measured. If the current and voltage are assumed to be harmonic, examine the frequency transfer function in terms of the applied frequency ω.

(*Hint*: the frequency transfer function is obtained by replacing the parameter s in the transfer function with $i\omega$, i being the imaginary number and not to be confused with the current i in the circuit.)

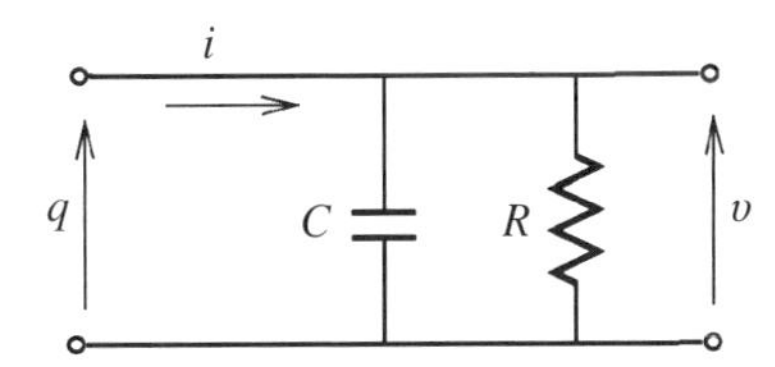

Figure 7Q2 Low frequency equivalent circuit of accelerometer with preamplifier

Q3. A moving-coil microphone as shown in Figure 7.5 has the following system parameters:

$$m_1 = 3 \times 10^{-4}\text{kg}, \quad m_2 = 10^{-4}\text{kg},$$

$$c_1 = 1.0\frac{\text{Ns}}{\text{m}}, \quad c_2 = 20\frac{\text{Ns}}{\text{m}}, \quad k_1 = 10^6\frac{\text{N}}{\text{m}}, \quad \text{and} \quad k_2 = 2 \times 10^6\frac{\text{N}}{\text{m}}.$$

Determine the transfer function of this dynamic microphone. Discuss the implications of an approximated transfer function for this microphone.

Q4. The voltage across the resistor for a condenser microphone shown in Figure 7.6 is given by Equation (7.18) as

$$v_L = \left(\frac{v_o C_1 R_L}{C_0}\right) \frac{\sin(\omega t + \varphi_1)}{\sqrt{\left(\frac{1}{\omega C_0}\right)^2 + R_L^2}}.$$

For a low-frequency response condenser microphone, evaluate the load resistance R_L and discuss its implications when $C_0 = 55$ pF and $C_1 = 1.4$ pF.

Q5. In applying a condenser microphone for measuring sound or dynamic acoustic pressure, a preamplifier is required as in the case of using a piezoelectric accelerometer. The microphone preamplifier performs the basic role of converting the high impedance output of the condenser microphone into a low impedance signal suitable for direct transmission to measuring and analyzing instrumentation. The low-frequency equivalent circuit of a piezoelectric condenser microphone with charge preamplifier is shown in Figure 7Q5. Derive the transfer function of the output voltage v_o to charge q of the condenser microphone with preamplifier system. The charge is generated by the dynamic acoustic pressure that impinges on the diaphragm of the condenser microphone. Find the low-frequency limit given that the input resistance $R = 6.0$ GΩ, preamplifier input capacitance $C_2 = 50$ pF, and the cartridge capacitance $C_1 = 1.4$ pF.

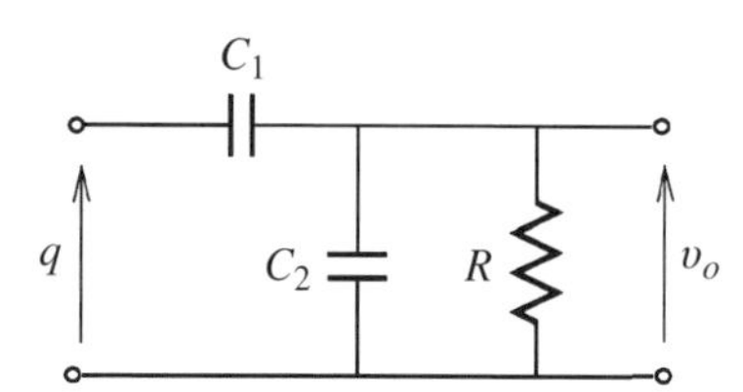

Figure 7Q5 Low-frequency equivalent circuit of condenser microphone with charge preamplifier

References

[1] Hauptmann, P. (1991). *Sensors: Principles and Applications*. Prentice-Hall, Englewood Cliffs, NJ.
[2] Seto, W.W. (1971). *Acoustics*. Schaum's Outline Series, McGraw-Hill, New York.
[3] Kinsler, L.E., and Frey, A.R. (1962). *Fundamentals of Acoustics*. John Wiley & Sons, Inc., New York.
[4] Doebelin, E.O. (1983). *Measurement Systems: Application and Design*, 3rd edn. McGraw-Hill, New York.
[5] Hassall, J.R., and Zaveri, K. (1979). *Acoustic Noise Measurements*, 4th edn. Brüel & Kjær, Copenhagen.

8

Fundamentals of Control Systems

- The term *control* refers to the process of *modifying* the *dynamic behavior* of a system in order to achieve some *desired outputs*.
- Control systems exist in many devices and fields. As pointed out in Chapter 1, the theory of control has been employed by economists, medical personnel, financial experts, political scientists, biologists, and engineers, to name but a few.
- Within the field of mechanical engineering, the heating system and water heater in a house, or the heating, ventilation, and air conditioning (HVAC) system in a modern building, employ control systems.
- In aerospace, the control of aircrafts, helicopters, satellites, and missiles requires very sophisticated advanced control systems [1].
- In shipbuilding industries, control systems are often employed [2].

The primary *objectives* of control system analysis are the determination of the degree or extent of system stability, the steady-state performance, and characteristics of transient response.

The general *procedure of analysis* of a control system may contain:

1. the derivation of the equation of motion or transfer function (TF) for every component;
2. selection of a representation of the system, such as the block diagram; and
3. determination of system characteristics.

The essential *aim of control system design* is to satisfy performance specifications, which are the constraints imposed on the mathematical functions governing the system characteristics. In practice, performance specifications may be stated in either the *time* domain or *frequency* domain. Either of these two forms has its advantages and disadvantages. The frequency domain

Introduction to Dynamics and Control in Mechanical Engineering Systems, First Edition. Cho W. S. To.

Companion website: www.wiley.com/go/to/dynamics

and time domain approaches will be studied in more detail in Chapters 9–12. Meanwhile, fundamentals of control systems are presented in the following sections. It may be appropriate to mention that contents in this and the remaining chapters are aimed at the introductory level, and therefore many applications are left to a second course in this subject.

8.1 Classification of Control Systems

There are many ways of classifying control systems. The following is adopted in the present book:

- *Process control or regulator systems*: the *controlled variable* (output from the system) is confined as closely as possible to a usually constant *desired value* (input).
- *Servomechanisms*: the *input to the system varies* and the *output is made to follow it* (the input) as closely as possible.

8.2 Representation of Control Systems

The overall linear control system is commonly represented by the block diagram, which is believed to be less abstract mathematically than the signal flow graph representation. Thus, in the present book the block diagram representation is adopted. Every block in the block diagram contains the TF of a component or element of the system.

There are two forms of control systems. These are the open-loop and closed-loop control systems, as shown in Figures 8.1a and b, respectively. Clearly, the open-loop control systems are special cases of closed-loop ones, and are relatively simple to deal with. This and subsequent chapters are therefore concerned with closed-loop control systems, unless stated otherwise.

8.3 Transfer Functions

The transfer function of a system is generally defined as the ratio of the Laplace transformed output to the Laplace transformed input. Thus, it can be considered as a mathematical operator. The TF of an open-loop system, $G(s)$, is shown in Figure 8.2.

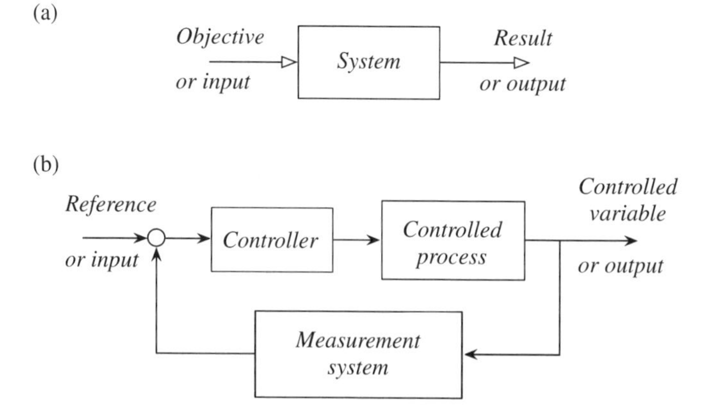

Figure 8.1 Control system forms: (a) open-loop system and (b) closed-loop system

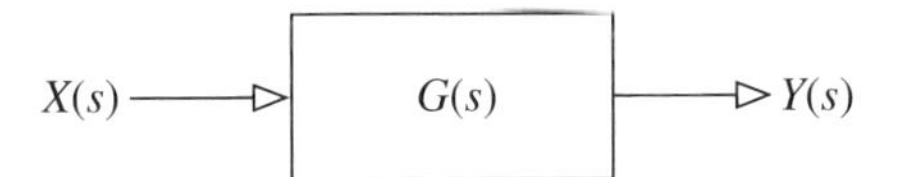

Figure 8.2 Transfer function of an element

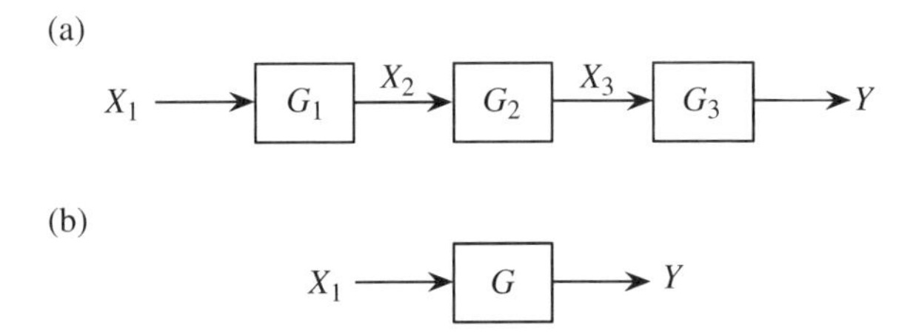

Figure 8.3 (a) Elements in series and (b) representation of elements in series

With reference to Figure 8.2, the TF of the element is defined as

$$G(s) = \frac{Y(s)}{X(s)} \quad \text{or} \quad G = \frac{Y}{X}. \tag{8.1}$$

8.3.1 *Transfer Function of Elements in Cascade Connection*

The elements are connected in series as shown in Figure 8.3a. The objective here is to find the TF for the system in Figure 8.3a such that it can be represented as that in Figure 8.3b.

Starting from applying the definition of TF for the system in Figure 8.3b, one has

$$G = \frac{Y}{X_1},$$

which can be written as

$$G = \frac{Y}{X_1} = \frac{Y}{X_3}\frac{X_3}{X_2}\frac{X_2}{X_1} = G_3 G_2 G_1.$$

Of course, one can generalize the above expression for n elements connected in series so that the TF becomes

$$G = G_1 G_2 G_3 \ldots = \prod\nolimits_{i=1}^{n} G_i. \tag{8.2}$$

8.3.2 *Transfer Function of Elements in Parallel Connection*

In this case, the elements are connected in parallel as shown in Figure 8.4a. The objective is to find the TF for the system in Figure 8.4a such that it can be represented as that in Figure 8.4b.

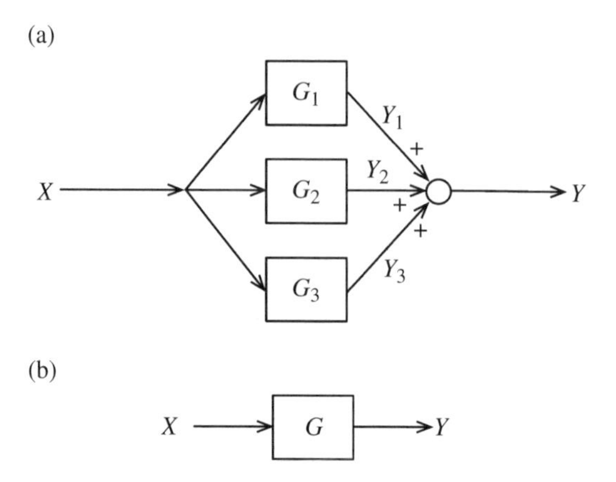

Figure 8.4 (a) Elements in parallel connection and (b) representation of elements in parallel connection

To find the TF for this system as shown in Figure 8.4b, one can apply the definition again so that

$$G = \frac{Y}{X}.$$

But the output signal is

$$Y = Y_1 + Y_2 + Y_3 = G_1X + G_2X + G_3X = (G_1 + G_2 + G_3)X$$

Therefore, upon generalization of the above expression for n elements connected in parallel, one has the TF of the system as

$$G = (G_1 + G_2 + G_3 + \ldots) = \sum_{i=1}^{n} G_i. \tag{8.3}$$

8.3.3 Remarks

1. A general engineering control system is usually made up of a combination of elements in series as well as elements in parallel connections.
2. Two basic assumptions for the TF representation and solution are *passivity* (that is, the individual element has no energy source) and *linearity* (such that the principle of superposition can be applied).

8.4 Closed-Loop Control Systems

In this section, two closed-loop control systems are considered. The objective of each case is to find the overall transfer function (OTF) of that system (for simplicity, the OTF shall be referred to as the TF).

8.4.1 Closed-Loop Transfer Functions and System Response

Consider the closed-loop control system shown in Figure 8.5, in which the following symbols are applied (all upper-case symbols denote the Laplace transforms of the corresponding lower cases):

- R is the reference or input;
- B is the feedback signal;
- E is the error or actuating signal;
- C is the output signal;
- G_c is the TF of the controller;
- G_a is the TF of the amplifier;
- G_p is the TF of the system to be controlled (commonly referred to as the TF of the plant);
- H is the TF of the feedback element;
- $G = C/E = G_cG_aG_p$ is known as the forward-loop transfer function (to be referred to as FLTF);
- GH is the open-loop transfer function (to be referred to as OLTF); and
- $C = C/R$ is the closed-loop TF of the system.

First, the error equation is given by $E = R - B$.
Also, $C = GE$, $B = HC$, and $G = G_cG_aG_p$.
Substituting for E and B, one has

$$\frac{C}{G} = R - B = R - HC.$$

Re-arranging the above equation, one obtains

$$\left(H + \frac{1}{G}\right)C = R.$$

Therefore, the closed-loop or overall TF becomes

$$\frac{C}{R} = \frac{G}{1 + GH}. \tag{8.4}$$

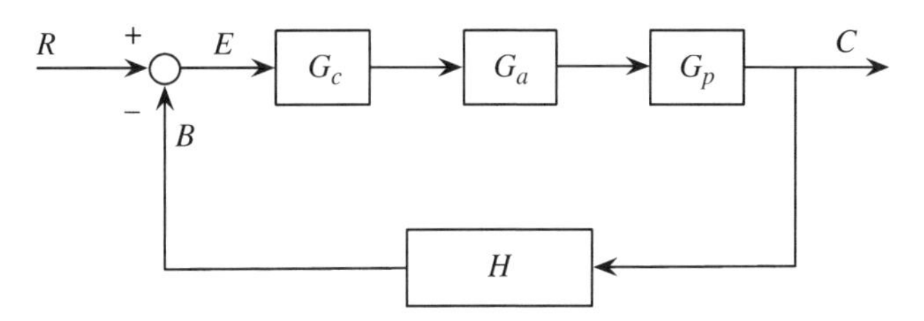

Figure 8.5 A feedback control system

Dividing both sides of Equation (8.4) by G,

$$\frac{C}{RG} = \frac{1}{1+GH}.$$

But by definition $C/G = E$, therefore:

$$\frac{E}{R} = \frac{1}{1+GH}. \tag{8.5}$$

Equation (8.5) is the error-input TF.

- When $H = 1$, then the system is referred to as the unity feedback control system.
- In general, H need not be unity and the control system may contain many feedback loops.
- The technique of obtaining the closed-loop or overall TF of multi-loop systems is identical to that presented above, except that additional error equations and substitution steps are necessary.

To further illustrate the derivation of the OTF, one additional control system with two inputs is studied here. The block diagram of the system is included in Figure 8.6.

First, it is assumed that the disturbance $U = 0$, then upon application of Equation (8.4), the output is

$$C_R = \left(\frac{G_1G_2G_3}{1+G_1G_2G_3H}\right)R. \tag{8.6a}$$

Second, it is assumed that the input $R = 0$, so that upon application of Equation (8.4) again, the output becomes

$$C_U = \left(\frac{G_3}{1+G_1G_2G_3H}\right)U. \tag{8.6b}$$

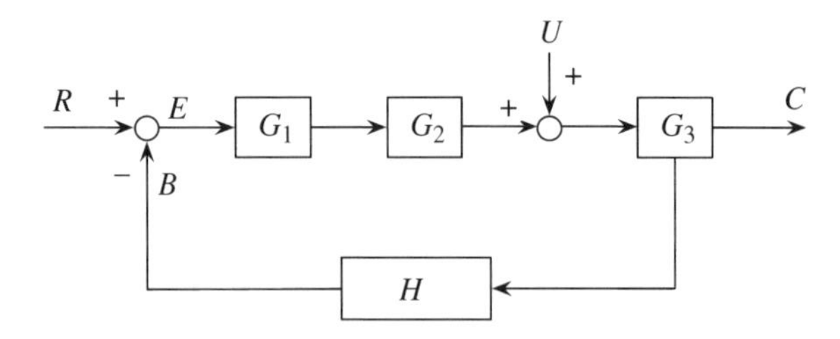

Figure 8.6 A feedback control system with two inputs

Third, by the principle of superposition (since it is assumed that the system is linear), the total response becomes

$$C = C_R + C_U = \left(\frac{G_1G_2G_3}{1+G_1G_2G_3H}\right)R + \left(\frac{G_3}{1+G_1G_2G_3H}\right)U. \tag{8.6c}$$

- Note that the denominator terms in Equations (8.6a) and (8.6b) are identical, implying that the system characteristics governed by the denominator term of a TF are independent of the inputs. This, in turn, implies that for linear systems their dynamic characteristics are not affected by their inputs.

8.4.2 Summary of Steps for Determination of Closed-Loop Transfer Functions

The steps for closed-loop TFs illustrated in the foregoing can be summarized in five steps. They are:

1. Write down error equations.
2. Write down remaining equations for *G*s and *H*s, and so on.
3. Apply the definition of a TF for a feedback control system, that is, Equation (8.4):

$$\frac{C}{R} = \frac{G}{1+GH}.$$

4. Use the relations in steps 2 and 3 above, and substitute into the error equations.
5. Re-arrange terms associated with *C* and *R* so that the ratio of *C* to *R* can be obtained.

8.5 Block Diagram Reduction

There are steps that can be applied to simplify the procedure of obtaining the closed-loop TF of a complicated control system. In the following, only two commonly used techniques are introduced and an example of block diagram reduction is subsequently included to demonstrate the steps required.

8.5.1 Moving Starting Points of Signals

In this technique, the starting point of a signal from behind of TF *G* is moved to the front. This is illustrated in Figure 8.7.

With reference to the block diagram on the lhs of Figure 8.7, the take-off point is behind the TF *G*. Then the signal $B = R$ and $R = C/G$. Therefore

$$B = \frac{C}{G}. \tag{8.7}$$

Figure 8.7 Equivalent block diagrams of moving a signal

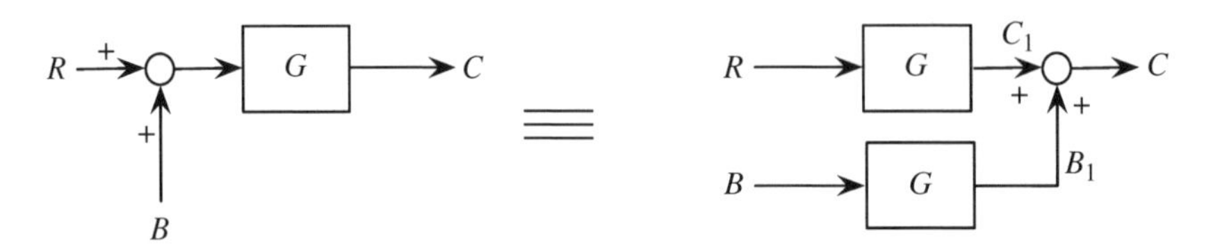

Figure 8.8 Equivalent block diagrams of moving a summing point

If now the take-off point is in front of TF G as indicated in the block diagram on the rhs of Figure 8.7, then the signal must go through a TF $1/G$ to yield B. In other words, the signal $B = C/G$ or $C = RG$. Substituting the latter into Equation (8.7), one has

$$B = \frac{RG}{G} = R.$$

Thus, the block diagram on the rhs of Figure 8.7 is equivalent to that on the lhs.

8.5.2 Moving Summing Points

This technique is illustrated in Figure 8.8.

With reference to the block diagram on the lhs of Figure 8.8, one has the output signal from the TF:

$$C = EG = (R + B)G = RG + BG.$$

This is equivalent to the block diagram on the rhs of Figure 8.8 since $C_1 = RG$, and $B_1 = BG$.

8.5.3 System Transfer Function by Block Diagram Reduction

It is understood that the TFs of many closed-loop control systems can be derived by applying the method of signal-flow graph. However, such an approach is believed to be mathematically more abstract than the block diagram reduction technique based on the block diagram representation. Thus, in this subsection the closed-loop TF of a control system is determined by applying the block diagram reduction technique. Such a system is shown in Figure 8.9.

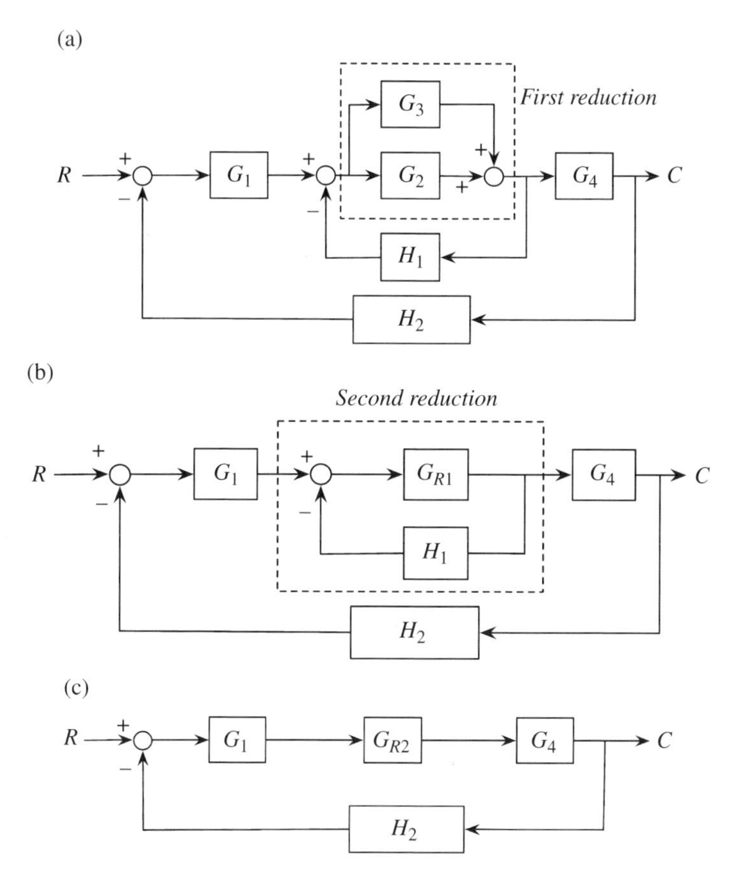

Figure 8.9 Control system showing (a) first block diagram reduction, (b) second block diagram reduction, and (c) final feedback control system

8.5.3.1 First Reduction

With reference to the control system shown in Figure 8.9a, the two TFs in parallel connections and boxed with a dotted line are first considered. By Equation (8.3), the TF of the first reduction is

$$G_{R1} = G_2 + G_3. \tag{8.8a}$$

8.5.3.2 Second Reduction

Applying Equation (8.4) to the inner closed-loop in Figure 8.9b, the TF is

$$G_{R2} = \frac{G_{R1}}{1 + G_{R1} H_1}. \tag{8.8b}$$

Then, the FLTF in Figure 8.9c becomes

$$G = G_1 G_{R2} G_4. \tag{8.8c}$$

8.5.3.3 Final Block Diagram Reduction

Applying Equation (8.4) to the feedback loop in Figure 8.9c, one has

$$\frac{C}{R}=\frac{G}{1+GH_2}. \tag{8.8d}$$

Backward substituting for Equations (8.8c), (8.8b), and (8.8a), one obtains

$$\frac{C}{R}=\frac{\left[\dfrac{G_1(G_2+G_3)G_4}{1+(G_2+G_3)H_1}\right]}{1+\left[\dfrac{G_1H_2(G_2+G_3)G_4}{1+(G_2+G_3)H_1}\right]}.$$

Simplifying the above expression, one obtains

$$\frac{C}{R}=\frac{G_1(G_2+G_3)G_4}{1+(G_2+G_3)H_1+G_1H_2(G_2+G_3)G_4}. \tag{8.8e}$$

This is the required closed-loop TF of the control system in Figure 8.9a.

8.6 Questions and Solutions

In this section, solutions to three questions are presented. The first two questions are concerned with the reduction of block diagram representations of closed-loop control systems. The third question deals with the construction of the block diagram for a water-level control system.

Example 1
A feedback control system with two inputs is represented by the block diagram shown in Figure 8E1a. In Figure 8E1b, Laplace transformed errors are added. Express the Laplace transformed output C in terms of the two inputs R and X, and the two transfer functions G_1 and G_2.

Solution:
Consider the error equations from the lhs to rhs of the block diagram in Figure 8E1b. Thus, the error equation at the first summing point on the lhs becomes

$$E_1=R-E_2, \tag{i}$$

$$E_2=X_1+X-C=E_1G_1+X-C.$$

Substituting for Equation (i), it gives

$$E_2=(R-E_2)G_1+X-C.$$

Therefore,

$$E_2(1+G_1)=RG_1+X-C. \tag{ii}$$

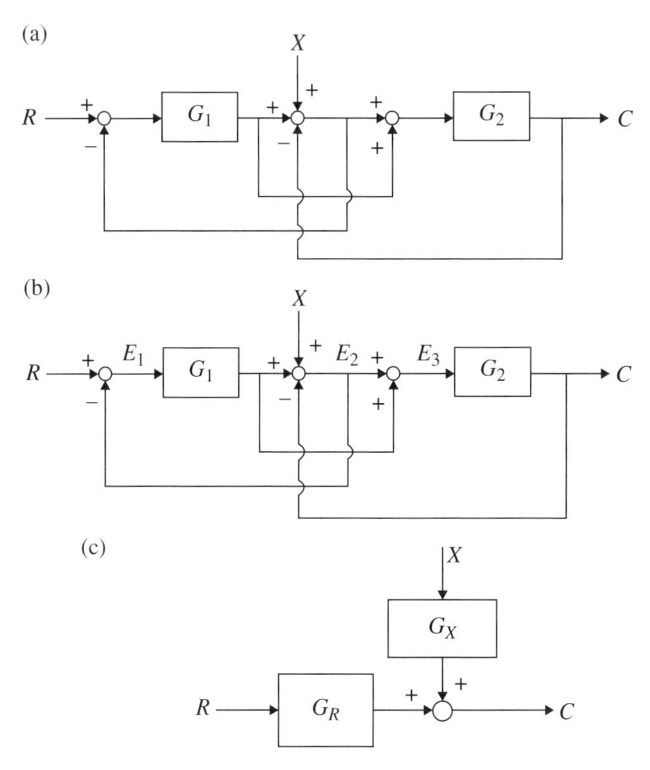

Figure 8E1 (a) A two-input feedback control system. (b) Two-input feedback control system with error signals. (c) Block diagram of the control system with two inputs

The third error equation becomes

$$E_3 = E_2 + X_1 = E_2 + E_1G_1.$$

Substituting for Equation (i), one has

$$E_3 = E_2 + E_1G_1 = E_2 + (R - E_2)G_1 = E_2(1 - G_1) + RG_1.$$

But $E_3 = \dfrac{C}{G_2}$. Substituting this into the lhs of the last equation, one obtains

$$\frac{C}{G_2} = E_2(1 - G_1) + RG_1. \qquad \text{(iii)}$$

Substituting Equation (ii) into Equation (iii), one has

$$\frac{C}{G_2} = \left(\frac{RG_1 + X - C}{1 + G_1}\right)(1 - G_1) + RG_1.$$

Moving G_2 to the rhs of the equation gives

$$C = \frac{(RG_1G_2 + XG_2 - CG_2)(1 - G_1) + RG_1G_2(1 + G_1)}{1 + G_1}.$$

Expanding,

$$C(1 + G_1) = RG_1G_2 + XG_2 - CG_2 - RG_1^2G_2 - XG_1G_2 + CG_1G_2 + RG_1G_2(1 + G_1).$$

Regrouping terms associated with the output C on one side and terms associated with the inputs R and X on the other side of the equality sign:

$$C(1 + G_1 + G_2 - G_1G_2) = (2G_1G_2)R + G_2(1 - G_1)X.$$

Upon rearranging, it becomes

$$C = G_R R + G_X X \tag{iv}$$

where

$$G_R = \frac{2G_1G_2}{1 + G_1 + G_2 - G_1G_2} \quad \text{and} \quad G_X = \frac{G_2(1 - G_1)}{1 + G_1 + G_2 - G_1G_2}.$$

Equation (iv) is the required result and can be represented by the block diagram in Figure 8E1c.

Example 2

A simplified block diagram for a linear control system of a large antenna is presented in Figure 8E2, in which

$$G_1 = \frac{1}{1 + \tau s}, \quad G_2 = k_p + \frac{k_i}{s}, \quad G_3 = \frac{k_1}{s},$$
$$G_4 = \frac{k_3}{s}, \quad G_5 = \frac{k_2}{s^2}.$$

Determine the closed-loop transfer function of the system.

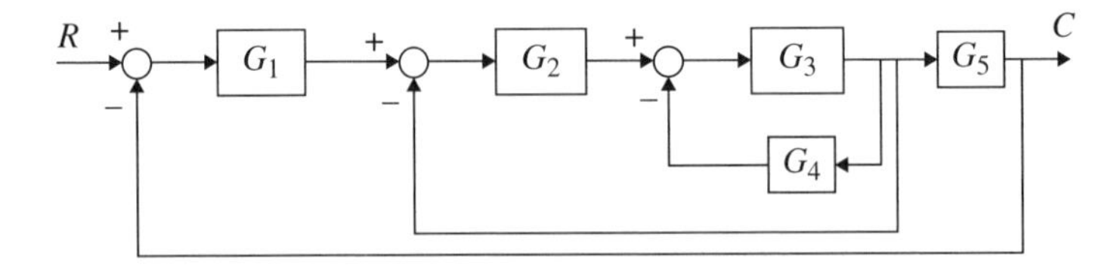

Figure 8E2 Simplified block diagram of a large antenna

Solution:
With reference to Figure 8E2, the transfer function of the innermost feedback loop is

$$G_{IMF} = \frac{G_3}{1 + G_3 G_4}. \tag{i}$$

Similarly, the transfer function of the inner feedback loop becomes

$$G_{IF} = \frac{G_2 G_{IMF}}{1 + G_2 G_{IMF}} = \frac{G_2 G_3}{1 + G_2 G_3 + G_3 G_4}. \tag{ii}$$

From the outer feedback loop, one has

$$\frac{C}{R} = \frac{G_1 G_{IF} G_5}{1 + G_1 G_{IF} G_5} = \frac{G_1 G_2 G_3 G_5}{1 + G_2 G_3 + G_3 G_4 + G_1 G_2 G_3 G_5}. \tag{iii}$$

Substituting the given transfer functions, one can show that

$$\frac{C}{R} = \frac{N}{D}, \tag{iv}$$

where the numerator term is given by $N = (k_1 k_2 k_p)s + k_1 k_2 k_i$, and the denominator term $D = a_5 s^5 + a_4 s^4 + a_3 s^3 + a_2 s^2 + a_1 s + a_0$, in which $a_5 = \tau$, $a_4 = 1 + k_1 k_p \tau$, $a_3 = k_1 k_p + k_1 k_i \tau + k_1 k_3 \tau$, $a_2 = k_1 k_i + k_1 k_3$, $a_1 = k_1 k_2 k_p$, $a_0 = k_1 k_2 k_i$.

Example 3
The schematic diagram of a water-level control system with a proportional pneumatic controller (P controller) is shown in Figure 8E3a, in which air and water supply pressures are assumed to be constant. Variations of nozzle back pressure p_0 are proportional to variations of the flapper-to-nozzle displacement x_f. Displacement x_b, at the left-hand end of the lever which is connected to the bellows by a vertical linkage, is proportional to changes of pressure p_0 in the bellows. For the pneumatically actuated water control valve, the simple lag transfer function $\frac{K}{(\tau s + 1)}$ can be assumed, with K and τ being constant. Note that the desired water level is set by screw adjustment of the float post. Introducing parameters as needed and writing the necessary equations, develop a block diagram for the water-level control system.

Solution:
The objective of the problem is to provide a block diagram representation for the output h and input h_{set} of the water-level control system.

The problem has two parts: the geometry and kinematics, and dynamic behaviors of various components (valve, tank, flapper-nozzle, and bellows).

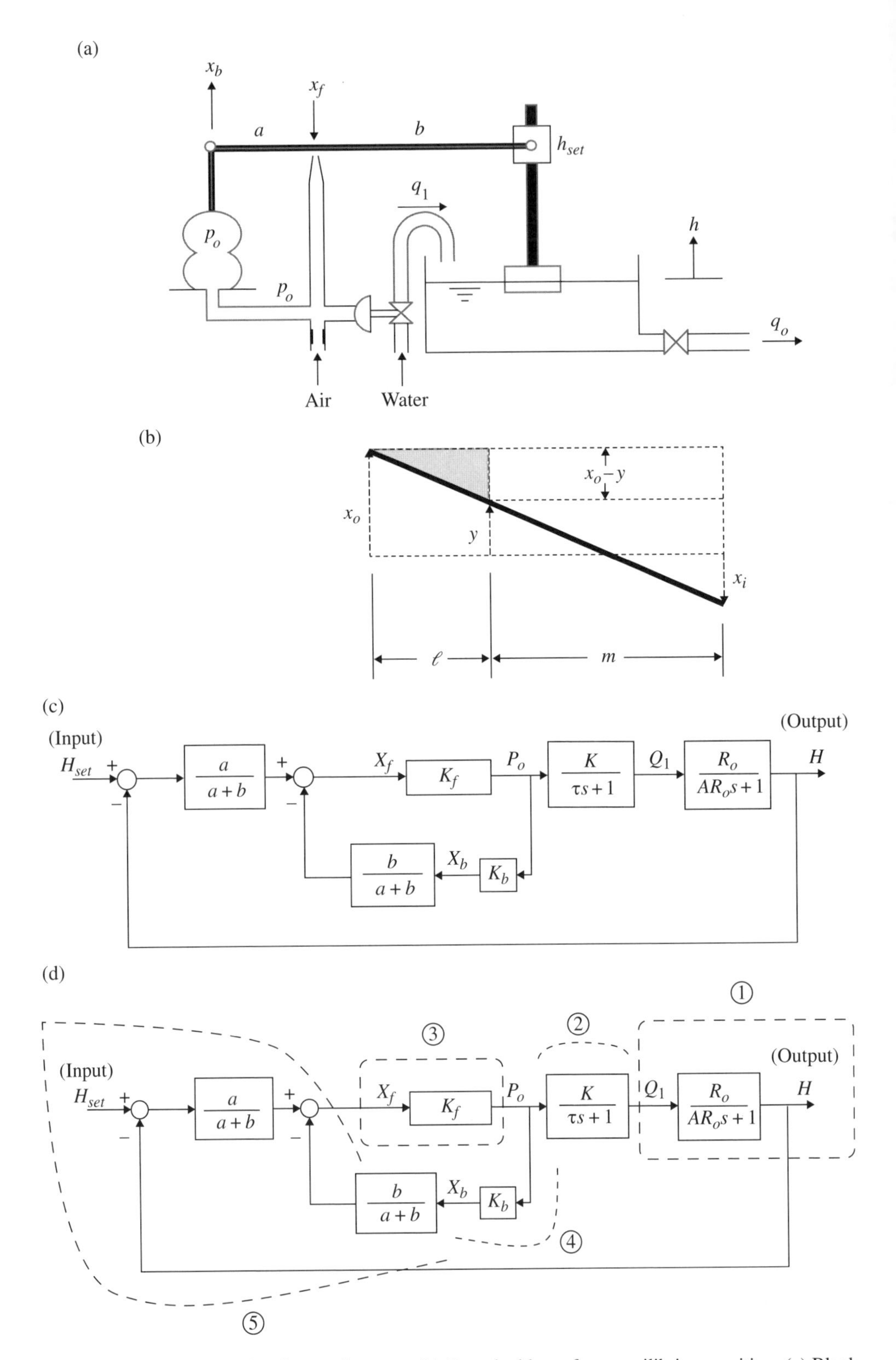

Figure 8E3 (a) A water-level control system. (b) Perturbed lever from equilibrium position. (c) Block diagram of water-level control system. (d) Steps in construction of block diagram in (c)

For the geometry and kinematics:

For the linkage or lever, by using similar triangles in Figure 8E3b, one can write

$$\frac{x_i + x_o}{\ell + m} = \frac{x_o - y}{\ell}$$

so that

$$y = x_o - \frac{(x_i + x_o)}{\ell + m}\ell. \tag{i}$$

Now, returning to the symbols used in the question, one identifies them as

$$y = -x_f, \quad x_i = h_{set} - h, \quad x_o = x_b, \quad m = b, \quad \ell = a,$$

such that Equation (i) becomes

$$x_f = \frac{a(h_{set} - h)}{a+b} - \frac{b\,x_b}{a+b}.$$

Taking the Laplace transform, one has

$$X_f(s) = \left(\frac{a}{a+b}\right)[H_{set}(s) - H(s)] - \left(\frac{b}{a+b}\right)X_b(s). \tag{ii}$$

For the dynamic behaviors of the valve:

Consider the pneumatically actuated valve. It is given that

$$\frac{Q_1(s)}{P_o(s)} = \frac{K}{\tau s + 1}. \tag{iii}$$

For the dynamic behaviors of the tank:

$$q_1 - q_o = A\frac{dh}{dt}, \quad q_o = \frac{h}{R_o},$$

where A is the uniform cross-sectional area of the tank.

Taking the Laplace transform and assuming the system starts from rest, one has

$$Q_1(s) - \frac{H(s)}{R_o} = AsH(s), \quad Q_1(s) = (1 + AR_o\,s)\frac{H(s)}{R_o}.$$

Therefore, the transfer function for the tank is given by

$$\frac{H(s)}{Q_1(s)} = \frac{R_o}{(AR_o)s + 1}. \tag{iv}$$

For the dynamic behaviors of the flapper-nozzle:

Since it is given that $p_0 \propto x_f$, then one can write $p_0 = K_f x_f$, where K_f is the proportionality constant. Taking the Laplace transform gives

$$P_0(s) = K_f X_f(s) \tag{v}$$

For the dynamic behaviors of bellows:

Since it is given that $x_b \propto p_0$, one can write $x_b = K_b p_0$, where K_b is the proportionality constant. Taking the Laplace transform, one obtains

$$X_b(s) = K_b P_0(s). \tag{vi}$$

By making use of Equations (i)–(vi) the block diagram may be constructed and presented in Figure 8E3c. The sequential steps in the construction of Figure 8E3c are included in Figure 8E3d, in which the encircled integer is the step number.

Exercise Questions

Q1. The block diagram for a model of a hydrogovernor-turbine is shown in Figure 8Q1, in which

$$G_1 = \frac{k_1}{(1+\tau_1 s)\tau_2}, \quad G_2 = \frac{k_2}{s},$$

$$G_3 = \frac{1-\tau_3 s}{1+\left(\frac{\tau_3}{2}\right)s}, \quad H_1 = K, \quad H_2 = \frac{\tau_5 \tau_4 s}{1+\tau_4 s},$$

where k_i, $i = 1, 2$; K; τ_n, $n = 1,2, \ldots, 5$ are constant. Find the closed-loop transfer function of the model in Figure 8Q1.

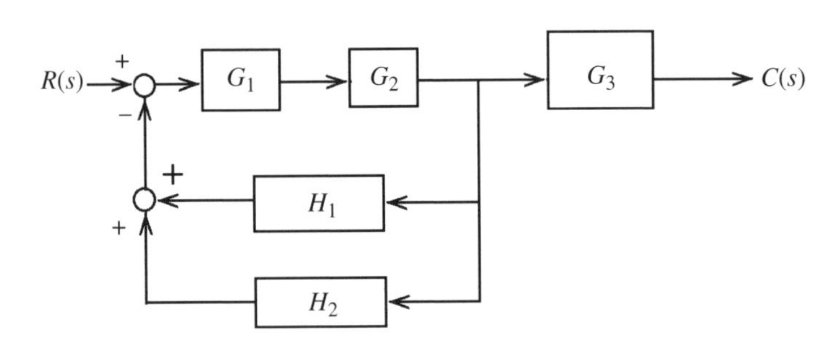

Figure 8Q1 Block diagram representation of a hydrogovernor-turbine model

Q2. The block diagram of a control scheme for a jet aircraft pitch-rate control mechanism is shown in Figure 8Q2, in which

$$G_1 = \frac{k_1}{s}, \quad G_2 = \frac{a}{s+a}, \quad G_3 = \frac{m(s+b)}{s^2 + 2\zeta\omega_n s + \omega_n^2},$$
$$H_1 = k_2 = H_2, \quad H_3 = \alpha + \beta s,$$

where k_i, $i = 1, 2$; a, b, α, β; m, ζ, and ω_n are constant. Find the closed-loop transfer function of the control mechanism.

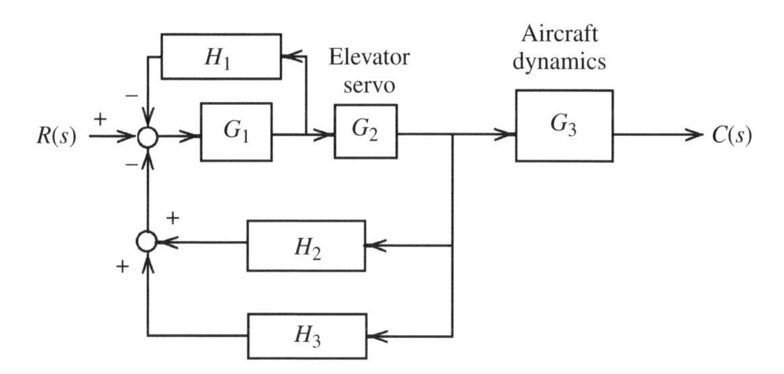

Figure 8Q2 Block diagram of a jet aircraft pitch-rate control mechanism

Q3. A displacement servo is defined as a control system whose output displacement follows the input displacement. The block diagram representation of a displacement servo is shown in Figure 8Q3 in which

$$G_1 = K_p, \quad G_2 = K_m, \quad G_3 = \frac{1}{ms+c}, \quad G_4 = K_a, \quad G_5 = \frac{1}{s}$$

and in such a control system the Laplace transformed input displacement, $X_i(s)$ is that of the spool valve, whereas the Laplace transformed output displacement, $X_o(s)$ is that of the piston. Note that in the figure, K_p, K_m, K_a, m, and c are positive and constant.

Determine the transfer function, in terms of the parameters, of the control system shown in Figure 8Q3.

It is known that the parameter K_a in the inner feedback loop is the cross-sectional area of the piston. What would be the effect of increasing K_a on the output $X_o(s)$, if all the other parameters in the figure are kept constant?

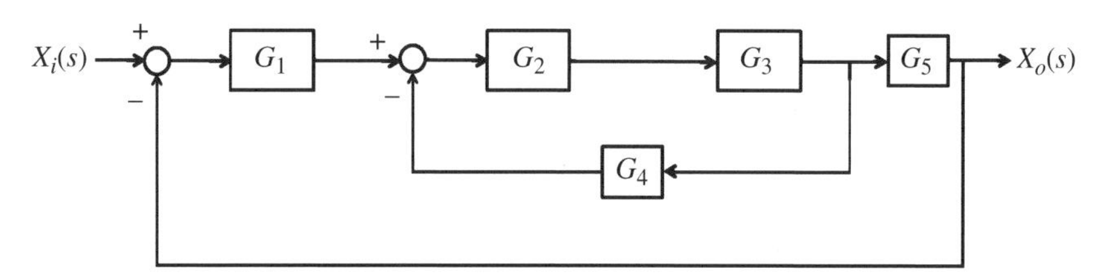

Figure 8Q3 Block diagram of a displacement servo

Q4. The block diagram of an electric servo-mechanism employed to control the angular position of a turntable is given in Figure 8Q4. Determine

a. the open-loop and closed-loop transfer functions of the system shown, and
b. the condition for the system to have oscillatory response.

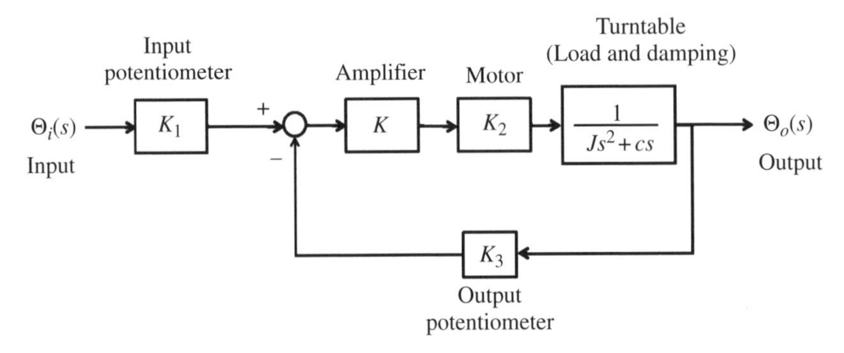

Figure 8Q4 Block diagram of turntable position control system

References

[1] Siouris, G.M. (1993). *Aerospace Avionics Systems: A Modern Synthesis.* Academic Press, New York.
[2] Fossen, T.I. (1994). *Guidance and Control of Ocean Vehicles.* John Wiley & Sons, Ltd., Chichester.

9

Analysis and Performance of Control Systems

This chapter is concerned with the analysis and performance of control systems. Section 9.1 deals with the analysis of response in the time domain. The transient responses as functions of closed-loop poles are presented in Section 9.2. Control system design based on transient responses is considered in Section 9.3. The analysis of the common control types, that is, the so-called proportional, integral, and derivative (PID) control systems are included in Section 9.4. The concepts of steady-state errors are addressed in Section 9.5. Performance criteria and sensitivity functions are introduced in Section 9.6, while problem examples are presented in Section 9.7.

9.1 Response in the Time Domain

Consider the second-order unity feedback control system shown in Figure 9.1. The output can be found as

$$C = \frac{KR}{As^2 + B_1 s + K}$$

where A, K, and B_1 are constant. One can define

$$\omega_n^2 = \frac{K}{A}, \quad 2\zeta\omega_n = \frac{B_1}{A},$$

in which ω_n is the natural frequency and ζ is the damping ratio of the system.

Introduction to Dynamics and Control in Mechanical Engineering Systems, First Edition. Cho W. S. To.

Companion website: www.wiley.com/go/to/dynamics

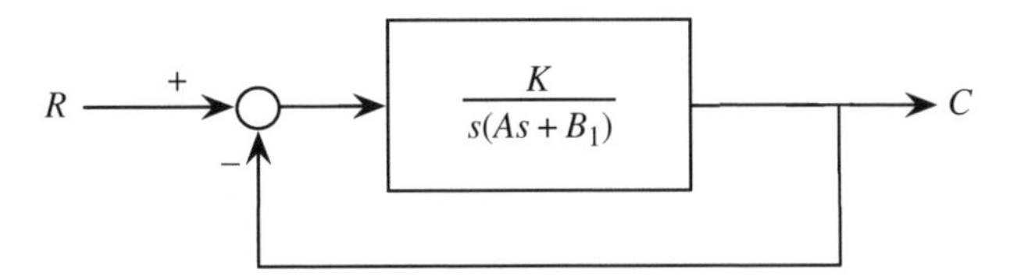

Figure 9.1 Block diagram of a unity feedback system

Substituting the above definitions the output becomes

$$C = \frac{\omega_n^2 R}{s^2 + 2\zeta\omega_n s + \omega_n^2}. \tag{9.1}$$

The transient response of the above control system depends on the roots of the characteristic equation (c.e.):

$$s^2 + 2\zeta\omega_n s + \omega_n^2 = 0. \tag{9.2}$$

That is, the c.e. is formed by setting the denominator term on the rhs of Equation (9.1) to zero. In general, control of dynamic systems assumes that the system is lightly damped or under-damped. That is, one assumes $1 > \zeta > 0$, so that the roots of the c.e. become

$$s_1, s_2 = -\zeta\,\omega_n \pm i\omega_n\sqrt{1-\zeta^2} \quad \text{or}$$

$$s_1, s_2 = -\zeta\omega_n \pm i\omega_d \tag{9.3}$$

where the damped natural frequency is defined as $\omega_d = \omega_n\sqrt{1-\zeta^2}$.

If the input to the system is a unit step, that is, $r(t) = 1.0$ so that its Laplace transform becomes $R = 1/s$, the response in the time domain becomes

$$c(t) = 1 - \frac{e^{-\zeta\omega_n t}}{\sqrt{1-\zeta^2}}\sin(\omega_d t + \phi), \tag{9.4}$$

where

$$\tan\phi = \frac{\sqrt{1-\zeta^2}}{\zeta}.$$

Explanation:

Since $0 < \zeta < 1.0$ and $R = 1/s$, then the Laplace transformed output according to Equation (9.1) becomes

$$C = \frac{\omega_n^2\left(\frac{1}{s}\right)}{s^2 + 2\zeta\omega_n s + \omega_n^2}. \tag{9.5}$$

The denominator on the rhs of Equation (9.5) can be written as

$$(s+\alpha)^2+\omega_d^2,\ \alpha=\zeta\omega_n$$

because

$$(s+\alpha)^2+\omega_d^2=s^2+2\zeta\omega_n s+\omega_n^2.$$

With the above results, Equation (9.5) becomes

$$C=\left(\frac{1}{s}\right)\frac{\omega_n^2}{(s+\alpha)^2+\omega_d^2}.$$

This equation can be expressed as

$$C=\frac{A_1}{s}+\frac{A_2 s+A_3}{(s+\alpha)^2+\omega_d^2}$$

where A_i (i = 1, 2, 3) are constants. By using the partial fraction method that was reviewed in Chapter 2, one can show that

$$C=\frac{1}{s}-\frac{s+\alpha+\alpha}{(s+\alpha)^2+\omega_d^2}. \tag{9.6}$$

Taking the inverse Laplace transform, one obtains Equation (9.4).

Note that ϕ is in the third quadrant in order to provide a positive angle. This positive angle is a required condition since the system has positive damping which, in turn, means that the response is delayed by the phase angle ϕ. Thus,

$$\tan\phi=\frac{-\omega_n\sqrt{1-\zeta^2}}{-\zeta\omega_n}$$

from Equation (9.3). That is, the tangent of the angle is equal to the negative imaginary part divided by the negative real part of the root of the c.e.

9.2 Transient Responses as Functions of Closed-Loop Poles

To provide a quick reference to the transient responses as functions of closed-loop poles on the s-plane, Figure 9.2 is presented. Note that all the transient responses in Figure 9.2 are based on a unit step input.

Remark 9.2.1 When the control system is stable, its closed-loop poles have to be *on the lhs of the s-plane.*

Remark 9.2.2 When the control system is stable, its closed-loop poles have to be *simple if they exist on the imaginary axis.*

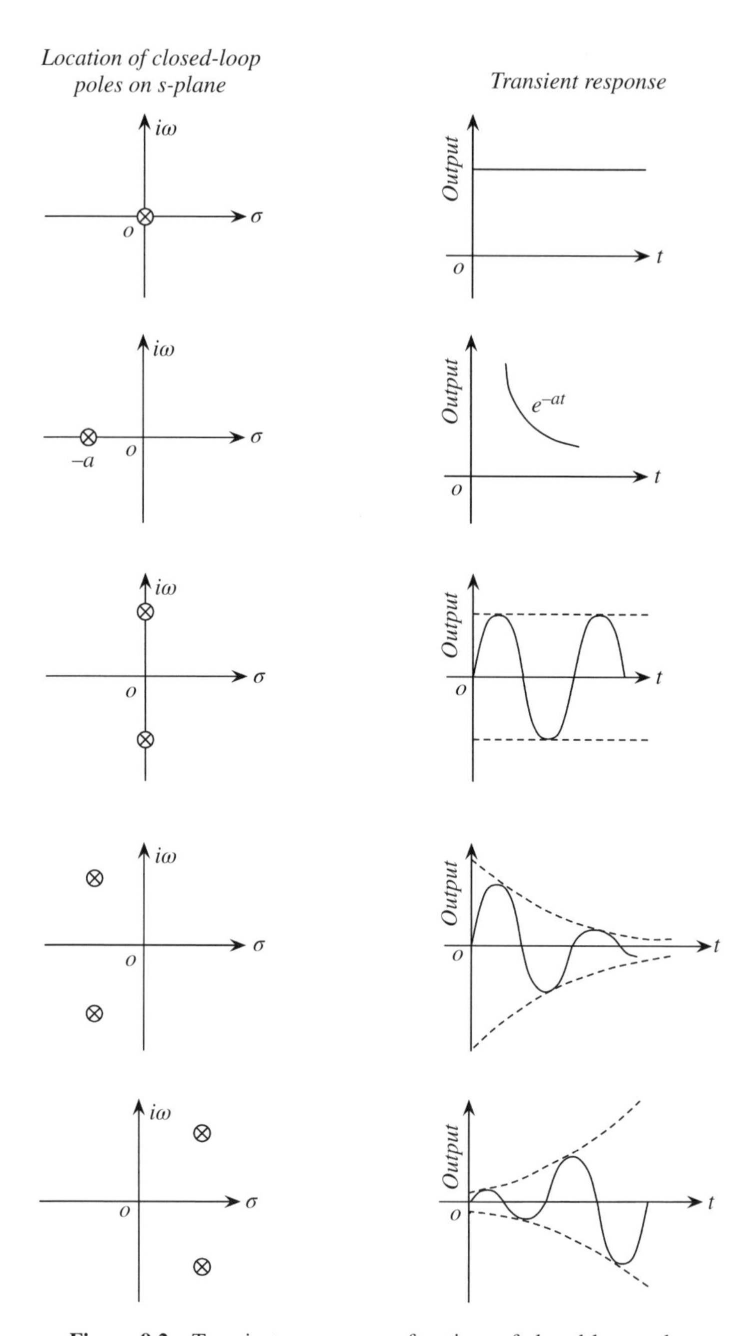

Figure 9.2 Transient responses as functions of closed-loop poles

9.3 Control System Design Based on Transient Responses

Transient responses are important in control system design. In particular, the amount of overshoot (of the response) is usually used as a design parameter. Other parameters used in the design include the rise time, t_r, the delay time, t_d, and the settling time, t_s. The response defined by Equation (9.4) is plotted in Figure 9.3. The rise time is the duration for the response to reach 90% (some other definitions can be found in the literature; the duration between 10% and 90% of the steady state value has also been used as the rise time) of its steady state value (for the plot in Figure 9.3, the steady-state value is 1.0). The delay time is the duration for the response to reach 50% of its steady-state value. The settling time is the duration required for the oscillation of the response to stay within some specified small percentage of the steady-state or final value. Common values for t_s are 2% and 5%. For the unity feedback control system considered above, if the settling time is defined within ±2% of the steady-state value of the response, one can show that it is

$$t_s = \frac{4}{\zeta\omega_n}. \tag{9.7}$$

Unless stated otherwise, this is the settling time adopted for the design in this book.

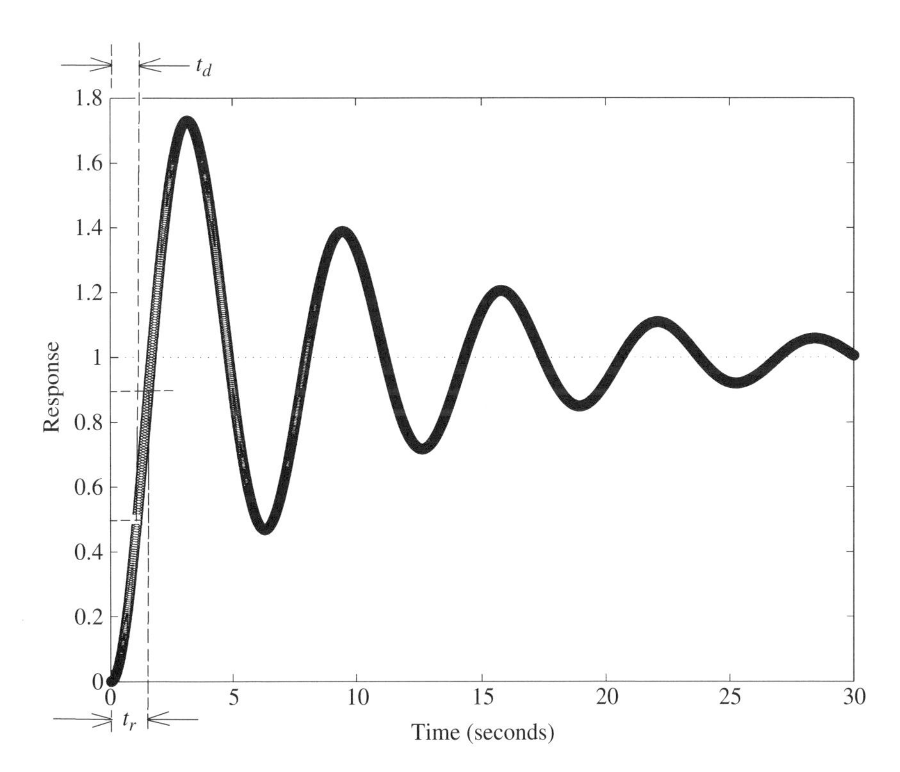

Figure 9.3 Response of dynamic system to unit step input (the settling time is not included in this figure because it is beyond the range plotted)

The amount of overshoot may be obtained by considering the fact that at the first and largest peak of the response its gradient is zero. Mathematically,

$$\frac{dc(t)}{dt} = 0.$$

In obtaining the above, one can simply make use of the Laplace transform rather than performing the relatively tedious differentiation in the time domain. Thus, applying the Laplace transform, one has

$$\mathcal{L}\left[\frac{dc(t)}{dt}\right] = sC(s).$$

But

$$C = \left(\frac{1}{s}\right)\frac{\omega_n^2}{(s+\alpha)^2 + \omega_d^2}$$

therefore

$$\mathcal{L}\left[\frac{dc(t)}{dt}\right] = sC(s) = \frac{\omega_n^2}{(s+\alpha)^2 + \omega_d^2}.$$

Taking the inverse Laplace transform,

$$\frac{dc(t)}{dt} = \mathcal{L}^{-1}[sC(s)] = \mathcal{L}^{-1}\left[\frac{\omega_n^2}{(s+\alpha)^2 + \omega_d^2}\right].$$

From the Laplace transform table, one obtains

$$\frac{dc(t)}{dt} = \frac{\omega_n^2}{\omega_d} e^{-\alpha t} \sin \omega_d t. \tag{9.8}$$

Then $dc(t)/dt = 0$ gives, from Equation (9.8), $\sin \omega_d t = 0$. This, in turn, gives $\omega_d t = 0, \pi, \ldots$ The first solution, 0, is trivial, and applying the second solution, one has $\omega_d t = \pi$, which gives

$$t = \frac{\pi}{\omega_d} = \frac{\pi}{\omega_n \sqrt{1-\zeta^2}}.$$

This is the time when the response is at its maximum value:

$$t_{max} = \frac{\pi}{\omega_n \sqrt{1-\zeta^2}}. \tag{9.9}$$

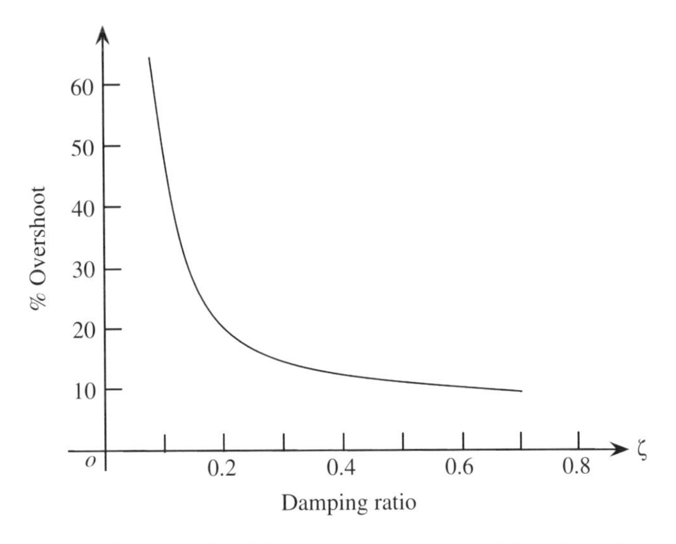

Figure 9.4 Relationship between overshoot and damping ratio

The overshoot is obtained by substituting Equation (9.9) into Equation (9.4) to give

$$c(t_{max}) = 1 - \frac{e^{-\zeta\omega_n\left(\frac{\pi}{\omega_n\sqrt{1-\zeta^2}}\right)}}{\sqrt{1-\zeta^2}} sin(\pi+\phi). \tag{9.10}$$

But $sin(\pi+\phi) = sin\pi\, cos\phi + cos\pi\, sin\phi = -sin\phi$.
Recall that the phase angle is defined by

$$\tan\phi = \frac{\sqrt{1-\zeta^2}}{\zeta}$$

which gives $\sin\phi = \sqrt{1-\zeta^2}$, $\cos\phi = \zeta$.
In addition, $\sin(\pi+\phi) = -\sin\phi = -\sqrt{1-\zeta^2}$.
Substituting the above results into Equation (9.10), one arrives at

$$c(t_{max}) = 1 + e^{-\zeta\left(\frac{\pi}{\sqrt{1-\zeta^2}}\right)}. \tag{9.11}$$

Equation (9.11) is evaluated for various values of the damping ratio and the results are plotted in Figure 9.4.

9.4 Control Types

Frequently when the transient and steady-state responses are used in the control system design, they may not satisfy the system requirements. In these cases, in order to optimize the performance of the system, one can add a *controller* into the control system. Such a system is shown in Figure 9.5.

With reference to Figure 9.5, the closed-loop transfer function (CLTF), by Equation (8.4), is given by

$$\frac{C}{R}=\frac{G_cG}{1+G_cGH}. \tag{9.12}$$

- This equation shows that the response C is affected by the controller transfer function (TF) G_c. As the designer is able to select G_c, s/he has some control over the response for a given input.

Fundamentally, there are three types of control the designer can select. These are the PID controls. In general, the designer can choose any combination of the PID controls.

9.4.1 *Proportional Control*

In this type of control, as shown in Figure 9.6, the actuating signal U is proportional to the error signal E, that is:

$$U=K_p(R-B)=K_pE. \tag{9.13}$$

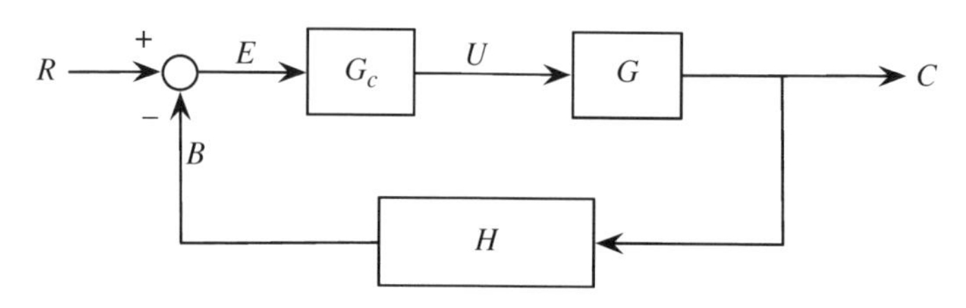

Figure 9.5 System with an added controller

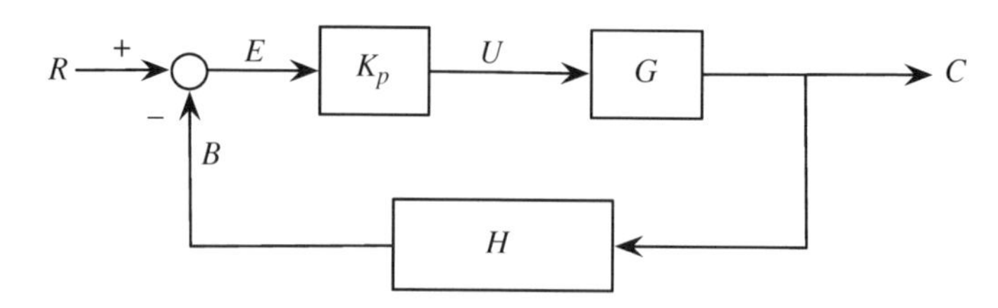

Figure 9.6 System including proportional control

- While this type of control is simple to implement, it can give excessive overshoot or even instability when an optimum gain K_p satisfies the steady-state error.

9.4.2 Integral Control

The block diagram of an integral control system is provided in Figure 9.7. With reference to the latter figure, the actuating signal is related to the error signal by the following equation

$$U = \left(1 + \frac{K_I}{s}\right) E, \tag{9.14}$$

and the CLTF becomes

$$\frac{C}{R} = \frac{s + K_I}{As^3 + B_1 s^2 + s + K_I}. \tag{9.15}$$

- This type of control changes the second-order system to a third-order one. The latter may not be stable over all ranges of gain K_I.

9.4.3 Derivative Control

A system including this type of control is sketched in Figure 9.8. The particular derivative control shown is defined by

$$U = K_d s E, \tag{9.16}$$

where K_d is the gain of the controller.

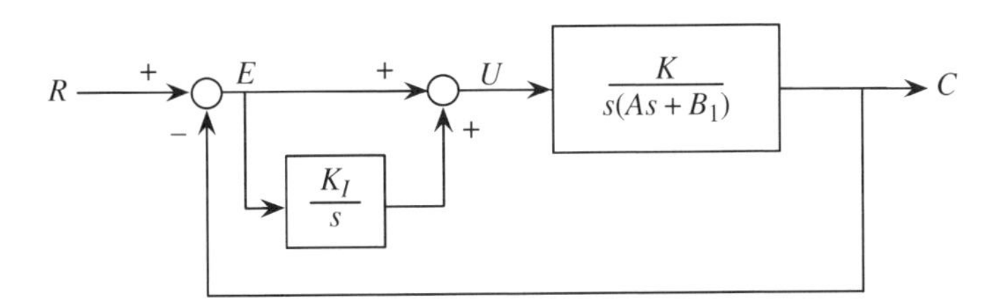

Figure 9.7 System including integral control

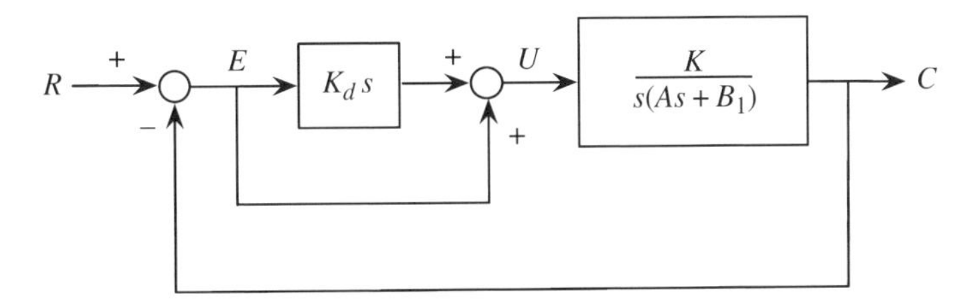

Figure 9.8 System including derivative control

With reference to the system in Figure 9.8, one has the CLTF

$$\frac{C}{R}=\frac{K_d s+1}{As^2+(B_1+K_d)s+1}. \tag{9.17}$$

- Equation (9.17) indicates that the system damping is increased and it must be applied together with other types of control.
- With the increased system damping shown in Equation (9.17), it enables the designer to control system overshoot as the latter is directly governed by the damping in the system (see Figure 9.4 for example).

9.5 Steady-State Errors

Aside from the stability and transient response analysis, the steady-state errors are also applied in the analysis, design, and performance of control systems.

The error expression in the time domain is

$$\varepsilon(t)=r(t)-b(t),$$

where $\varepsilon(t)$ is the error signal in the time domain, $r(t)$ is the input or reference signal, and $b(t)$ is the feedback signal. Note that $\varepsilon(t)$ is used henceforth instead of $e(t)$ in order not to confuse it with the exponential function.

The steady-state error becomes

$$\varepsilon_{ss}(t)=\lim_{t\to\infty}\varepsilon(t).$$

Application of the final-value theorem (see Chapter 2) gives

$$\varepsilon_{ss}(t)=\lim_{t\to\infty}\varepsilon(t)=\lim_{s\to 0}sE(s), \tag{9.18}$$

where $E(s)$ or simply E is the Laplace transformed error signal and it can be obtained by using the error-input TF in Equation (8.5). Equation (9.18) gives the steady-state error in terms of the Laplace transformed error signal instead of the time domain error signal. The implication of this is that much less algebraic manipulation is required in obtaining the steady-state error.

9.5.1 *Unit Step Input*

Applying the error TF function defined by Equation (8.5), the steady-state error due to a unit step input becomes

$$\varepsilon_{ss}(t)=\lim_{s\to 0}sE(s)=\lim_{s\to 0}\left[\frac{s\left(\frac{1}{s}\right)}{1+GH}\right]. \tag{9.19}$$

By defining

$$K_p = \lim_{s \to 0} GH \tag{9.20}$$

where K_p is called the *step error constant* or *positional error constant.*

Equation (9.19) can be written as

$$\varepsilon_{ss}(t) = \frac{1}{1 + K_p}. \tag{9.21}$$

9.5.2 Unit Ramp Input

Since the Laplace transform of a unit ramp is $1/s^2$, the steady-state error due to a unit ramp is given by

$$\begin{aligned} \varepsilon_{ss}(t) &= \lim_{s \to 0} sE(s) = \lim_{s \to 0} \left[s\left(\frac{1}{s^2}\right)/(1 + GH) \right] \quad \text{or} \\ \varepsilon_{ss}(t) &= \lim_{s \to 0} \left(\frac{1}{sGH}\right) = \frac{1}{\lim_{s \to 0} sGH}. \end{aligned} \tag{9.22}$$

By defining

$$K_v = \lim_{s \to 0} sGH \tag{9.23}$$

then Equation (9.22) becomes

$$\varepsilon_{ss}(t) = 1/K_v, \tag{9.24}$$

where K_v is called the *ramp error constant* or *velocity error constant.*

9.5.3 Unit Parabolic Input

Since the unit parabolic input $r(t) = t^2$ and its Laplace transform is $R = 1/s^3$, then the steady-state error becomes

$$\begin{aligned} \varepsilon_{ss}(t) &= \lim_{s \to 0} sE(s) = \lim_{s \to 0} \left[s\left(\frac{1}{s^3}\right)/(1 + GH) \right] \\ \text{or} \quad \varepsilon_{ss}(t) &= \lim_{s \to 0} \left(\frac{1}{s^2 GH}\right) = \frac{1}{\lim_{s \to 0} s^2 GH}. \end{aligned} \tag{9.25}$$

Upon defining

$$K_a = \lim_{s \to 0} s^2 GH \tag{9.26}$$

then Equation (9.25) becomes

$$\varepsilon_{ss}(t) = 1/K_a, \tag{9.27}$$

where K_a is called the *parabolic error constant* or *acceleration error constant.*

9.6 Performance Indices and Sensitivity Functions

When a control system is designed for an application, it has to satisfy certain requirements. These latter are called *design or performance specifications.*

The performance of a control system generally consists of three components.

- The first component has to do with the definition of performance specifications, which are normally related to the *system response.*
- The second component is the so-called *performance index*, which is a quantitative measure of the performance of a system. It is governed by the specific criterion that one applies.
- The third component is the issue of system error due to parameter variations.

The system response has been dealt with in Sections 9.1–9.3; this section is concerned with performance indices and sensitivity functions.

9.6.1 *Performance Indices*

In general, the performance index I_i is expressed as an integral

$$I_i = \int_0^T f_i(p)dt \quad i = 1, 2, 3, \ldots \tag{9.28}$$

where p is a system parameter and may be the system error, output, input, or some combinations of the above. T, the upper integration limit, is generally chosen to be the time required for the system response to reach its steady-state value. It is often chosen to be the settling time t_s.

- To provide a useful measure, the integrand in Equation (9.28) must be positive or zero.
- If $i = 1$, $f_1(p)$ in Equation (9.28) is defined as

$$f_1(p) = \varepsilon^2(t). \tag{9.29a}$$

This is known as the *integral of square of the error* (ISE).

- If $i = 2$, $f_2(p)$ in Equation (9.28) is defined as

$$f_2(p) = |\varepsilon(t)|. \tag{9.29b}$$

  This is known as the *integral of the absolute error* (IAE).
- If $i = 3$, $f_3(p)$ in Equation (9.28) is defined as

$$f_3(p) = t|\varepsilon(t)|. \tag{9.29c}$$

  This is known as the *integral of time multiplied by the absolute error* (ITAE).
- If $i = 4$, $f_4(p)$ in Equation (9.28) is defined as

$$f_4(p) = t\varepsilon^2(t). \tag{9.29d}$$

  This is known as the *integral of time multiplied by the square of the error* (ITSE).

In closing this subsection, it may be noted that in theory there are many ways of defining the performance indices. But it is not pursued further here as the foregoing performance indices have basically covered all of the frequently applied cases.

9.6.2 Sensitivity Functions

The dynamic behavior of a feedback control system changes as its parameter or input varies. Thus, the output from the system is affected and therefore the overall performance of the system is influenced.

If a parameter K changes, causing a change in the TF F, then the percentage change in F, due to K, divided by the percentage change in K is defined as the *sensitivity function*:

$$S_{FK} = \frac{\Delta F/F}{\Delta K/K} = \left(\frac{K}{F}\right)\frac{\Delta F}{\Delta K} \tag{9.30}$$

in which Δ denotes the incremental change. Thus, in the limit (that is, when $\Delta \to 0$) Equation (9.30) becomes

$$S_{FK} = \left(\frac{K}{F}\right)\frac{\partial F}{\partial K}. \tag{9.31}$$

9.7 Questions and Solutions

In this section, two questions are solved. The first example is concerned with a speed control system for passenger cars on a smart highway. The second example deals with sensitivity functions of an oil-level control system.

Example 1

A speed control system is proposed for application to passenger cars traveling on the smart highways of the future. The block diagram representation of the proposed control system is presented in Figure 9E1 in which

$$G_1(s) = \frac{K_1}{\tau_1 s + 1}, \quad \text{and} \quad G(s) = \frac{K_e}{\tau_e s + 1}.$$

The load disturbance (that is, the load torque in the present case) due to percentage grade is in general represented as $\Delta D(s)$. The engine gain K_e changes within the range of 10–1000 for various models of automobiles. The engine time constant τ_e is 20 s. The transfer function of the tachometer is $K_t = 1$.

a. Determine the effect of the load torque on the speed v(t) of the car.
b. Determine the constant percentage grade $\Delta D(s) = \dfrac{\Delta d}{s}$, in terms of the gain factor, for which the engine stalls. Note that since the grade is constant, the steady-state solution is sufficient for the present problem. Assume the setting speed $r(t)$ is 30 km/h so that its Laplace transform $R(s) = \dfrac{30}{s}$ km/h and $K_e K_1 \gg 1$.
c. When $\dfrac{K_g}{K_1} = 2$ what percent grade Δd would cause the automobile to stall?

Solution:

a. With reference to Figure 9E1, the effect of load torque on the speed is

$$V(s) = \frac{G_1(s)G(s)R(s)}{1 + G_1(s)G(s)K_t} + \frac{-K_g(s)G(s)\,\Delta D(s)}{1 + G_1(s)G(s)K_t}.$$

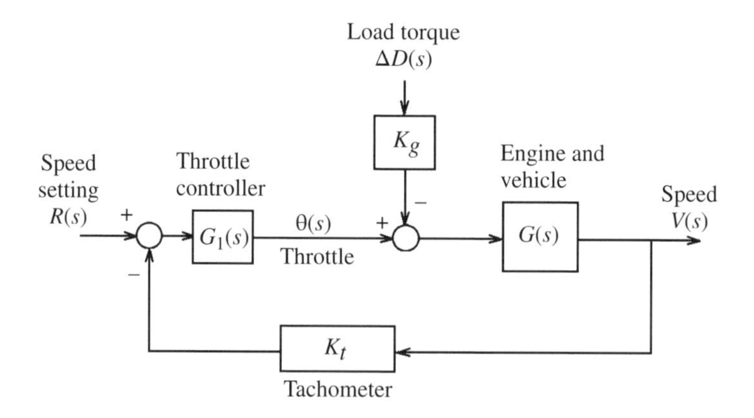

Figure 9E1 Speed control system

Since it is given that $K_t = 1$, therefore

$$V(s) = \frac{G_1(s)G(s)R(s)}{1+G_1(s)G(s)} - \frac{K_g(s)G(s)\,\Delta D(s)}{1+G_1(s)G(s)}.$$

b. The steady-state solution of the speed is

$$\mathrm{v}_{ss}(t) = \mathrm{v}(\infty) = \lim_{s\to 0}\left[s\mathrm{V}(s)\right].$$

When the engine stalls it means $\mathrm{v}_{ss}(t) = 0$. Therefore, applying the equation in (a) above, one has

$$\mathrm{v}_{ss}(t) = \mathrm{v}(\infty) = 0 = \lim_{s\to 0}\left\{ s\left[\frac{G_1(s)G(s)R(s)}{1+G_1(s)G(s)} - \frac{K_g(s)G(s)\,\Delta D(s)}{1+G_1(s)G(s)}\right]\right\}.$$

$$0 = \lim_{s\to 0}\left\{ s\left[\left(\frac{30}{s}\right)\frac{G_1(s)G(s)}{1+G_1(s)G(s)} - \left(\frac{\Delta d}{s}\right)\frac{K_g(s)G(s)}{1+G_1(s)G(s)}\right]\right\}.$$

Substituting for the transfer functions, this becomes

$$0 = 30\left(\frac{K_1K_e}{1+K_1K_e}\right) - (\Delta d)\left(\frac{K_gK_e}{1+K_1K_e}\right).$$

With $K_eK_1 \gg 1$, this equation reduces to

$$0 = 30 - (\Delta d)\left(\frac{K_g}{K_1}\right).$$

Therefore, the constant percentage grade for which the engine stalls is

$$\Delta d = 30\left(\frac{K_1}{K_g}\right).$$

c. When $\frac{K_g}{K_1} = 2$, one substitutes this value into the last equation so that

$$\Delta d = 30\left(\frac{1}{2}\right) = 15.$$

That is, the automobile stalls at a grade of 15 m rise vertically per 100 m horizontally.

Example 2

Consider an oil-level control system, whose block diagram representation is shown in Figure 9E2a, in which

$$G_a(s) = \frac{A}{\tau_a s + 1}, \quad \text{and} \quad G(s) = \frac{R_v}{(AR_v)s + 1}.$$

Suppose that the controller is represented as a gain, and the Laplace transformed supply pressure disturbances $P(s)$ and the time constant τ_a of the actuator are negligibly small, such that the system can be simplified to that in Figure 9E2b. In this simplified model, the valve flow gain K_g, incorporated into K, and the outflow valve resistance R_v of the tank are linearized gains that change with operating point. Generally, they are not precisely known. The gain K may vary due to gain variations of the valve actuator. The level sensor gain H may change from its perfect operating value of 1. For this simplified system represented in Figure 9E2b:

a. Derive the sensitivity functions to changes in K, R_v, and H.
b. Determine the static sensitivities and hence comment on the effects of K, R_v, and H under static conditions.
c. Compare the sensitivity to R_v with that for open-loop control, and hence discuss the effects of feedback.

Solution:

a. With reference to Figure 9E2b, the CLTF is

$$F_c(s) = \frac{KR_v}{(AR_v)s + 1 + KR_v H}.$$

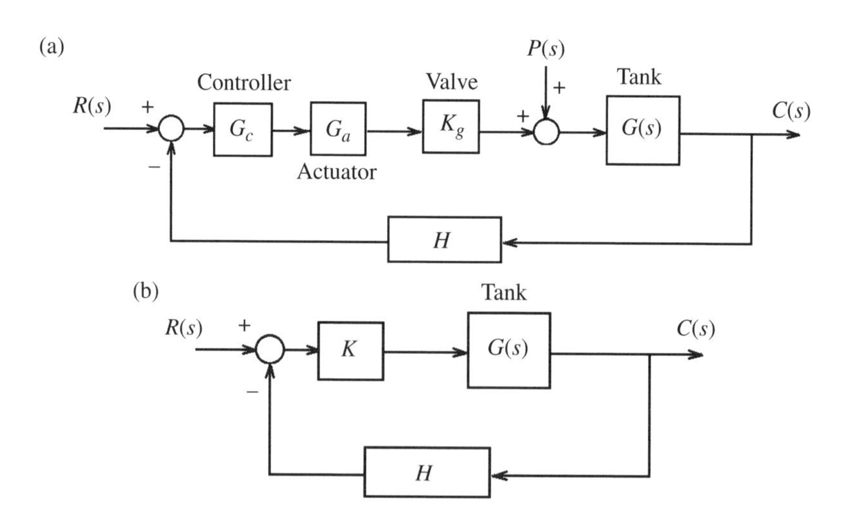

Figure 9E2 (a) Block diagram representation of oil-level control system; (b) simplified oil-level control system

Applying Equation (9.31), the sensitivity function to change in R_v becomes

$$S_{FR_v} = \left(\frac{R_v}{F}\right)\frac{\partial F}{\partial R_v} = \frac{\left(\frac{R_v}{F}\right)\left[D\left(\frac{\partial N}{\partial R_v}\right) - N\left(\frac{\partial D}{\partial R_v}\right)\right]}{D^2},$$

in which $F = F_c(s) = \frac{N}{D}$, $N = KR_v$, and $D = (AR_v)s + 1 + KR_vH$. Thus,

$$S_{FR_v} = \left(\frac{R_v}{F}\right)\frac{K}{D^2} = \frac{N}{\left(\frac{N}{D}\right)D^2} = \frac{1}{D} = \frac{1}{(AR_v)s + 1 + KR_vH}.$$

Similarly, the sensitivity function to change in K,

$$S_{FK} = \left(\frac{K}{F}\right)\frac{\partial F}{\partial K} = \left(\frac{K}{F}\right)\frac{\left[D\left(\frac{\partial N}{\partial K}\right) - N\left(\frac{\partial D}{\partial K}\right)\right]}{D^2}.$$

Simplifying, one obtains

$$S_{FK} = \frac{(AR_v)s + 1}{(AR_v)s + 1 + KR_vH}.$$

The sensitivity function to change in H,

$$S_{FH} = \left(\frac{H}{F}\right)\frac{\partial F}{\partial H} = \left(\frac{H}{F}\right)\left[D\left(\frac{\partial N}{\partial H}\right) - N\left(\frac{\partial D}{\partial H}\right)\right]/D^2.$$

Therefore,

$$S_{FH} = \left(\frac{H}{\frac{N}{D}}\right)\left[\frac{-N(KR_v)}{D^2}\right] = \frac{-H(KR_v)}{(AR_v)s + 1 + KR_vH}.$$

b. For static sensitivities, it means setting $s = 0$ so that

$$S_{FR_v} = \frac{1}{1 + KR_vH} = S_{FK}, \quad \text{and} \quad S_{FH} = \frac{-HKR_v}{1 + HKR_v}.$$

Thus, increasing the loop gain HKR_v reduces the static sensitivities S_{FR_v} and S_{FK},, but the static sensitivity S_{FH} approaches −1.

That is, for large loop gain HKR_v, the change of the CLTF is as large as the change of the feedback loop transfer function.

c. The open-loop transfer function (OLTF) of the system is given by

$$F_o(s) = \frac{HKR_v}{(AR_v)s + 1}.$$

Therefore, the sensitivity function due to R_v for the OLTF or the loop gain function is

$$S_{FR_v} = \left(\frac{R_v}{F}\right)\frac{\partial F}{\partial R_v} = \left(\frac{R_v}{F}\right)\left[\frac{D\left(\frac{\partial N}{\partial R_v}\right) - N\left(\frac{\partial D}{\partial R_v}\right)}{D^2}\right],$$

where now $F = F_o(s) = \frac{N}{D}$, $N = HKR_v$ and $D = (AR_v)s + 1$. Thus,

$$S_{FR_v} = \left(\frac{R_v}{\frac{N}{D}}\right)\left[\frac{D(HK) - N(As)}{D^2}\right] = \left(\frac{R_v}{N}\right)\left[\frac{D(HK) - N(As)}{D}\right].$$

Upon substitution of the parameters and simplification, the sensitivity function due to R_v for the OLTF is

$$S_{FR_v} = \frac{1}{D} = \frac{1}{(AR_v)s + 1}.$$

The static sensitivity function for this case $S_{FR_v} = 1$.

Comparing with the closed-loop transfer function case, the static sensitivity for the closed-loop system reduces greatly with increasing loop gain.

Exercise Questions

Q1. An important problem for mast antenna structures that support very high frequency electronic devices on board frigates and destroyers is the loss of transmission and reception of signals during the launching of surface-to-surface or surface-to-air missiles. A system, whose block diagram is shown in Figure 9Q1, has been proposed to reduce or eliminate the effects of signal loss in these electronic devices during the launching of missiles. In this figure, the rate gyro and the motor transfer functions are given, respectively, as:

$$G_1(s) = \frac{K_g}{\tau_g s + 1}, \quad \text{and} \quad G(s) = \frac{K_m}{\tau_m s + 1}.$$

A maximum antenna support speed of 25°/s is expected. (Note that s denotes seconds, and is not to be confused with the *s*-parameter of the Laplace transform.) If $K_t = K_g = 1$ and τ_g is negligible:

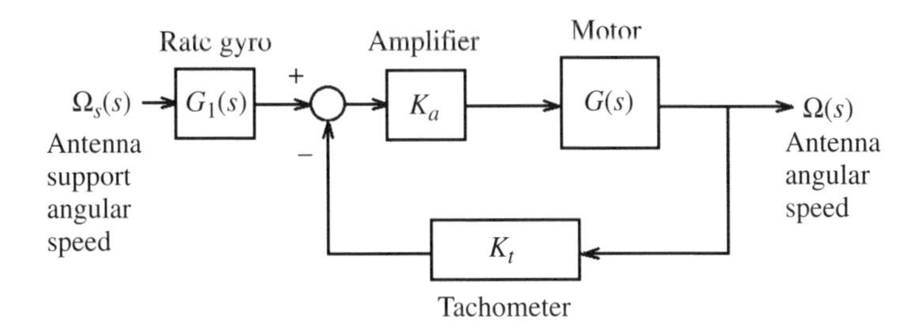

Figure 9Q1 A system for reducing or eliminating signal loss in mast antennae

a. determine the error of the system, $E(s)$; and
b. determine the necessary loop gain, $K_a\,K_m\,K_t$, when a 1°/s steady-state error is allowable.

Q2. The block diagram of a simplified microprocessor-controlled robot is shown in Figure 9Q2, in which

$$G_1(s) = \frac{k_1}{s}, \quad G(s) = \frac{k_2}{\tau s + 1},$$

$H(s) = k_3 + k_4 s$, and the load torque $t_L(t) = D$, with the latter being constant such that the Laplace transform of this load torque is $T_L(s) = \frac{D}{s}$.

a. If the index position is such that $R(s) = 0$, determine the effect of the load torque $T_L(s)$ on the output $C(s)$.
b. Find the sensitivity of the closed-loop transfer function $F(s)$ to the gain k_2, $S_{FK}(s)$, and its static value.
c. When the load torque $T_L(s) = 0$ and the index position now is $R(s) = 1/s$, determine the steady-state error.

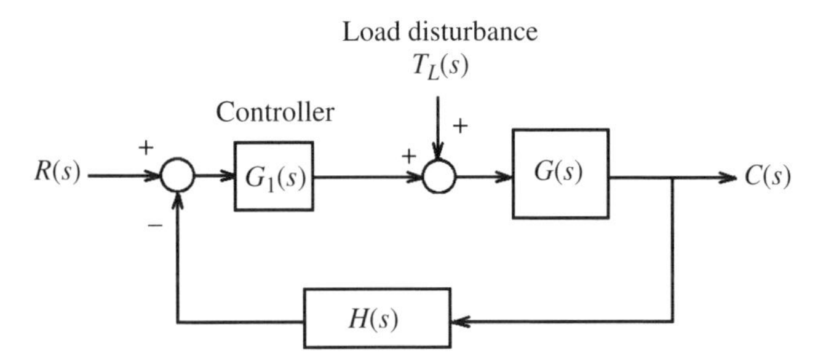

Figure 9Q2 Block diagram of a simplified robot

Q3. The block diagram of a proposed pacemaker with a rate measurement sensor is shown in Figure 9Q3, in which the transfer functions of the pacemaker, heart, and rate measurement sensor are given, respectively as

$$G_1(s) = \frac{10K}{s + 10}, \quad G_2(s) = \frac{1}{s}, \quad \text{and} \quad H(s) = K_m = 1.$$

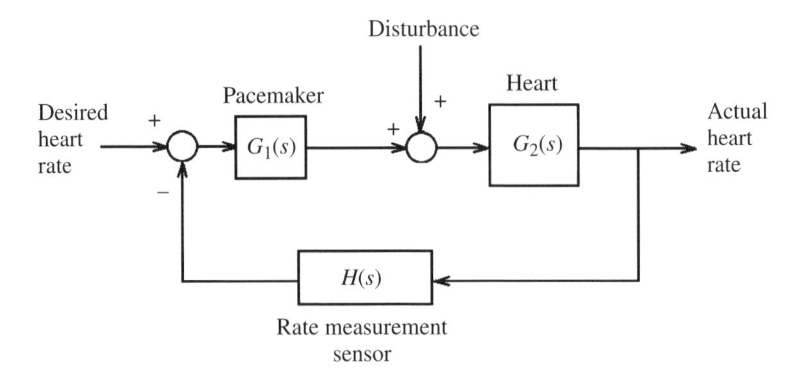

Figure 9Q3 Block diagram of the heart pump and pacemaker

With reference to the latter figure, the transfer function of the heart pump and the pacemaker is therefore obtained as:

$$G(s) = \frac{10K}{s^2 + 10s}.$$

For the design, it is desired to have an amplifier gain to achieve a tightly controlled system with a settling time to a step disturbance of less than 1 s. The overshoot to a step input in desired heart rate should be less than 15% (this corresponds to a damping ratio of approximately $\zeta = 0.5$).

a. If $K = 4$, determine the sensitivity of the system to small changes in K.
b. Find the sensitivity under DC (direct current); that is, by setting $s = 0$ in the resulting sensitivity.
c. Find the magnitude of the sensitivity at the normal heart rate of 60 beats/min (that is, 60 cycles/min; note that s in the sensitivity is replaced by $i\omega$ where i is the imaginary number and ω is in rad/s).

Q4. Most ship stabilization systems employ fins or hydrofoils extending into water in order to generate a stabilization torque on the ship. The block diagram of a simplified ship stabilization system is shown in Figure 9Q4b. The rolling motion of the ship can conceptually be considered as an oscillating pendulum with an angular displacement θ from the vertical, with reference to Figure 9Q4a. A typical period of oscillation for the ship is around 3 s and its transfer function is given by

$$G(s) = \frac{\omega_n^2}{s^2 + 2\zeta\omega_n s + \omega_n^2},$$

where the natural frequency $\omega_n = 2\pi f_n = 2\pi/T = 2$ rad/s, period of oscillation $T = 3.14$ s, and the damping ratio $\zeta = 0.10$ or 10% critical. With this relatively low damping ratio,

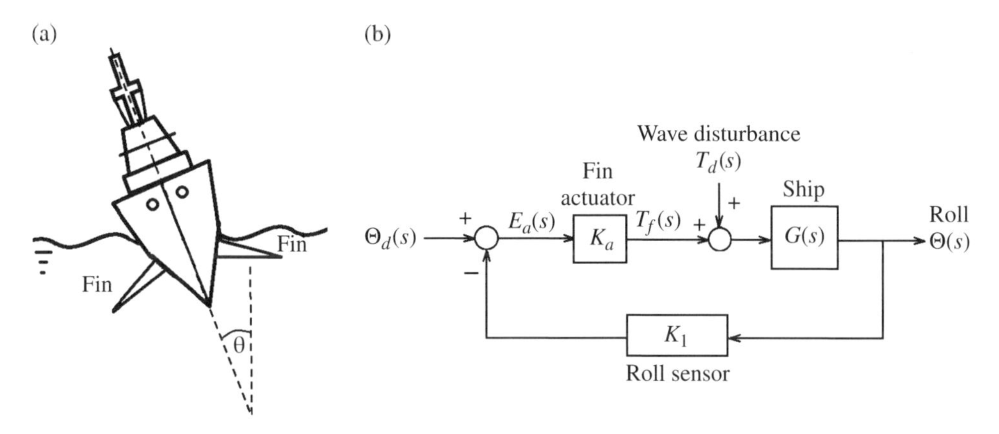

Figure 9Q4 A ship stabilization system: (a) oscillation of ship and (b) block diagram of the stabilization system

the oscillations of the ship persist for several cycles and the rolling amplitude can reach approximately 20° for the expected amplitude of waves in a normal sea.

Determine and compare the open-loop and closed-loop transfer functions of the system shown in Figure 9Q4b for its ability to reduce the effects of the disturbance $T_d(s)$ of the waves. Note that the desired roll $\Theta_d(s)$ is zero degrees.

Q5. For a second-order system it was found that its error is given by

$$\varepsilon(t) = \frac{e^{-\zeta\omega_n t} sin(\omega_d t + \varphi)}{\sqrt{1-\zeta^2}}$$

in which $\tan\varphi = \frac{\sqrt{1-\zeta^2}}{\zeta}$. Plot the performance index I_1 for this system when its natural frequency ω_n = 1 rad/s and the damping ratio ζ is between 0 and 1.0, where $I_1 = \int_0^T \varepsilon^2(t)dt$. Determine the minimum value of I_1.

10

Stability Analysis of Control Systems

The stability analysis in this chapter is confined to *linear* control systems. The concept of stability is introduced in Section 10.1, and the Routh-Hurwitz criterion [1, 2] is presented in Section 10.2, in which examples illustrating its use are included. Problem examples in stability analysis and steady-state errors of systems are provided in Section 10.3. Three questions and solutions are included in Section 10.4. It should be noted that no other methods of stability analysis are considered in this chapter. Further stability analyses of control systems are presented in Chapter 11.

10.1 Concept of Stability in Linear Control Systems

In control system analysis and design, the issue of stability is a primary one. Generally, a system is said to be stable if its response is bounded. In other words, if the response of the system grows with time and is unbounded, the system is unstable.

Recall that in Section 9.2 it was remarked that when the control system is stable the closed-loop poles have to be on the lhs of the s-plane. Thus, *a necessary and sufficient condition for a feedback control system to be stable is that all poles of the system closed-loop transfer function (TF) have negative real parts.*

10.2 Routh–Hurwitz Stability Criterion

From the foregoing section, it appears that in order to determine the stability or instability of a feedback control, one has to find the poles of the closed-loop TF (that is, the roots of the characteristic equation (c.e.)). The Routh-Hurwitz criterion is a rigorous means for determining the stability of the system without actually solving for the roots of the c.e.

- The Routh-Hurwitz stability criterion is a necessary and sufficient condition for the stability of linear systems.

Introduction to Dynamics and Control in Mechanical Engineering Systems, First Edition. Cho W. S. To.

Companion website: www.wiley.com/go/to/dynamics

The c.e. is formed by setting to zero the sum of the product of the numerator terms and product of the denominator terms of the forward loop and feedback loop TF of the system. In other words, if one writes

$$G=\frac{P}{Q},\quad H=\frac{A}{B},$$

then the c.e. is formed as

$$PA+BQ=0. \tag{10.1}$$

Now, after using Equation (10.1) and simplifying, one can obtain the c.e. of the feedback control system as

$$a_n s^n + a_{n-1}s^{n-1} + a_{n-2}s^{n-2} + \ldots + a_0 = 0 \tag{10.2}$$

where n is a positive integer and each a is a coefficient of the polynomial in s.

- From the theory of equations, it is known that if any coefficient a_r is negative, then there must be roots with positive real parts.
- Also, if any coefficient except a_0 is zero, then one either has roots with positive real parts or roots that exist on the imaginary axis of the s-plane.

Before the Routh-Hurwitz criterion can be applied, the Routh array or schedule has to be obtained. The Routh array corresponding to Equation (10.2) may be constructed as

$$\begin{array}{c|cccc}
s^n & a_n & a_{n-2} & \ldots & \\
s^{n-1} & a_{n-1} & a_{n-3} & \ldots a_0 & \\
\hline
s^{n-2} & b_1 & b_2 & \ldots & \\
s^{n-3} & c_1 & c_2 & \ldots & \\
s^{n-4} & d_1 & d_2 & \ldots & \\
. & . & . & . & \\
. & . & . & . & \\
s^0 & a_0 & & &
\end{array}$$

where

$$b_1=\frac{-\begin{vmatrix} a_n & a_{n-2} \\ a_{n-1} & a_{n-3}\end{vmatrix}}{a_{n-1}},\quad b_2=\frac{-\begin{vmatrix} a_n & a_{n-4} \\ a_{n-1} & a_{n-5}\end{vmatrix}}{a_{n-1}},\quad b_3=\frac{-\begin{vmatrix} a_n & a_{n-6} \\ a_{n-1} & a_{n-7}\end{vmatrix}}{a_{n-1}},$$

$$c_1=\frac{-\begin{vmatrix} a_{n-1} & a_{n-3} \\ b_1 & b_2\end{vmatrix}}{b_1},\quad c_2=\frac{-\begin{vmatrix} a_{n-1} & a_{n-5} \\ b_1 & b_3\end{vmatrix}}{b_1},$$

$$d_1=\frac{-\begin{vmatrix} b_1 & b_2 \\ c_1 & c_2\end{vmatrix}}{c_1},\quad d_2=\frac{-\begin{vmatrix} b_1 & b_3 \\ c_1 & c_3\end{vmatrix}}{c_1},$$

and so on.

- The coefficients are evaluated until a zero is obtained in every row. Once the table is completed, the Routh-Hurwitz criterion can be applied. It states that *the number of zeros (roots) of the c.e. with positive real parts is equal to the number of sign changes in the first column (that is, the one on the rhs of the vertical line in the array) of the array.*
- Thus, the criterion means whenever there are sign changes in the first column of the array, there are positive real parts of roots of the c.e. This, in turn, means that the system is unstable. For a stable system, there should be no sign change in the first column of the array.

10.3 Applications of Routh–Hurwitz Stability Criterion

The Routh Hurwitz stability criterion is applied to determine the stabilities of systems. The following examples illustrate its use and the steps involved in the stability analysis.

Example 1
Consider a feedback control system whose c.e. is given by:

$$s^3 + 12s^2 + 6s + 8 = 0.$$

Determine whether or not this system is stable.

Solution:
The Routh array becomes

s^3	1	6
s^2	12	8
s^1	16/3	0
s^0	8	0

In the foregoing array,

$$b_1 = \frac{-\begin{vmatrix} a_n & a_{n-2} \\ a_{n-1} & a_{n-3} \end{vmatrix}}{a_{n-1}} = \frac{-\begin{vmatrix} 1 & 6 \\ 12 & 8 \end{vmatrix}}{12} = \frac{72-8}{12} = \frac{16}{3},$$

$$b_2 = \frac{-\begin{vmatrix} a_n & a_{n-4} \\ a_{n-1} & a_{n-5} \end{vmatrix}}{a_{n-1}} = \frac{-\begin{vmatrix} 1 & 0 \\ 12 & 0 \end{vmatrix}}{12} = \frac{0}{12} = 0, \quad c_1 = 8, \quad c_2 = 0.$$

Since there is no change of sign in the first column, there is no root having positive real parts. This means that the above feedback control system is stable.

Example 2
Consider the following c.e. of a control system:

$$s^3 + 4s^2 + 4s + 20 = 0$$

Determine whether or not this system is stable.

Solution:

The Routh array becomes

s^3	1	4
s^2	4	20
s^1	−1	0
s^0	20	0

In the foregoing array,

$$b_1 = \frac{-\begin{vmatrix} a_n & a_{n-2} \\ a_{n-1} & a_{n-3} \end{vmatrix}}{a_{n-1}} = \frac{-\begin{vmatrix} 1 & 4 \\ 4 & 20 \end{vmatrix}}{4} = \frac{16-20}{4} = -1,$$

$$b_2 = \frac{-\begin{vmatrix} a_n & a_{n-4} \\ a_{n-1} & a_{n-5} \end{vmatrix}}{a_{n-1}} = \frac{-\begin{vmatrix} 1 & 0 \\ 4 & 0 \end{vmatrix}}{4} = \frac{0}{4} = 0, \quad c_1 = 20, \quad c_2 = 0.$$

Since there are two sign changes (that is, changing from positive 4 to negative 1 and then from negative 1 to positive 20) in the first column, there are two roots having positive real parts. This means that the above feedback control system is unstable.

Example 3

Consider a unity feedback control system having the block diagram as shown in Figure 10.1. Find the gain K that ensures stability of the system.

Solution:

The forward loop TF of the given system is

$$G = \left(\frac{K}{s+100}\right)\left(\frac{200}{s^2+80s}\right),$$

and the feedback loop TF is $H = 1$.

Applying Equation (10.1), the c.e. of the given system becomes

$$s^3 + 180s^2 + 8000s + 200K = 0.$$

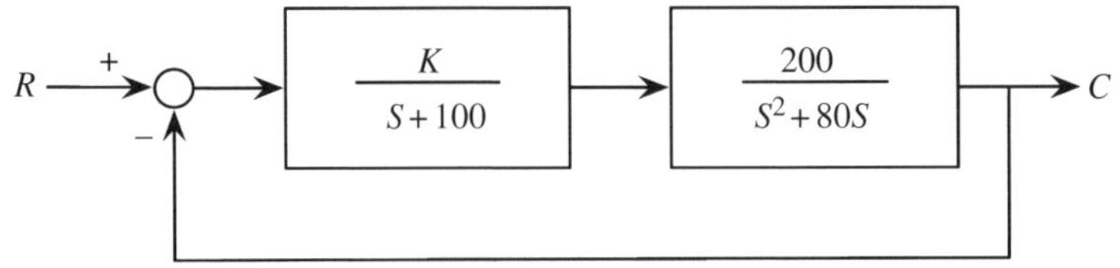

Figure 10.1 Unity feedback control system

The Routh array for this system becomes

$$\begin{array}{c|cc} s^3 & 1 & 8000 \\ s^2 & 180 & 200K \\ \hline s^1 & b_1 & b_2 \\ s^0 & c_1 & 0 \end{array}$$

in which

$$b_1 = \frac{-\begin{vmatrix} a_n & a_{n-2} \\ a_{n-1} & a_{n-3} \end{vmatrix}}{a_{n-1}} = \frac{-\begin{vmatrix} 1 & 8000 \\ 180 & 200K \end{vmatrix}}{180},$$

$$b_2 = 0, \quad c_1 = 200K, \quad c_2 = 0.$$

Since b_1 and c_1 must be positive in order to maintain no change of sign in the first column for stability. Then K must be positive and

$$(180)(8000) - 200K > 0, \quad \text{or} \quad 7200 > K.$$

Therefore, in order to ensure stability of the system, the gain K must satisfy the following relation

$$7200 > K > 0.$$

Example 4

Consider a feedback control system that has the following c.e.:

$$s^4 + 2s^3 + 4s^2 + 8s + 20 = 0.$$

Determine whether or not the system is stable.

Solution:

The Routh array for the given system is

$$\begin{array}{c|ccc} s^4 & 1 & 4 & 20 \\ s^3 & 2 & 8 & 0 \\ \hline s^2 & b_1 & b_2 & \\ s^1 & c_1 & c_2 & \\ s^0 & 20 & & \end{array}$$

in which

$$b_1 = \frac{-\begin{vmatrix} a_n & a_{n-2} \\ a_{n-1} & a_{n-3} \end{vmatrix}}{a_{n-1}} = \frac{-\begin{vmatrix} 1 & 4 \\ 2 & 8 \end{vmatrix}}{2} = 0,$$

$$b_2 = \frac{-\begin{vmatrix} a_n & a_{n-4} \\ a_{n-1} & a_{n-5} \end{vmatrix}}{a_{n-1}} = \frac{-\begin{vmatrix} 1 & 20 \\ 2 & 0 \end{vmatrix}}{2} = 20.$$

Since $b_1 = 0$, a zero appears in the first column of the array. This requires the replacement of b_1 and b_2 with ε and b_2, where $\varepsilon \to 0$. Then the above array is replaced by the following new scheme:

s^4	1	4	20
s^3	2	8	0
s^2	ε	20	
s^1	c_1	c_2	
s^0	20		

where

$$c_1 = \frac{-\begin{vmatrix} a_{n-1} & a_{n-3} \\ b_1 & b_2 \end{vmatrix}}{b_1} = \frac{-\begin{vmatrix} 2 & 8 \\ \varepsilon & 20 \end{vmatrix}}{\varepsilon} = \frac{8\varepsilon - 40}{\varepsilon}, \quad c_2 = 0.$$

As ε approaches zero, $c_1 \to -\infty$. Therefore, the first column of the array has two sign changes. This means that there are two roots having positive real parts and thus, the above feedback control system is unstable.

- Note that the above method of replacing the coefficient cannot be applied if all coefficients of a row are zeros.
- If all the coefficients or elements in any row of the Routh array are zero, it indicates that there are roots of equal magnitude lying opposite in the s-plane – that is, two real roots with equal magnitudes and opposite signs and/or two conjugate imaginary roots.

In such a case, the evaluation of the rest of the array can be continued by forming an auxiliary polynomial with the coefficients of the last row and by using the coefficients of the derivative of this polynomial in the next row. Such roots with equal magnitudes and lying opposite in the s-plane can be found by solving the *auxiliary* polynomial, which is always even. Thus, for a $2n$-degree auxiliary polynomial, there are n pairs of equal and opposite roots.

Example 5
Consider the following characteristic equation

$$s^5 + 2s^4 + 24s^3 + 48s^2 - 25s - 50 = 0.$$

Solve for the stability of this system.

Solution:

It should be mentioned that since the above c.e. has two negative coefficients, according to the theory of equations there must be roots with positive real parts. This means that the system is unstable. The following study is simply made to show (a) that indeed the system is unstable and (b) how to use the Routh array approach when all elements or entries in an entire row are zeros.

The partial Routh array becomes

s^5	1	24	−25	
s^4	2	48	−50	← Auxiliary polynomial $P(s)$
s^3	0	0		

The terms in the s^3 row are all zero because

$$b_1 = \frac{-\begin{vmatrix} a_n & a_{n-2} \\ a_{n-1} & a_{n-3} \end{vmatrix}}{a_{n-1}} = \frac{-\begin{vmatrix} 1 & 24 \\ 2 & 48 \end{vmatrix}}{2} = 0,$$

$$b_2 = \frac{-\begin{vmatrix} a_n & a_{n-4} \\ a_{n-1} & a_{n-5} \end{vmatrix}}{a_{n-1}} = \frac{-\begin{vmatrix} 1 & -25 \\ 2 & -50 \end{vmatrix}}{2} = 0.$$

The auxiliary polynomial is then formed from the coefficients of the s^4 row. In other words, the auxiliary polynomial $P(s)$ is

$$P(s) = 2s^4 + 48s^2 - 50$$

which indicates that there are two pairs of roots of equal magnitude and opposite sign. These pairs are obtained by solving the auxiliary polynomial equation $P(s) = 0$. The derivative of $P(s)$ with respect to s is

$$\frac{dP(s)}{ds} = 8s^3 + 96s.$$

The terms in the s^3 row are now replaced by the coefficients of the last equation, that is, 8 and 96. The new array of coefficients then becomes

s^5	1	24	−25	
s^4	2	48	−50	← Auxiliary polynomial $P(s)$
s^3	8	96		
s^2	24	−50		
s^1	112.7	0		
s^0	−50			

It is observed that there is one change in sign in the first column of the new array. Therefore, the original equation has one root with a positive real part.

By solving for roots of the auxiliary polynomial equation $2s^4 + 48s^2 - 50 = 0$, one obtains

$$(s+1)(s-1)(s+i5)(s-i5)=0.$$

These two pairs of roots are a part of the roots of the original equation. In fact, the original equation can be written in factored form as

$$(s+1)(s-1)(s+i5)(s-i5)(s+2)=0.$$

This shows that the original equation has one root with a positive real part giving by the factor $(s-1)$ and therefore the system is not stable.

10.4 Questions and Solutions

In this section the Routh-Hurwitz stability criterion is employed to analyse the stability conditions of three control systems. The steady-state errors are considered in the first and third systems. The first control system is a model of a phase detector, the second one is a model of a motorized wheelchair, and the third is concerned with the system that does not contain any moving mechanical parts for control of speed of a steam turbine. These three control systems with some changes are drawn from the book by Dorf [3], in which many excellent, realistic, and practical questions are presented.

Example 1

The stability feature of a linear model of a phase detector (phase-lock loop) is considered and its block diagram representation is included in Figure 10E1. Phase detectors are frequently applied in space telemetry, color television, and missile tracking. They are designed to eliminate phase difference between the input carrier signal and a local voltage-controlled oscillator. In Figure 10E1 the filter for a particular application has been selected as

$$G(s)=\frac{s+10}{\left(s+\frac{1}{20}\right)(s+50)}.$$

It is required to minimize the steady-state error of the system for a ramp change in the phase information signal.

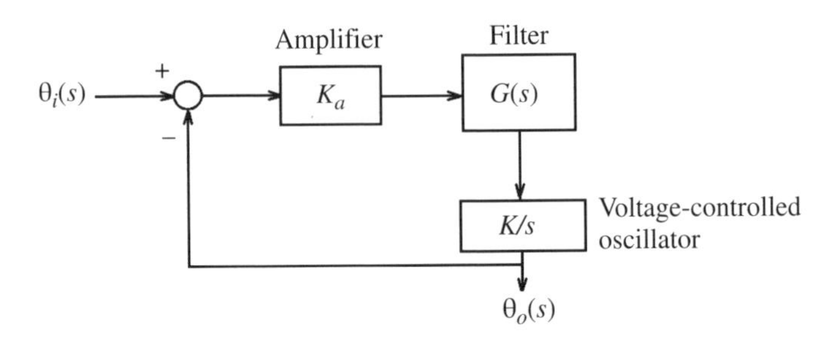

Figure 10E1 Block diagram representation of a phase detector

a. Determine the limiting value of gain $K_v = K_aK$ in order to maintain the stability of the system.
b. It has been decided that a steady-state error equal to 0.5° is acceptable for a ramp signal of 100 rad/s. For that value of gain K_v, locate the roots of the system in the s-plane and their implication for the stability of the system.

Solution:

a. Since $K_v = K_aK$ and $G_o = \left(\frac{K_aK}{s}\right)\frac{(s+10)}{\left(s+\frac{1}{20}\right)(s+50)} = \frac{20K_aK(s+10)}{s(20s+1)(s+50)}$, by applying Equation (10.1) one obtains the c.e. as

$$20s^3 + 1001s^2 + (20K_v + 50)s + 200\,K_v = 0$$

The Routh array for this system becomes

s^3	20	$20K_v + 50$
s^2	1001	$200K_v$
s^1	b_1	b_2
s^0	c_1	0

in which

$$b_1 = \frac{-\begin{vmatrix} a_n & a_{n-2} \\ a_{n-1} & a_{n-3} \end{vmatrix}}{a_{n-1}} = \frac{-\begin{vmatrix} 20 & 20K_v+50 \\ 1001 & 200K_v \end{vmatrix}}{1001} = \frac{16020K_v + 50050}{1001},$$

$$b_2 = 0, \qquad c_1 = 200K_v.$$

b_1 and c_1 must be positive in order to maintain no change of sign in the first column for stability, that is, when $K_v > 0$ and $16020K_v + 50050 > 0$.
Therefore, the system is always stable for all value of $K_v > 0$.

b. The steady-state error,

$$\varepsilon_{ss} = \varepsilon(\infty)\lim_{s\to 0} s\left[\frac{R(s)}{1+\frac{20K_aK(s+10)}{s(20s+1)(s+50)}}\right],$$

where $R(s) = \mathcal{L}\{At\} = \frac{A}{s^2}$ in which $A = 100$ rad/s, is the amplitude of the ramp for the present problem. Thus,

$$\varepsilon_{ss} = \lim_{s\to 0}\left[\frac{s\left(\frac{A}{s^2}\right)}{1+\frac{20K_aK(s+10)}{s(20s+1)(s+50)}}\right] = \frac{5A}{20K_v}.$$

But the given condition is that $\varepsilon_{ss} = \dfrac{0.5°}{\left(\frac{360°}{2\pi\ \text{rad}}\right)} = 0.008727\ \text{rad}.$

Equating $0.008727 = \dfrac{5A}{20K_v} = \dfrac{5(100)}{20K_v}$, gives $K_v = 2864.67$, which is the required gain. Sub stituting this gain value into the c.e.,

$$20s^3 + 1001s^2 + [(20)(2864.67) + 50]s + (200)(2864.67) = 0.$$

$$20s^3 + 1001s^2 + 57343.4s + 572934 = 0.$$

The roots of this c.e., which is a cubic polynomial in s, are

$$s_1 = -11.8666, \ s_2 = -19.0917 + i45.2722, \ \text{and}$$

$$s_3 = -19.0917 + i45.2722,$$

where s_2 and s_3 are complex conjugates to each other.

With reference to the above roots of the c.e., one can conclude that the system is stable since all roots are on the left-hand half s-plane. This simply confirms that found in (a) of the present question.

Example 2

A velocity control system for a motorized wheelchair has been designed for a person paralysed from the neck down. In this system, velocity sensors are mounted in a headgear and its block diagram representation is shown in Figure 10E2. The headgear sensor provides an output signal proportional to the magnitude of the head movement. The sensors are mounted at 90° intervals so that forward, left, right, or backward movements can be determined. The transfer functions in the figure are given by

$$G_1(s) = \frac{K_1}{\tau_1 s + 1}, \quad G_2(s) = K_2, \quad G_3(s) = \frac{K_3}{\tau_3 s + 1}, \quad G_4(s) = \frac{1}{s}, \quad H(s) = \frac{K_4}{\tau_4 s + 1}.$$

Through a series of tests it was found that typical values for the time constants are $\tau_1 = \tau_1 = 0.5$ s, $\tau_3 = 1.0$ s, and $\tau_4 = \frac{1}{3}$ s. For this system,

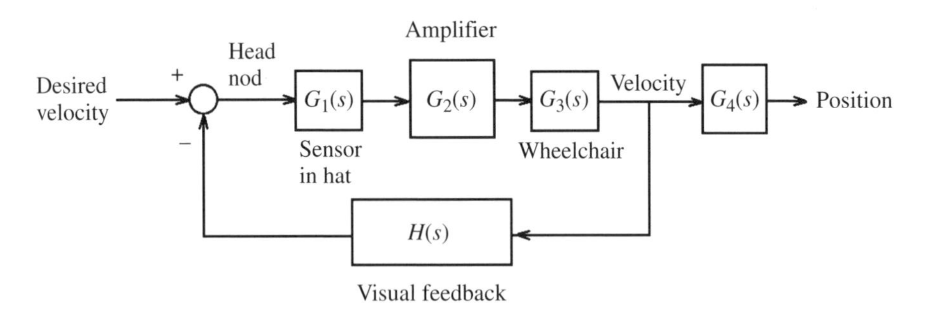

Figure 10E2 Velocity control system for a motorized wheelchair

a. determine the limiting gain $K = K_1K_2K_3K_4$; and
b. if the gain K is equal to a third of the upper limiting value, find whether the settling time of the system is less than 4.0 s. Assume the settling time $t_s = \frac{4}{\zeta\omega_n}$.

Solution:

a. The c.e. is given by $1 + G(s)H(s) = 0$. Thus

$$1+\frac{K_1K_2K_3K_4}{\left(\frac{s}{2}+1\right)(s+1)\left(\frac{s}{3}+1\right)}=0.$$

Simplifying, one has

$$s^3+6s^2+11s+6(1+K)=0.$$

The Routh array for this system becomes

s^3	1	11
s^2	6	$6(1+K)$
s^1	b_1	b_2
s^0	c_1	

in which

$$b_1=\frac{-\begin{vmatrix} a_n & a_{n-2} \\ a_{n-1} & a_{n-3}\end{vmatrix}}{a_{n-1}}=\frac{-\begin{vmatrix} 1 & 11 \\ 6 & 6(1+K)\end{vmatrix}}{6} \quad \text{or}$$

$$b_1=11-(1+K)$$

$$b_2=0, \quad c_1=6(1+K).$$

b_1 and c_1 must be positive in order to maintain no change of sign in the first column for stability. This implies $11 - (1 + K) \geq 0$ and $6(1 + K) \geq 0$.
Therefore, for stability, $-1 \leq K \leq 10$.

b. $K = 10/3$, which is a third of the upper limiting value as found in (a). To proceed, one substitutes this value into the c.e. to give:

$$s^3+6s^2+11s+26=0.$$

Solving for the roots of this equation gives

$$s_1=-\ 4.8371, \quad s_2=-0.5814+i2.2443, \quad s_3=-0.5814-i2.2443.$$

That is, all roots of the c.e. are in the left-hand half s-plane, and therefore the system is stable. Also, s_2 and s_3 are complex conjugates to each other. They are the roots of the second-order system associated with the c.e.:

$$(s - s_2)(s - s_3) = 0.$$

That is, $(s - 0.5814 + i2.2443)(s - 0.5814 - i2.2443) = 0$, or $s^2 + 2(0.5814)s + 5.3749 = 0$ which corresponds to

$$s^2 + 2(\zeta\omega_n)s + \omega_n^2 = 0.$$

Therefore, $\zeta\omega_n = 0.5814$ for the present system.
The settling time is defined by

$$t_s = \frac{4}{\zeta\omega_n} = \frac{4}{0.5814}\,\text{s} = 6.8799\ \text{s}$$

which is longer than 4.0 s.

Example 3

A pure fluid control system that does not have any moving mechanical parts is designed for the control of speed of a steam turbine. Fluid control systems have many advantages in practice because they are insensitive and reliable over a wide range of electromagnetic and nuclear radiation, temperature, acceleration, and vibration. The designed system maintains the speed within 0.5% of the desired speed by employing a tuning fork reference and a valve actuator. The amplification within the system is achieved by applying a fluid jet deflection amplifier. The present system is designed for a 500 kW steam turbine with an operating speed of 12,000 rpm. The simplified block diagram representation for this system is shown in Figure 10E3 where

$G_1(s) = \dfrac{s+1}{s+11}$ is the TF of the filter,
$G_2(s) = 1/s$ is the TF of the valve actuator,
$G_3(s) = \dfrac{1}{Js + f}$ is the TF of the turbine, and
$H(s) = K_1$ is the TF of the tuning fork and error detector.

It is assumed that the friction f of the large inertia of the turbine is negligible, so in Figure 10E3, $f = 0$. It is also assumed that the disturbance to the system is zero, so $T_s(s) = 0$.

With $J = K_1 = 10^6$, determine the closed-loop transfer function of the system in Figure 10E3. With the given parameters above:

a. find the characteristic equation (c.e.) of the system, set up the Routh-Hurwitz array, and determine whether or not the system is stable;
b. evaluate the steady-state error of the system response to a unit step input.

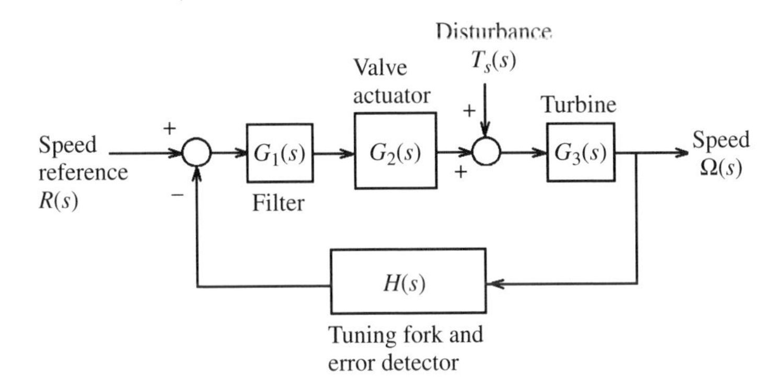

Figure 10E3 A pure fluid speed control for a 500 kW steam turbine

Solution:

The closed-loop transfer function of the given system is defined by

$$\frac{\Omega(s)}{R(s)} = \frac{G_1(s)G_2(s)G_3(s)}{1 + G_1(s)G_2(s)G_3(s)H(s)}$$

where the transfer functions are given above.

Upon substituting the TF of the individual components in the system, the closed-loop transfer function becomes

$$\frac{\Omega(s)}{R(s)} = \frac{s + 1}{s(s + 11)(Js + f) + (s + 1)K_1}.$$

Substituting for the given parameters and simplifying, one has the TF as

$$\frac{\Omega(s)}{R(s)} = \frac{s + 1}{(10^6)s^3 + (11 \times 10^6)s^2 + (10^6)s + 10^6}.$$

a. With reference to the denominator term of the last equation, the c.e. is

$$(10^6)s^3 + (11 \times 10^6)s^2 + (10^6)s + 10^6 = 0.$$

The Routh array for this system becomes

s^3	10^6	10^6
s^2	11×10^6	10^6
s^1	b_1	b_2
s^0	c_1	

in which

$$b_1 = \frac{-\begin{vmatrix} a_n & a_{n-2} \\ a_{n-1} & a_{n-3} \end{vmatrix}}{a_{n-1}} = \frac{-\begin{vmatrix} 10^6 & 10^6 \\ 11\times 10^6 & 10^6 \end{vmatrix}}{11\times 10^6} \quad \text{or}$$

$$b_1 = 9.0909\times 10^5, \quad b_2 = 0, \quad c_1 = 10^6.$$

Since b_1 and c_1 are positive, there is no sign change in the first column of the Routh array. This means that the real parts of the roots of the c.e. are negative, and therefore the system is stable.

b. The input is a unit step function and therefore $R(s) = 1/s$.

The steady-state error of the system response is

$$\omega(\infty) = \lim_{s\to 0} s\,\Omega(s) = \lim_{s\to 0} \frac{s\left(\frac{1}{s}\right)(s+1)}{(10^6)s^3 + (11\times 10^6)s^2 + (10^6)s + 10^6}.$$

Therefore, the steady-state error of the system response is

$$\omega(\infty) = \frac{1}{10^6} = 10^{-6}.$$

Exercise Questions

Q1. A control system has an open-loop transfer function given by

$$G(s)H(s) = \frac{K(s+2)}{s(s-2)}.$$

a. Find the value of the gain K when the damping ratio ζ of the closed-loop roots is equal to $1/\sqrt{2}$.
b. Find the value of the gain K when the closed-loop system has two roots on the imaginary axis.

Q2. A conceptual design of a small vertical-takeoff fighter jet airplane which is invisible to radar has a direction control system as shown in Figure 10Q2 in which

$$G_1 = K, \quad \text{and} \quad G_2(s) = \frac{s+20}{s^3 + 20s^2 + 100s}.$$

Determine the maximum gain K of the direction control system for stable operation.

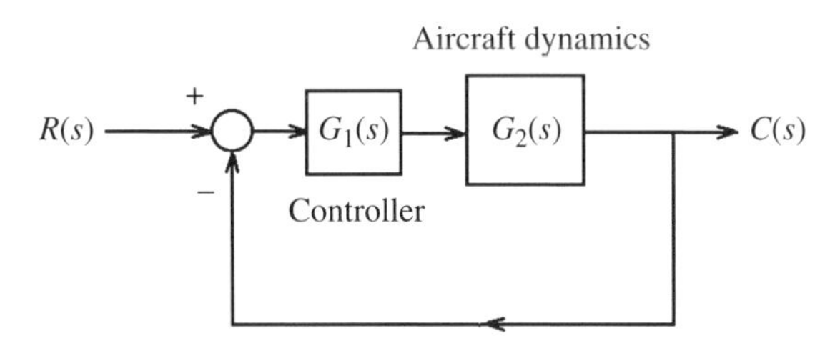

Figure 10Q2 Direction control system

Q3. Determine the stability of the following characteristic equations by using the Routh-Hurwitz criterion:

a. $s^3 + 4s^2 + 5s + 6 = 0$,
b. $s^4 + s^3 + 2s^2 + 6s + 8 = 0$.

For the above two cases, find the number of roots, if any, in the right-hand s-plane.

Q4. It is generally believed that one of the most important areas of application for industrial robots is arc welding. In many welding situations, uncertainties in geometry of the joint, the welding process itself, and dimensions of the part call for the use of sensors in order to maintain weld quality. Vision devices are frequently employed to measure the geometry of the puddle of melted metal. One such device applying a constant rate of feeding the wire to be melted has the block diagram shown in Figure 10Q4, in which

$$G_1(s) = \frac{K}{s+1}, \quad G_2(s) = \frac{1}{s+1}, \quad H(s) = \frac{2}{s+2}.$$

a. Find the limiting value or values for K of the vision device that will result in a stable response.
b. Applying the above limiting value or values for K, determine the steady-state error of puddle diameter if the desired diameter can be considered as a unit step input.

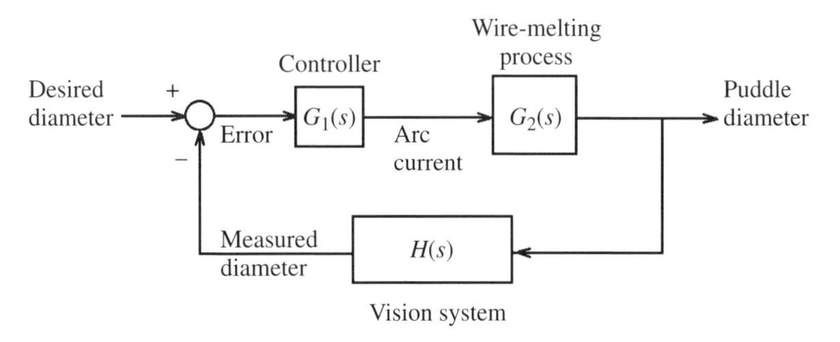

Figure 10Q4 Block diagram of a vision device used in arc welding

Q5. The block diagram representation of a roll stabilizer system for a ship is shown in Figure 10Q5, in which

$$G(s) = \frac{5}{10s^2 + 2s + 10}.$$

a. Find the transfer function of the effect of wave torque $T_d(s)$ on ship roll $C(s)$.
b. Find the characteristic equation of the control system in Figure 10Q5.
c. If the gain of the amplifier $K_a = K$, $K_1 = 8K_a$ and the system critical damping ratio $\zeta = 0.25$ determine whether or not the system will oscillate. If the answer is affirmative, what is the natural frequency of the system?

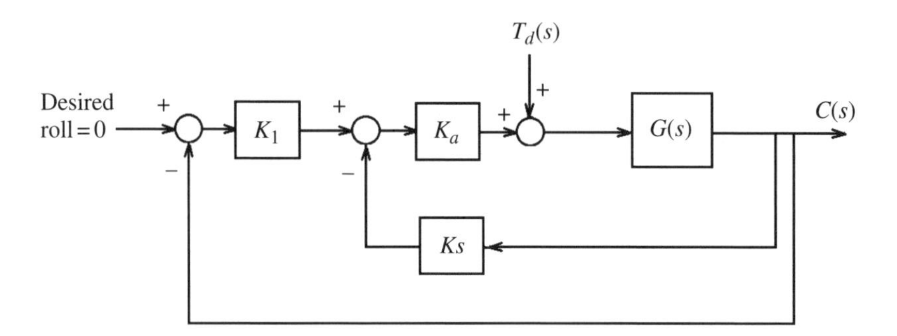

Figure 10Q5 Block diagram of a roll stabilizer for a ship

References

[1] Routh, E.J. (1877). *A Treatise on the Stability of a Given State of Motion*. Macmillan and Co., London.
[2] Hurwitz, A. (1895). On the conditions under which an equation has only roots with negative real parts. *Mathematische Annalen*, **46**, 273–284.
[3] Dorf, R.C. (1986). *Modern Control Systems*, 4th edn. Addison-Wesley, Reading, MA.

11

Graphical Methods for Control Systems

Up to this stage, the methods and techniques applied to the feedback control systems are analytical and in both the time and s-parameter domains. In many situations when the systems are linear, graphical methods may prove to be very useful to reveal system characteristics and performance. In a sense, these graphical methods compress a great deal of information in the time domain into the parameter domain in a single figure. The parameter domains considered in this chapter are the s-parameter domain of the Laplace transformation, and the ω-parameter domain of the frequency response function. The frequency response function can simply be obtained by replacing the s-parameter of the system transfer function with $i\omega$.

In this chapter the root locus method and root locus plots are presented in Section 11.1, while in Section 11.2 polar and Bode plots are included. Section 11.3 is concerned with the Nyquist stability criterion (NSC) and Nyquist diagrams. Gain and phase margins in relative stability analysis are introduced in Section 11.4. Contours of magnitude and phase of system frequency response, the so-called M and N circles, are presented in Sections 11.5 and 11.6, respectively. In Section 11.7, Nichols charts are introduced. Finally, Section 11.8 includes various examples using MATLAB [1, 2].

11.1 Root Locus Method and Root Locus Plots

In previous chapters it has been shown analytically that the transient response of a linear system under a deterministic forcing function depends on the roots of the characteristic equation (c.e.). In parallel to the analytical approaches, a graphical method is introduced in this section for the determination of the actual values of the roots of the c.e. when the system parameter is varied. This method is called the *root locus method* [3], and the resulting plot in the complex s-plane as the parameter varies is known as the *root locus plot*.

Introduction to Dynamics and Control in Mechanical Engineering Systems, First Edition. Cho W. S. To.

Companion website: www.wiley.com/go/to/dynamics

Consider the closed-loop transfer function in Chapter 8. The open-loop transfer function is defined by

$$G(s)H(s) = KG_o(s) = \frac{K(s-z_1)(s-z_2)\ldots(s-z_m)}{(s-p_1)(s-p_2)\ldots(s-p_n)}, \tag{11.1}$$

where p_i and z_j are the poles and zeros, respectively, and K is known as the system gain constant or the open-loop gain factor. In this equation it is assumed that $n > m$.

While in principle the root locus plot in the complex s-plane may be constructed by the trial-and-error method, in order to reduce the time for such a construction the following rules or steps are presented.

11.1.1 Rules for Root Locus Plots of Negative Feedback Control Systems

In order to provide a set of simple rules for the construction of root loci for negative feedback control systems, it is convenient to make use of the open-loop transfer function given by Equation (11.1). The following rules, without proof for brevity, are based on the latter equation.

Rule 1: Number of loci

The number of branches or loci is equal to the number of open-loop poles, n, or degree of the c.e.

Rule 2: Symmetry of loci

The root loci of a real c.e. are symmetrical with respect to the real axis, since complex roots of a real c.e. appear in conjugate pairs.

Rule 3: Poles of $G_o(s)$

These poles of $G_o(s)$ lie on the root loci and correspond to $K = 0$.

Rule 4: Zeros of $G_o(s)$

The zeros of $G_o(s)$ lie on the root loci and correspond to $K = \pm\infty$.

Rule 5: Asymptotes of loci

If $G_o(s)$ has $\delta = (n - m)$ more poles than zeros, the root loci are asymptotic to δ straight lines, making angles $(2k + 1)\pi/\delta$ with the real axis for positive K where $k = 0, 1, 2, \ldots, \delta - 1$. The root loci are also asymptotic to δ straight lines, making angles $2k\pi/\delta$ with the real axis for negative K.

The root loci approach the asymptotic lines of angles $(2k + 1)\pi/\delta$ as $K \to +\infty$ and approach those asymptotic lines of angles $2k\pi/\delta$ as $K \to -\infty$.

Rule 6: Intersection points of asymptotes

Both sets of asymptotes intersect on the real axis at a point with abscissa

$$\sigma_a = \frac{\sum_{i=1}^{n} p_i - \sum_{j=1}^{m} z_j}{n-m}, \tag{11.2}$$

which is sometimes referred to as the *center of gravity* or *center of asymptotes*.

Rule 7: Real axis loci

The root locus on the real axis in the s-plane is determined by counting the total number of the finite poles and zeros of $G_o(s)$ to the rhs of the points in question.

For positive K, points of the root locus on the real axis lie to the left of an *odd* number of finite poles and zeros.

For negative K, points of the root locus on the real axis lie to the left of an *even* number of finite poles and zeros.

Rule 8: Breakaway points

Breakaway points imply the existence of multiple characteristic roots and appear at those values of s which satisfy

$$\frac{dK}{ds} = 0. \tag{11.3}$$

The corresponding loci leave or enter the real axis at angles of $\pm\pi/2$.

Rule 9: Intersections of root loci with imaginary axis

These intersections with the imaginary axis can be obtained by determining the values of the gain K and angular frequency ω of the c.e. in which the parameter s is replaced with $i\omega$. These values of K and ω are evaluated by equating separately the real part and the imaginary part of the resulting c.e. to zero. The obtained angular frequencies are those at which the root loci cross the imaginary axis. The corresponding value of K to every crossing frequency ω is the gain at that crossing point.

Rule 10: Departure and arrival angles

The *departure angle* of the root locus from a complex pole is defined as

$$\theta_d = 180^\circ + \angle(GH)_1, \tag{11.4}$$

and the *arrival angle* of the root locus at a complex zero is defined as

$$\theta_a = 180^\circ - \angle(GH)_2, \tag{11.5}$$

where $\angle(\mathrm{GH})_1$ is the phase angle in degrees of the open-loop transfer function GH evaluated at the complex pole, but disregarding the contribution of that particular pole, and $\angle(\mathrm{GH})_2$ is the phase angle of GH determined at the complex zero, but ignoring the contribution of that zero.

Rule 11: Determination of system gain K on the root loci

The absolute magnitude of the system gain K corresponding to any point s_K on a root locus can be obtained by measuring the lengths of the vectors from the poles and zeros of $G_o(s)$ to s_K. Thus,

$$|K| = \frac{|s_K - p_1||s_K - p_2||s_K - p_3|\ldots}{|s_K - z_1||s_K - z_2||s_K - z_3|\ldots}. \tag{11.6}$$

11.1.2 Construction of Root Loci

Consider the feedback control system having the open-loop transfer function,

$$GH = KG_o(s) = \frac{K(s+2)}{s(s+3)(s+2-i)(s+2+i)} = \frac{K(s+2)}{s(s+3)(s^2+4s+5)},$$

where K is positive. Note that the c.e. becomes

$$s^4 + 7s^3 + 17s^2 + (15+K)s + 2K = 0$$

From the foregoing,

$$G_o(s) = \frac{s+2}{s(s+3)(s+2-i)(s+2+i)} = \frac{s+2}{s^4 + 7s^3 + 17s^2 + 15s}$$

This latter expression will be used in the root locus construction. To this end, one applies the rules in Section 11.1.1 as in the following.

Rule 1: Since there are four poles, therefore there are four loci.

Rule 2: For real K, loci are symmetrical about the real axis.

Rule 3: In order to determine the poles of $G_o(s)$, one sets the denominator term of $G_o(s)$ to 0. Thus, there are four poles and they are $s = 0, -3, -2 \pm i$. Root loci pass through poles at $K = 0$.

Rule 4: The zero is the numerator term of the open-loop transfer function being set to zero. That is, $s + 2 = 0$. Thus, the zero is $s = -2$. The root locus passes through this zero at $K = \infty$.

Rule 5: There are $n - m = 4 - 1 = 3$ more poles than zeros. Thus, $\delta = 3$ and these asymptotic lines make angles of $\pi/3$, π, and $5\pi/3$ with the real axis for $K \to \infty$.

Rule 6: The asymptotes intersect at $\sigma_a = -5/3$. This is because, according to Equation (11.2), one obtains

$$\frac{(0-3-2+i-2-i)-(-2)}{3} = -\frac{5}{3}$$

Rule 7: Since all the points on the real axis between 0 and −2 lie to the lhs of an odd number of finite poles and zeros, these points are on the root locus for positive K. The segment of the real axis between −3 and $-\infty$ lies to the lhs of an odd number of finite poles and zeros, so that the points in this segment are also on the root locus for positive K. The loci on the real axis for the present system are indicated in Figure 11.1.

Rule 8: Two loci are on the real axis and two other loci are symmetrical with respect to the real axis. There are no multiple characteristic roots and therefore, there is no breakaway point.

Rule 9: Replace s by $i\omega$ in the c.e.

$$s^4 + 7s^3 + 17s^2 + (15+K)s + 2K = 0$$

Substituting $i\omega$ for s in this c.e. gives

$$(i\omega)^4 + 7(i\omega)^3 + 17(i\omega)^2 + (15+K)i\omega + 2K = 0$$

which upon equating the real part and imaginary part to zero, one obtains

$$\omega^4 - 17\omega^2 + 2K = 0, \quad \text{and} \quad 7(i\omega)^3 + (15+K)i\omega = 0.$$

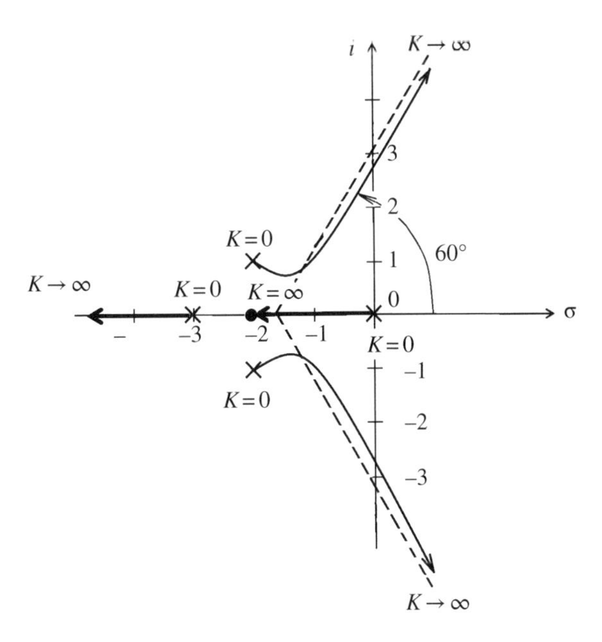

Figure 11.1 Root locus diagram of $G_o(s)$

The last equation becomes $-7\omega^2 + 15 + K = 0$ which gives $\omega^2 = \dfrac{15+K}{7}$.

Substituting this value into the above equation associated with the real part, one has

$$\left(\frac{15+K}{7}\right)^2 - 17\left(\frac{15+K}{7}\right) + 2K = 0$$

Upon expanding and simplifying, it reduces to $K^2 + 9K - 1560 = 0$.

Therefore,

$$K = \frac{-9+\sqrt{6321}}{2} = 35.25 \text{ which gives } \omega = 2.6793 \text{ rad/s.}$$

Rule 10: By Equation (11.4), the departure angle of the complex pole at $(-2 + i)$ is:

$$\theta_d = 180^\circ + 90^\circ - 45^\circ - 90^\circ - 153.43^\circ = -18.43^\circ$$

The root locus sketch is included in Figure 11.1.

11.2 Polar and Bode Plots

As mentioned in the introduction to this chapter, the frequency response function can be simply obtained by replacing the s-parameter in the system transfer function with $i\omega$. The frequency response function of a system can conveniently be displayed on an Argand diagram, which is called the polar plot by control engineers. Such a plot shows the variation of magnitude and phase of the output on polar coordinates, for a constant input, as the angular frequency varies from zero to infinity.

For example, the transfer function of a first-order system or *simple lag* is given by

$$G(s) = \frac{1}{1+\tau s}$$

The corresponding frequency response function is obtained by replacing the parameter s with $i\omega$:

$$G(i\omega) = \frac{1}{1+\tau(i\omega)}$$

Therefore, by rationalization, the frequency response function of the first-order system becomes

$$G(i\omega) = \frac{1}{1+\tau(i\omega)} = \frac{1-\tau(i\omega)}{1+(\tau\omega)^2} = \frac{1}{1+(\tau\omega)^2} - i\frac{\tau\omega}{1+(\tau\omega)^2} \tag{11.7}$$

This can be written as

$$G(i\omega) = |G(i\omega)| \angle G(i\omega) = \frac{1}{\sqrt{1+(\tau\omega)^2}} \angle\left(-\tan^{-1}\tau\omega\right). \tag{11.8}$$

With reference to Equation (11.8), one has

$$\begin{aligned} |G(i\omega)| &= 1 \text{ and } \angle G(i\omega) = 0 \quad \text{for } \omega = 0, \\ |G(i\omega)| &= 0 \text{ and } \angle G(i\omega) = -90^\circ \quad \text{for } \omega = \infty. \end{aligned}$$

The polar plot of this first-order system is a semicircle, as presented in Figure 11.2. It is a semicircle since one can start from Equation (11.7) in which the real part and the imaginary part on the rhs may be written, respectively, as

$$x = \frac{1}{1+(\tau\omega)^2} \text{ and } y = -\tau\omega x.$$

Eliminating the angular frequency ω, one obtains

$$y^2 = (\tau\omega x)^2 = x^2\left(\frac{1-x}{x}\right) = x - x^2$$

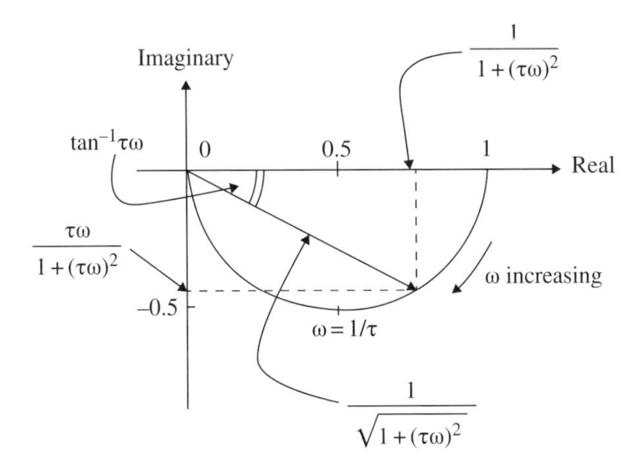

Figure 11.2 Polar plot of a simple lag system

Adding $(\frac{1}{2})^2$ to both sides and rearranging leads to

$$y^2 + \left(x - \frac{1}{2}\right)^2 = \left(\frac{1}{2}\right)^2$$

This shows that the polar plot of a simple lag system is a semicircle of radius 0.5 and center at (0.5, 0).

When the system transfer function $G(s)$ is available in the factorized form, an alternative method for obtaining and presenting system frequency response $G(i\omega)$ data is the Bode plot [4]. It is also known as the *corner plot* or *logarithmic plot*. It comprises two plots, one for the magnitude $|G(i\omega)|$ and the other for the phase $\angle G(i\omega)$. The magnitude is generally plotted on a logarithmic scale and expressed in decibels or dB. This is an advantage in that the overall magnitude and phase can be obtained by simply adding the component parts graphically. Another advantage of the Bode plot is that some approximations can be made by using straight-line constructions. These constructions can be achieved rapidly and the approximations are frequently of sufficient accuracy.

For illustration, two systems are included in the following. The first has the transfer function of a first-order system or *simple lag* given by

$$G(s) = \frac{1}{1 + \tau s}$$

This is the system considered for the polar plot presented in the foregoing.

The corresponding frequency response function is obtained by replacing the parameter s with $i\omega$:

$$G(i\omega) = \frac{1}{1 + \tau(i\omega)}$$

The magnitude of this transfer function is

$$20\ \log_{10}|G(i\omega)| = 20\ \log_{10}\left(\frac{1}{\sqrt{1+(\tau\omega)^2}}\right) = -20\ \log_{10}\sqrt{1+(\tau\omega)^2}\ \text{dB}.$$

With asymptotic approximation, one has

$$-20\ \log_{10}\sqrt{1+(\tau\omega)^2}\ \text{dB} = 0 \qquad \text{for } \omega\tau \ll 1 \text{ and}$$
$$-20\ \log_{10}\sqrt{1+(\tau\omega)^2}\ \text{dB} = -20\ \log_{10}\tau\omega\,\text{dB} \quad \text{for } \omega\tau \gg 1$$

The Bode plot of this first-order system is shown in Figure 11.3a, in which the straight line with a negative slope of 20 dB per decade of frequency intersects with the horizontal

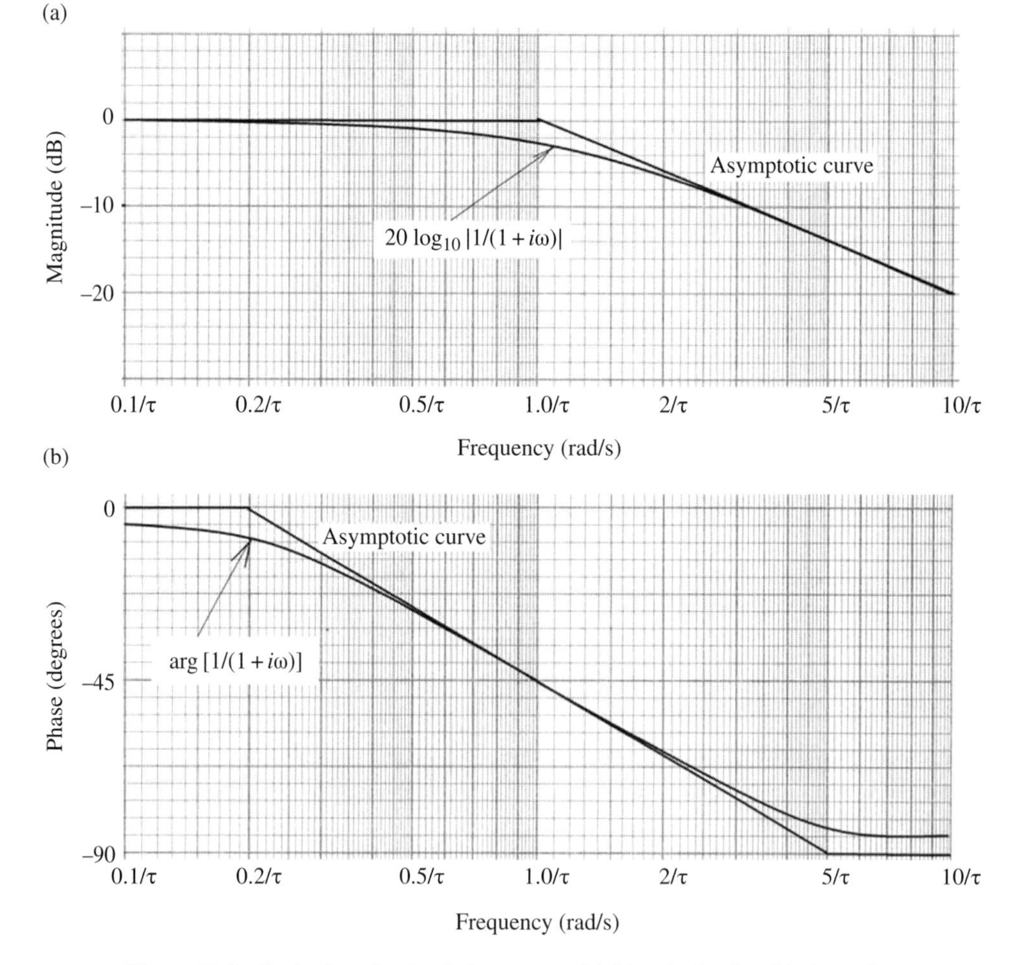

Figure 11.3 Bode plot of a simple lag system. (a) Magnitude plot; (b) phase plot

straight line of zero dB at $\omega = \frac{1}{\tau}$. The latter is called the *break point* or *corner frequency*. The phase is given by

$$\angle G(i\omega) = -\tan^{-1}\tau\omega$$

The exact solution and asymptotic approximation are included in Figure 11.3b.

It may be appropriate to note that the maximum error at the corner frequency between the exact plot and the asymptotic approximation for the magnitude, as indicated in Figure 11.3a, is 3 dB, while the maximum error at the corner frequency between the exact plot and the asymptotic approximation for the phase, as shown in Figure 11.3b, is zero. This is because the lag is exactly 45° at the corner frequency.

The second system has the following transfer function:

$$\begin{aligned} G(s) &= \frac{5}{(1+10s)(s^2+3s+25)} \\ &= \frac{5}{10s^3+31s^2+253s+25} \end{aligned} \tag{11.9}$$

The objective is to draw a Bode diagram for this transfer function. Thus, one replaces s with $i\omega$ such that Equation (11.9) becomes

$$\begin{aligned} G(i\omega) &= \frac{5}{(1+10i\omega)\left[(i\omega)^2+3i\omega+25\right]} \\ &= \frac{1/5}{(1+10i\omega)\left[1-\left(\frac{\omega}{5}\right)^2+i0.6\left(\frac{\omega}{5}\right)\right]} \end{aligned} \tag{11.10}$$

From Equation (11.10) one notes that there are three components in the Bode plot. These three components are listed in the following.

1. The first component is the constant gain (that is, by setting $i\omega = 0$) which is equal to 5/25 = 0.2 and has a constant magnitude of $20 \log_{10} 0.2 = -14$ dB at all frequencies. Since it is a constant, it has no effect on the phase.
2. The second component consists of a simple lag whose time constant is 10 s, and thus the break point is centered at 0.10 rad/s.
3. The third component is a quadratic lag with natural frequency $\omega_n = \sqrt{25} = 5$ and damping ratio $\zeta = \dfrac{3}{2\omega_n} = 0.3$, indicating the system is lightly damped. For this part the straight-line approximation is at 0 dB to the corner frequency 5 rad/s, and falling at 40 dB per decade beyond this. Falling at 40 dB per decade is because at $\omega/5 \gg 1$ the magnitude

$$|G(i\omega)| \simeq -20 \log_{10}\left(\frac{\omega}{5}\right)^2 = -40 \log_{10}\left(\frac{\omega}{5}\right)$$

At $\omega = \omega_n = 5$ the phase curve passes steeply through 90°.

(a)

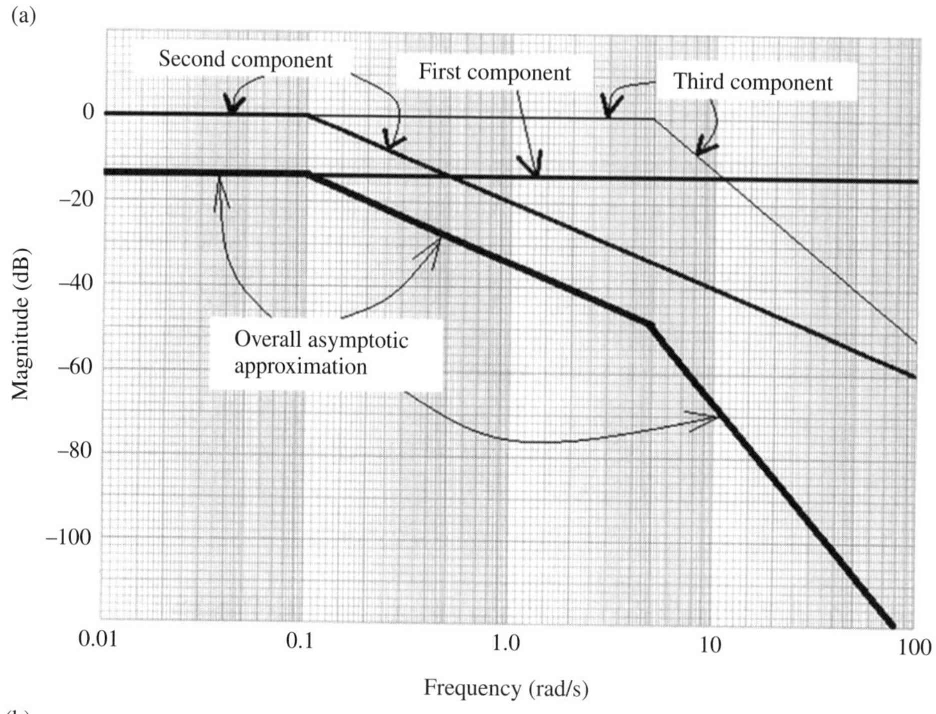

(b)

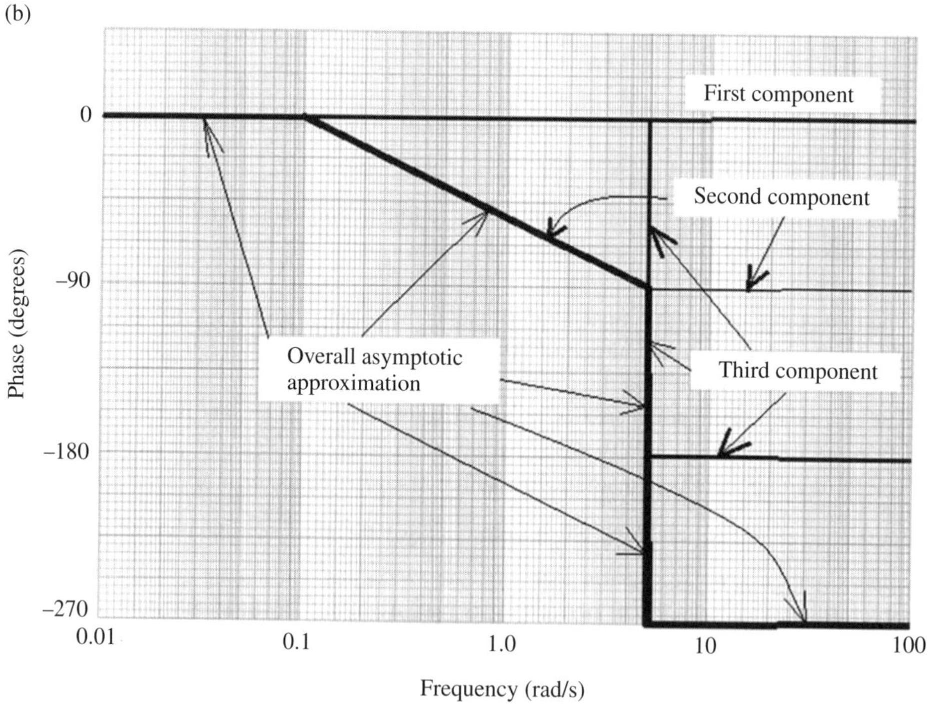

Figure 11.4 Asymptotic Bode plot of a system with $G(s) = \dfrac{5}{(1+10s)(s^2+3s+25)}$. (a) Magnitude plot; (b) phase plot

The asymptotic Bode plot of Equation (11.10) is shown in Figure 11.4. The approximate overall magnitude and phase of the system is determined simply by adding all the components.

11.3 Nyquist Plots and Stability Criterion

The Routh-Hurwitz criterion is applied to determine system stability in a binary sense, in that it indicates whether the system is stable or unstable. It does not provide the degree of stability or relative stability of a system.

To enable one to determine the relative stability of a system, the Nyquist stability criterion (NSC) [5] is often employed. However, the NSC in its most comprehensive form is rather complex, and to understand it calls for familiarity with the mathematical process of *conformal mapping*, to be explained later in this section. The full criterion is required to determine the relative stability of the system. Meanwhile, the simplified version of the NSC is introduced first in the following; the more elaborated one together with relevant material will be presented in Sections 11.3.1 and 11.3.2.

When the open-loop system is stable, a simplified form of the criterion can be employed. The simplified Nyquist stability criterion (SNSC) states that *if an open-loop system is stable then the system with the loop closed is also stable, provided that the open-loop locus on the polar plot does not enclose the* $(-1, i0)$ *point in the s-plane*. An illustration of this concept is given in Figure 11.5. Note that if the locus passes through the critical point $(-1, i0)$, which corresponds to a system with a pair of purely imaginary roots, the system is said to possess marginal stability.

The *Nyquist plot* or *Nyquist diagram* is in fact a polar plot of the open-loop transfer function of a system in the frequency domain. The polar plot of $G(i\omega)H(i\omega)$ enables one to find whether or not the roots have positive real parts without actually determining them from the c.e. It also provides the designer a means to measure the proximity of a root to instability of the system. Thus, it enables the designer to improve the system stability.

11.3.1 Conformal Mapping and Cauchy's Theorem

For completeness and before the introduction to the NSC [5], conformal mapping and Cauchy's theorem are presented in this subsection. This is because the stability criterion is based on

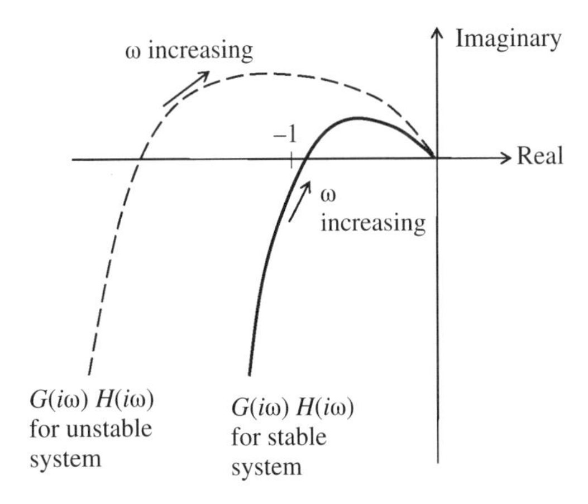

Figure 11.5 Nyquist plot of a system

Cauchy's theorem of *mapping* or *transformation* of contours in the s-plane. Central to this stability criterion is that the stability of a control system may be studied by investigating the c.e.:

$$F(s) = 1 + P(s) = 0 \tag{11.11}$$

$$\text{where } P(s) = G(s)H(s) = \frac{P_N(s)}{P_D(s)}$$

in which $P_N(s)$ and $P_D(s)$ are the polynomials in s in the numerator and denominator, respectively. In other words, $P(s)$ is a rational function of s. For stable response of the control system it is required that zeros of $F(s)$ or roots of the c.e. do not situate in the right-hand s-plane.

Before proceeding further, the definitions illustrated in Figure 11.6a,b are in order. All points to the right of a contour as it is traversed in the direction shown in Figure 11.6a are said to be *enclosed* by it. A clockwise (CW) traverse around a contour is said to be positive as indicated in Figure 11.6b. It may be appropriate to point out that in the literature, contour traverses in the opposite direction to that shown in Figure 11.6a,b have been employed. For example, in [6] such a different contour has been applied. In such a definition all points to the left of a contour as it is traversed are regarded as being enclosed by it.

A *closed contour* (by definition, a *closed contour* in a complex plane is a continuous curve beginning and ending at the same point) in the $P(s)$-plane is considered to make N *positive encirclements,* N being an integer, of the origin if an imaginary radial line drawn from the origin to a point on the $P(s)$ curve rotates in a CW direction completely through $360N$ degrees. If the

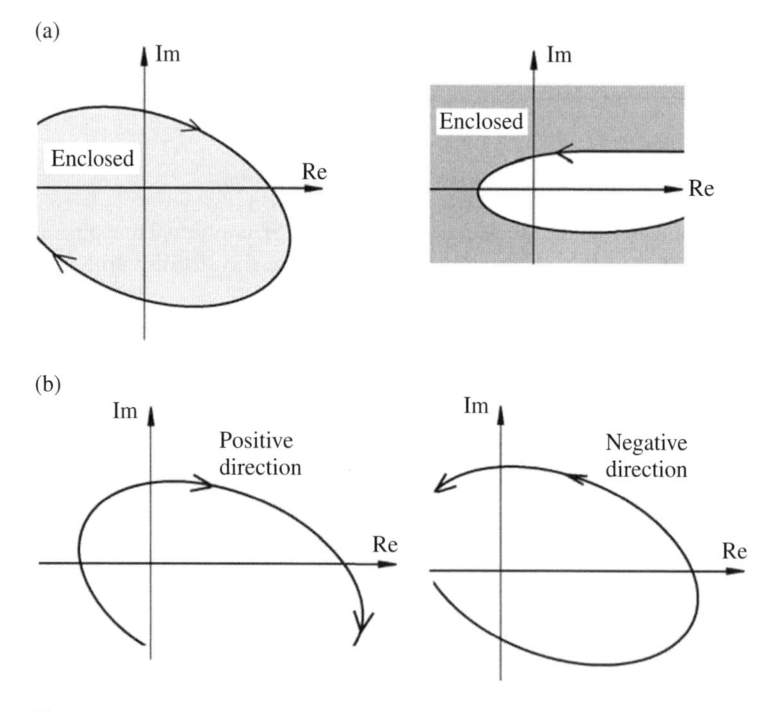

Figure 11.6 Enclosure and encirclement: (a) enclosure and (b) encirclement

traverse of a point on the $P(s)$ curve is performed in the opposite direction, a *negative encirclement* is achieved. The total number of encirclements N is equal to the CW minus the counterclockwise (CCW) encirclements.

In addition, the following assumptions are required:

1. the function $P(s)$ in Equation (11.11) is singled-valued and a rational function;
2. the function $P(s)$ is *analytic* in the s-plane except at some finite number of points (poles); and
3. for every value s_i on the closed contour Γ_s one wants to compute and plot a corresponding point $P(s_i)$ on the $P(s)$-plane.

The resulting contour on the $P(s)$-plane is *closed* and denoted by Γ_p. The transformation of Γ_s to Γ_p is called *conformal mapping*.

In the foregoing, the word *analytic* means that if the derivative of $P(s)$ at s_i defined by

$$\left.\frac{dP(s)}{ds}\right|_{s_i} \equiv \lim_{s \to s_i}\left[\frac{P(s)-P(s_i)}{s-s_i}\right]$$

exists at all points in a region of the s-plane, meaning if the limit is finite and unique, then $P(s)$ is *analytic* in that region. Thus, a point at which $P(s)$ is not analytic is a *singular point* or *singularity* of $P(s)$. A *pole* of $P(s)$ is a singular point and a *zero* of $P(s)$ is also a singular point. The encirclement of the zeros and poles of $P(s)$ can be related to the encirclement of the origin in the $P(s)$-plane by *Cauchy's theorem*.

Cauchy's theorem, *also known as the* ***principle of the argument*** [7] *in complex variable theory, states that if the contour Γ_s encircles N_z zeros and N_P poles of $P(s)$ on the s-plane and does not pass through any poles or zeros of $P(s)$ as it traverses in the CW direction along the contour, then the corresponding contour Γ_p encircles the origin on the $P(s)$-plane $N = N_z - N_P$ times in the same CW direction.*

Clearly, if Γ_s encircles more zeros than poles, then $N > 0$. If Γ_s encircles more poles than zeros, then $N < 0$. This means that Γ_p encircles the origin N times in a direction opposite to that of Γ_s on the s-plane.

To better understand the use of Cauchy's theorem, one can consider a simple example in which the poles and zeros as well as the contour Γ_s are shown in Figure 11.7a. The contour encircles and encloses three zeros and one pole, so that according to Cauchy's theorem, one has $N = 3 - 1 = 2$, and Γ_p completes two CW encirclements of the origin in the $P(s)$-plane as shown in Figure 11.7b. For the zero and pole pattern as well as the contour Γ_s shown in Figure 11.7c, in which one pole is encircled and no zeros are encircled, one finds $N = 0 - 1 = -1$, and therefore the encirclement of the origin by the contour Γ_p is in the CCW direction as shown in Figure 11.7d.

11.3.2 Nyquist Method and Stability Criterion

The Nyquist method is concerned with the mapping of $F(s)$:

$$F(s) = 1 + G(s)H(s) = 1 + \frac{K\prod_{j=1}^{m}(s+z_j)}{\prod_{i=1}^{n}(s+p_i)} \tag{11.12}$$

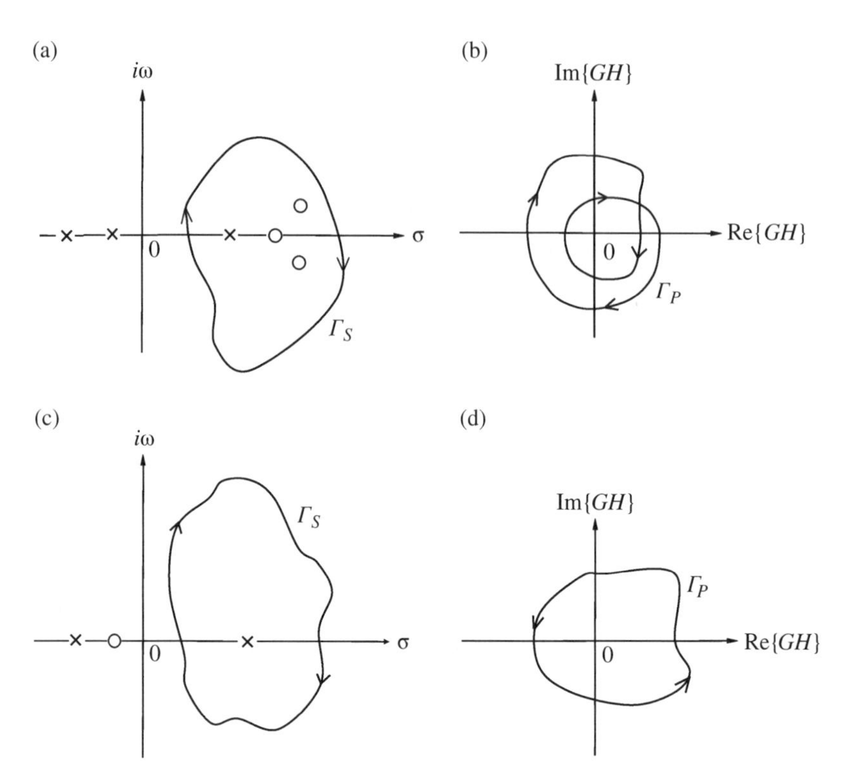

Figure 11.7 Use of Cauchy's theorem: (a) singularities and Γ_s; (b) encirclement and Γ_p of (a); (c) singularities and Γ_s, and (d) encirclement and Γ_p of (c)

and the encirclement of the origin on the $F(s)$-plane. Note that in the foregoing z_j are the zeros of the function $G(s)H(s)$, that is, the open-loop transfer function, and p_i are the poles of $G(s)H(s)$. From Equation (11.11), one can write

$$P(s) = F(s) - 1 = G(s)H(s) \tag{11.13}$$

so that the encirclement of the origin on the $F(s)$-plane is the same as the encirclement of the point (−1, 0) on the $G(s)H(s)$-plane or $P(s)$-plane. Thus, the NSC may be stated as follows:

A feedback control system is stable if and only if the contour Γ_p in the $P(s)$-plane does not encircle the (−1, 0) *point as the number of poles of $P(s)$ in the right-hand s-plane is zero.*

When the number of poles N_P in the right-hand s-plane is not zero, the NSC may be stated as:

A feedback control system is stable if and only if, for the contour Γ_p, the number of CCW encirclements of the (−1, 0) *point is equal to the number of poles of $P(s)$ with positive real parts.*

The basis on which the aforementioned two statements were made is the fact that for $P(s) = G(s)H(s)$ mapping, the number of roots (that is, zeros) of the c.e., $1 + P(s) = 0$, in the right-hand s-plane is represented by the relation

$$N_z = N + N_P \tag{11.14}$$

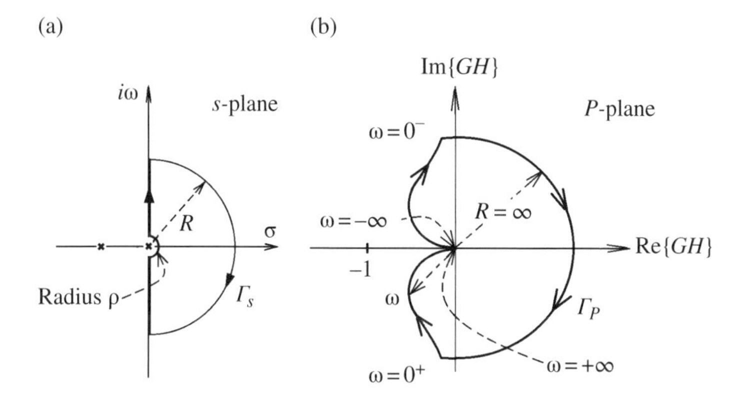

Figure 11.8 Nyquist plot and mapping for $P(s)$: (a) Nyquist plot in the s-plane; (b) mapping for $P(s)$

The following simple example is presented to illustrate the use of the NSC.

Consider the Nyquist stability plot of a feedback control system whose open-loop transfer function is:

$$G(s)H(s) = \frac{1}{s(s+1)}$$

In this transfer function there is one *pole* at the origin. The *Nyquist path* (that is, a closed contour in the s-plane which completely encloses the entire right half of the s-plane) is presented in Figure 11.8a. Note that a Nyquist path should not pass through any pole in accordance with Cauchy's theorem such that a small semi-circle (whose radius, mathematically, is interpreted as approaching zero in the limit) along the imaginary axis is required in the path.

Consider in details of the contour Γ_s in the s-plane and the corresponding contour Γ_p in the $G(s)H(s)$-plane or $P(s)$-plane.

a. *Origin of s-plane*:

The small semi-circular detour around the pole at the origin in the s-plane is written as $s = \lim_{\rho\to 0}\rho e^{i\varphi}$ for $-90° \le \varphi \le 90°$ and the mapping $P(s)$ is

$$\lim_{\rho\to 0} P\left(\rho\, e^{i\varphi}\right) = \lim_{\rho\to 0}\frac{1}{\rho e^{i\varphi}\left(\rho\, e^{i\varphi}+1\right)} = \lim_{\rho\to 0}\frac{1}{\rho e^{i\varphi}} = \infty\, e^{-i\varphi} = \infty\angle(-\varphi)$$

in which $\rho e^{i\varphi} + 1 \to 1$ as $\rho \to 0$ has been applied. Thus, the angle of the contour Γ_p in the $P(s)$-plane changes from 90° at $\omega = 0^-$ to $-90°$ at $\omega = 0^+$, passing through 0° at $\omega = 0$. The radius of this part of the contour in the $P(s)$-plane is infinite, as shown in Figure 11.8b.

b. *Portion of contour from* $\omega = 0^+$ *to* $\omega = +\infty$:

The portion of the contour Γ_s in the s-plane from $\omega = 0^+$ to $\omega = +\infty$ is mapped by $P(s)$ in the $P(s)$-plane such that

$$G(i\omega)H(i\omega) = \frac{1}{i\omega(i\omega+1)} = \frac{1}{\omega\sqrt{\omega^2+1}}\angle\left(-90° - \tan^{-1}\omega\right)$$

In the limits of ω, one has

$$\lim_{\omega\to 0} G(i\omega)H(i\omega) = \lim_{\omega\to 0}\frac{1}{i\omega(i\omega+1)} = \infty \angle(-90^\circ),$$

$$\lim_{\omega\to\infty} G(i\omega)H(i\omega) = 0\angle(-180^\circ).$$

In other words, when ω increases in the interval $0 < \omega < \infty$, $|G(i\omega)H(i\omega)|$ decreases from ∞ to 0 and the phase angle decreases steadily from −90° to −180°. Therefore, the contour does not cross the negative real axis, but approaches it from below as shown in Figure 11.8b.

c. *Portion of contour from* $\omega = +\infty$ *to* $\omega = -\infty$:
The portion of Γ_s in the s-plane from $\omega = +\infty$ to $\omega = -\infty$ is mapped into the point zero at the origin of the $P(s)$-plane. That is,

$$\lim_{R\to\infty} G(s)H(s)|_{s=Re^{i\varphi}} = \lim_{R\to\infty}\frac{1}{R^2}e^{-2i\varphi}$$

as the angle φ changes from +90° at $\omega = +\infty$ to −90° at $\omega = -\infty$ in the s-plane. Therefore, in the $P(s)$-plane the contour Γ_p moves from an angle of −180° at $\omega = +\infty$ to an angle of +180° at $\omega = -\infty$ The magnitude of the $P(s)$ contour as the radius R is infinite is always zero or a constant.

d. *Portion of contour from* $\omega = -\infty$ *to* $\omega = 0^-$:
The portion of Γ_s in the s-plane from $\omega = -\infty$ *to* $\omega = 0^-$ is mapped by $P(s)$ as

$$G(-i\omega)H(-i\omega) = \frac{1}{-i\omega(-i\omega+1)}$$

This is the complex conjugate to $G(i\omega)H(i\omega) = \dfrac{1}{i\omega(i\omega+1)}$. Thus, the plot for this portion is symmetrical to that in (b) as shown in Figure 11.8b.
With reference to Figure 11.8b and the NSC, one observes that the system is always stable.

Before leaving this subsection, it should be pointed out that Nyquist plots for many control systems can be constructed similarly as in the foregoing example. However, it is not pursued in the present chapter for conciseness.

11.4 Gain Margin and Phase Margin

In the foregoing section the significance of the Nyquist plot for the study of system stability has been demonstrated. In order to quantify the degree of stability or relative stability of a system, measures are applied to determine the closeness or proximity of the Nyquist plot to the critical point. These measures are the *gain margin* and the *phase margin*. The gain margin M_G is

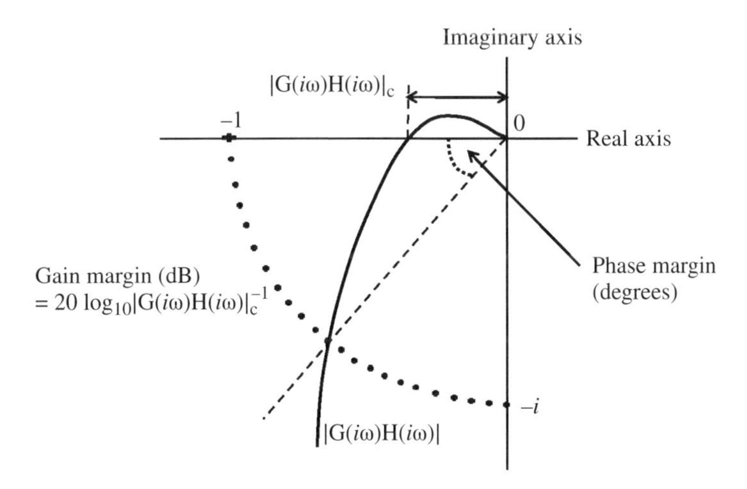

Figure 11.9 Gain margin and phase margin

defined as the amount by which the system gain can be increased before instability occurs. Generally, it is expressed as

$$M_G = 20\ \log_{10} \frac{1}{|G(i\omega)H(i\omega)|_c}\ \text{dB} \tag{11.15}$$

where $|G(i\omega)H(i\omega)|_c$ is the magnitude of the open-loop transfer function at the crossover point on the negative real axis. The magnitude corresponds to a phase lag of 180°.

The phase margin M_ϕ is defined as the angle through which the Nyquist locus must be rotated in order that the unity magnitude point on the locus passes through the critical point. That is,

$$M_\phi = \pi - \angle G(i\omega_c)H(i\omega_c) \tag{11.16}$$

where ω_c is the gain crossover frequency.

Both the gain margin and phase margin are illustrated in Figure 11.9, in which the plot crosses at about 0.35 and the gain margin $M_G = 20\ \log_{10}\frac{1}{0.35}\ \text{dB} = 9.12\ \text{dB}$.

The gain margin and phase margin can be obtained directly from a Bode plot. For example, the Bode plot in Figure 11.10 gives the gain margin, which is the attenuation at the *phase crossover frequency*, whereas the phase margin is the phase lag at the *gain crossover frequency* subtracted from 180°.

Example

Determine the gain margin M_G and phase margin M_ϕ as well as the crossover frequencies for the third-order system described by

$$G(s)H(s) = \frac{24.7}{(s+1)(s+2)(s+3)}$$

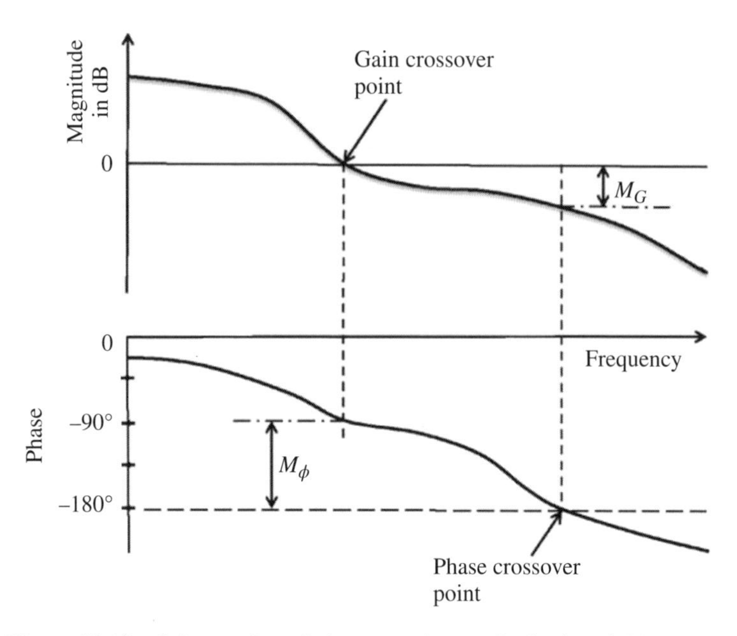

Figure 11.10 Gain margin and phase margin on a Bode plot of $G(i\omega)H(i\omega)$

Solution:
To obtain the phase crossover frequency ω_ϕ, one separates the real and imaginary parts in the denominator of $G(s)H(s)$:

$$G(i\omega)H(i\omega) = \frac{24.7}{(i\omega+1)(i\omega+2)(i\omega+3)}$$

$$= \frac{24.7}{(i\omega)^3 + 6(i\omega)^2 + 11(i\omega) + 6}$$

$$= \frac{24.7}{(6-6\omega^2) + i(11\omega - \omega^3)}$$

Equate the imaginary part to zero such that $11\omega = \omega^3$, which gives $\omega = \pm\sqrt{11}$. That is, $\omega_\phi = \pm\sqrt{11}$. Thus,

$$|G(i\omega)H(i\omega)|_{\omega_\phi} = |GH|_{\omega_\phi} = \left|G(i\omega_\phi)H(i\omega_\phi)\right| = \left|\frac{24.7}{6-6\omega_\phi^2}\right|$$

This gives $|GH|_{\omega_\phi} = \dfrac{24.7}{60}$ such that the gain margin

$$M_G = 20\log_{10}\frac{1}{|GH|_{\omega_\phi}}\,\text{dB} = 20\log_{10}\frac{60}{24.7}\,\text{dB} = 7.71\,\text{dB}$$

which means that the system gain may be increased by 7.71 dB (factor of 60/24.7 = 2.43) before becoming marginally stable.

To obtain the gain crossover frequency ω_c one applies $|G(i\omega)H(i\omega)| = 1$.

Therefore,

$$\left| \frac{24.7}{(i\omega+1)(i\omega+2)(i\omega+3)} \right| = 1$$

This gives

$$24.7 = |(i\omega+1)||(i\omega+2)||i\omega+3|$$

$$24.7 = \sqrt{\omega^2+1}\sqrt{\omega^2+4}\sqrt{\omega^2+9}.$$

Squaring both sides gives:

$$24.7^2 = \left(\omega^2+1\right)\left(\omega^2+4\right)\left(\omega^2+9\right).$$

Solving this equation, one finds $\omega = 2.10$ and therefore, $\omega_c = 2.10$ rad/s.

The phase margin

$$M_\phi = \pi - \angle G(i\omega_c)H(i\omega_c)$$

$$= \pi - \left[\tan^{-1}\left(\frac{\omega_c}{1}\right) + \tan^{-1}\left(\frac{\omega_c}{2}\right) + \tan^{-1}\left(\frac{\omega_c}{3}\right)\right]$$

$$M_\phi = \pi - \left[\tan^{-1}(2.1) + \tan^{-1}(1.05) + \tan^{-1}(0.70)\right] = 34.1^\circ.$$

The latter implies that the phase can be increased by 34.1° before the system becomes marginally stable.

Before leaving this example, it should be noted that as a general rule a phase margin $M_\phi > 30^\circ$ and a gain margin $M_G \approx 6$ dB are recommended for a good transient response of the system.

11.5 Lines of Constant Magnitude: *M* Circles

The magnitude of the closed-loop frequency response of a unity feedback control system can be evaluated directly from the polar plot of $G(i\omega)$.

Consider the overall transfer function of a unity feedback control system

$$\frac{C(s)}{R(s)} = \frac{G(s)}{1+G(s)}.$$

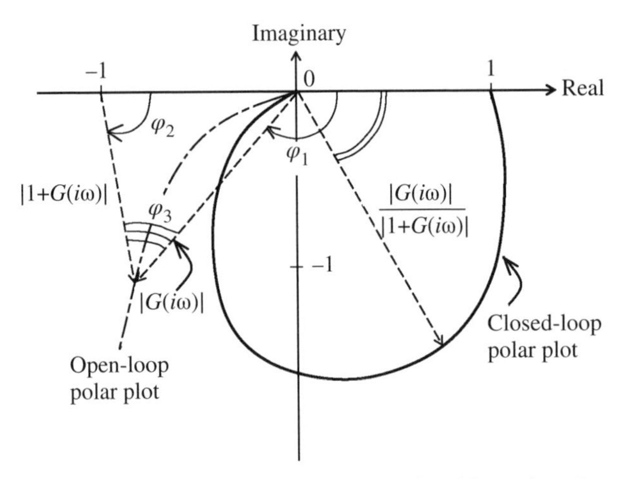

Figure 11.11 Open- and closed-loop polar plots of $G(s) = \dfrac{1}{s(1+\tau s)}$.

In the frequency domain,

$$\frac{C(i\omega)}{R(i\omega)} = \frac{G(i\omega)}{1+G(i\omega)}$$

The magnification M is defined by

$$M = \frac{|C(i\omega)|}{|R(i\omega)|} = \frac{|G(i\omega)|}{|1+G(i\omega)|}$$

For any given frequency the magnitudes $|G(i\omega)|$ and $|1+G(i\omega)|$ can be measured from the plot as indicated in Figure 11.11, while the corresponding magnification M can be determined by using these measured magnitudes. The closed-loop phase can be measured directly from the plot since

$$\angle C(i\omega) - \angle R(i\omega) = \angle\left[\frac{C(i\omega)}{R(i\omega)}\right] = \angle G(i\omega) - \angle[1+G(i\omega)].$$

Suppose one writes $G(i\omega) = x + iy$, where x and y are real. Then the magnification

$$M = \frac{|G(i\omega)|}{|1+G(i\omega)|} = \frac{|x+iy|}{|1+x+iy|} = \frac{\sqrt{x^2+y^2}}{\sqrt{(1+x)^2+y^2}}.$$

Squaring on both sides,

$$M^2\left[(1+x)^2+y^2\right] = x^2+y^2. \tag{11.17}$$

Expanding and rearranging, one has

$$\left(1-M^2\right)x^2-2M^2x+\left(1-M^2\right)y^2=M^2.$$

Dividing through by $(1-M^2)$, one obtains

$$x^2-\frac{2M^2}{1-M^2}x+y^2=\frac{M^2}{1-M^2}.$$

Adding the term $\left(\frac{M^2}{1-M^2}\right)^2$ to both sides gives

$$x^2-\frac{2M^2}{1-M^2}x+y^2+\left(\frac{M^2}{1-M^2}\right)^2=\frac{M^2}{1-M^2}+\left(\frac{M^2}{1-M^2}\right)^2.$$

Simplifying, it results in

$$\left(x-\frac{M^2}{1-M^2}\right)^2+y^2=\left(\frac{M}{1-M^2}\right)^2. \tag{11.18}$$

Equation (11.18) is the equation of a circle with center $\left(\frac{M^2}{1-M^2},\ 0\right)$ and radius $\frac{M}{1-M^2}$. Note that when $M = 1$ the above circle equation is invalid and one has to apply Equation (11.17) to find the value of x. For $M = 1$, the latter equation gives $x = -½$, which is a straight line parallel to the imaginary axis and passing through the point $(-½, 0)$ in the $G(i\omega)$-plane. A typical example of constant M circles is presented in Figure 11.12a. These circles are also called constant M contours or loci.

It is interesting to note that while the approaches presented in this chapter are primarily applied to the analysis and design of control systems in the frequency domain, the M contours or circles may be related to the response in the time domain. Specifically, these circles may be used to find the resonance peak M_P of the closed-loop response from the plot of $G(i\omega)$. In other words, these circles may be used to design a system with a specified M_P. The resonance peak M_P is the largest value of M of the M circle(s) tangent to the polar plot of $G(i\omega)$. The relationship between M contours and system response in the time domain is presented in Figure 11.12b.

In passing, it may be noted that damping of the system can be correlated to the phase margin M_ϕ of the system, while M_ϕ may be related to the resonance peak M_P of a second-order system. For example, if the open-loop transfer function of a unity feedback control system is given by $G(s)=\frac{\omega_n^2}{s^2+(2\zeta\omega_n)s}$, then the open-loop frequency response becomes $G(i\omega)=\frac{\omega_n^2}{(i\omega)^2+(2\zeta\omega_n)i\omega}$.

Writing $\mu = \omega_n/\omega$, one finds

$$G(i\omega)=\frac{\mu^2}{i(i+2\zeta\mu)}.$$

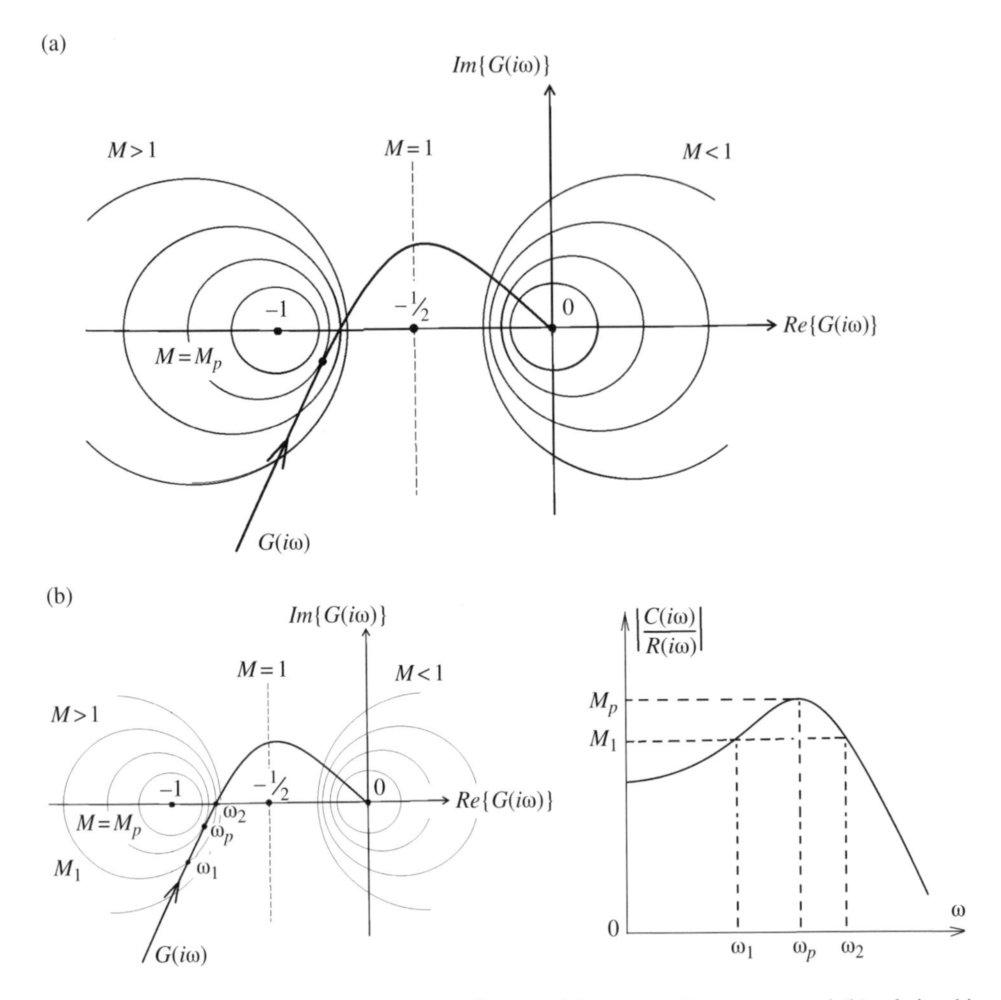

Figure 11.12 Constant M-circles on a polar diagram: (a) constant M contours; and (b) relationship between M contours and response of system

The phase margin M_ϕ occurs at the gain crossover frequency ω_c, in which

$$|G(i\omega)| = 1 = \frac{\mu_c^2}{\sqrt{1 + (2\zeta\mu_c)^2}}, \quad \mu_c = \frac{\omega_n}{\omega_c}.$$

Upon squaring both sides and rearranging, the resulting equation becomes

$$\mu_c^4 - (2\zeta\mu_c)^2 - 1 = 0 \tag{11.19}$$

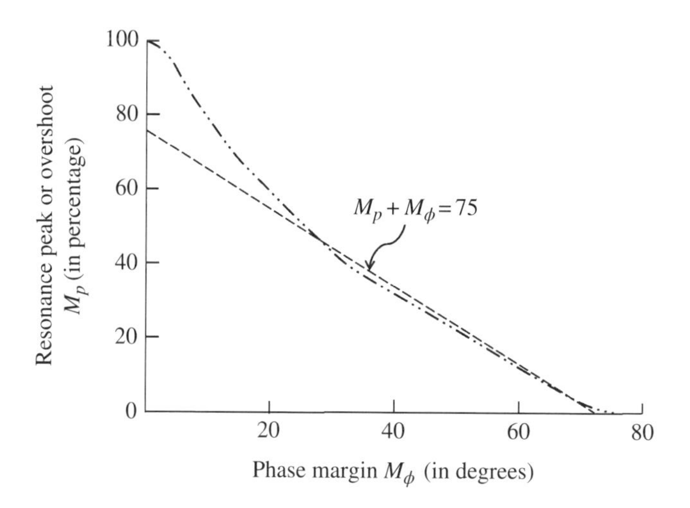

Figure 11.13 Relationship between phase margin and resonance peak of a second-order system

This is the equation relating the damping ratio to the gain crossover frequency. The corresponding phase margin M_ϕ is defined by

$$M_\phi = \pi + \angle G(i\omega_c) = \pi - \left[\tan^{-1}\left(\frac{1}{0}\right) + \tan^{-1}\left(\frac{1}{2\zeta\mu_c}\right)\right] \tag{11.20}$$

In addition to the foregoing relationship between phase margin and damping ratio of the second-order system, one can relate the phase margin to the resonance peak or peak overshoot of the response M_P.

The resonance peak is given by the second term on the rhs of Equation (9.11) as

$$M_P = e^{-\frac{\pi\zeta}{\sqrt{1-\zeta^2}}}. \tag{11.21}$$

Since the phase margin is related to the damping ratio by Equation (11.20), one may obtain an expression between the phase margin and resonance peak. The plot of the relationship between phase margin and resonance peak or overshoot is included in Figure 11.13, in which for $M_\phi \geq 30°$ the relationship may be approximated as a straight line such that

$$M_\phi(°) = 75(\%) - M_P(\%). \tag{11.22}$$

11.6 Lines of Constant Phase: *N* Circles

In a similar way, contours of constant phase shift constitute another family of circles, the so-called *N*-circles or loci.

Recall that $G(i\omega) = x + iy$ and $\dfrac{C(i\omega)}{R(i\omega)} = \dfrac{G(i\omega)}{1+G(i\omega)} = \dfrac{x+iy}{1+x+iy}$. Therefore,

$$\varphi_N = \tan^{-1}\left(\frac{y}{x}\right) - \tan^{-1}\left(\frac{y}{1+x}\right) = \varphi_1 - \varphi_2$$

or simply write $\varphi_N = \varphi_1 - \varphi_2$.

Taking the tangent on both sides, one has

$$\tan\varphi_N = \tan(\varphi_1 - \varphi_2) = \frac{\tan\varphi_1 - \tan\varphi_2}{1 + \tan\varphi_1 \tan\varphi_2}.$$

Since $\tan\varphi_1 = \dfrac{y}{x}$ and $\tan\varphi_2 = \dfrac{y}{1+x}$, substituting these into the last equation gives:

$$\tan\varphi_N = \frac{y}{x^2 + x + y^2}.$$

Writing $N = \tan\varphi_N$, then the last equation becomes

$$x^2 + x + y^2 - \frac{y}{N} = 0.$$

Adding the term $\dfrac{1}{4}\left(1 + \dfrac{1}{N^2}\right)$ to both sides, one has

$$x^2 + x + y^2 - \frac{y}{N} + \frac{1}{4}\left(1 + \frac{1}{N^2}\right) = \frac{1}{4}\left(1 + \frac{1}{N^2}\right).$$

Rearranging, one obtains

$$\left(x + \frac{1}{2}\right)^2 + \left(y - \frac{1}{2N}\right)^2 = \frac{1}{4}\left(1 + \frac{1}{N^2}\right) \tag{11.23}$$

This equation represents a circle with its center at $\left(-\dfrac{1}{2}, \dfrac{1}{2N}\right)$ and its radius is $\sqrt{\dfrac{N^2+1}{4N^2}}$. Typical contours of N-circles are presented in Figure 11.14, in which $G(i\omega)H(i\omega) = G(i\omega)$, since in the present closed-loop feedback system $H(i\omega) = 1$.

11.7 Nichols Charts

Generally, for analysis and design of control systems, it is more concise to plot the M loci in the polar coordinates and expressed in magnitude and phase coordinates. The resulting loci are known as a *Nichols chart* [8]. A typical Nichols chart or plot is presented in Figure 11.15,

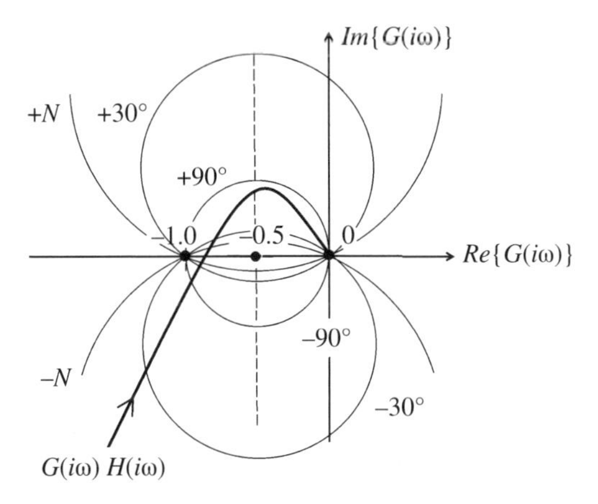

Figure 11.14 Constant N-circles on a polar diagram

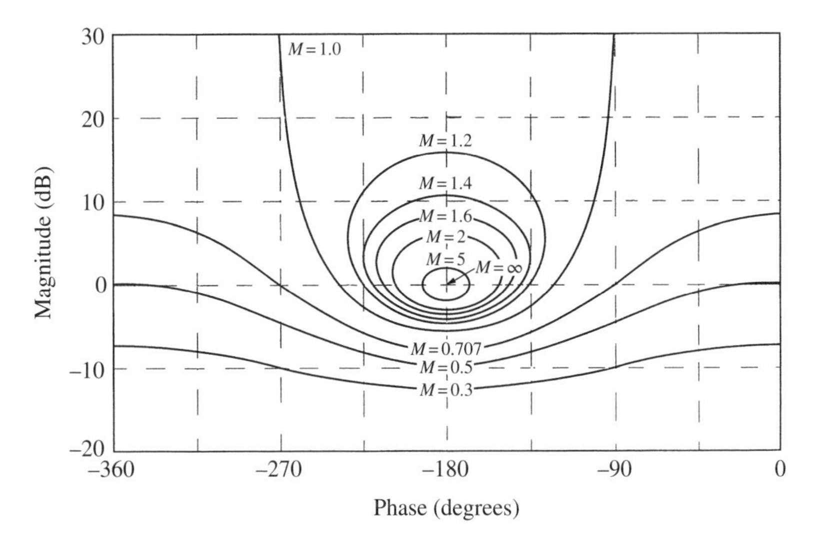

Figure 11.15 Nichols chart

in which the coordinates are open-loop magnitudes expressed in dB and the open-loop phases in degrees.

It may be appropriate to note that the Nichols chart and the M circles considered in the foregoing are confined to closed-loop systems with unity feedback. When the feedback is not of unity, the closed-loop transfer function is given by

$$\frac{C(s)}{R(s)} = \frac{G(s)}{1 + G(s)H(s)}$$

Multiplying both sides by $H(s)$:

$$F(s) = \frac{C(s)H(s)}{R(s)} = \frac{G(s)H(s)}{1 + G(s)H(s)} \tag{11.24}$$

where the frequency response $F(i\omega)$ can be obtained by plotting the function $G(i\omega)H(i\omega)$ in the amplitude and phase coordinates together with the Nichols chart. When this plot is completed, the frequency response may be computed as in Section 11.5 so that

$$M(i\omega) = \frac{|C(i\omega)|}{|R(i\omega)|} = \frac{|F(i\omega)|}{|H(i\omega)|}$$

In general, the magnitude and phase are expressed in dB and degrees. Thus,

$$|M(i\omega)| \text{ dB} = (|F(i\omega)| - |H(i\omega)|)\text{ dB}, \quad \text{and} \tag{11.25a}$$

$$\varphi_M = \angle M(i\omega) = \angle F(i\omega) - \angle H(i\omega). \tag{11.25b}$$

11.8 Applications of MATLAB for Graphical Constructions

Applications of MATLAB [1] and one of its modules, SIMULINK, have been popular among control engineers. This is mainly due to its efficiency, easy availability, and computing power. Many undergraduate and graduate textbooks have included programs and graphical illustrations by making use of MATLAB [2,9–11], for example. One of the early books exclusively making use of MATLAB for control system design is that by Shahian and Hassul [2].

In this section, the root locus plots, Bode diagrams, and Nyquist plots are presented in Sections 11.8.1–11.8.3, respectively. Several illustrative examples are provided in each subsection. They are intended to show the power and pitfalls of MATLAB for the graphical constructions of various control systems. While it is understood that these illustrative examples are not meant to replace the graphical and analytical procedures presented in the foregoing sections, it is emphasized that examples using MATLAB can be applied to verify the computed results and constructed diagrams.

11.8.1 Root Locus Plots

In this subsection, the root locus plots of five different control systems are presented. These include two third-order and three fourth-order systems. It is believed that, aside from the first system, the root locus plots of the remaining four systems are studied for the first time.

Example 1

By employing MATLAB, perform the root locus plot of the system considered earlier in Section 11.1.2; that is, the system has the open-loop transfer function

$$G(s)H(s) = KG_o(s) = \frac{K(s+2)}{s^4 + 7s^3 + 17s^2 + 15s}$$

Solution:
Given the equation above, then

$$G_o(s) = \frac{s + 2}{s^4 + 7s^3 + 17s^2 + 15s}$$

will be used for the root locus construction.

The input to and output from MATLAB are included in Program Listing 11.1 and Figure 11E1.

It may be appropriate to point out that, according to the rules in Section 11.1.1, the angle enclosed by the asymptotic line, in the first quadrant, or the region enclosed by the positive imaginary and real axes, and the real axis is 60°. On the other hand, the corresponding angle in Figure 11E1 output from MATLAB looks far less than 60°. The cause of this pitfall is that in Figure 11E1 the scales for the vertical or imaginary axis and that for the horizontal or real axis are different. This type of pitfall from graphs plotted with the latest version of MATLAB should be noted.

Program Listing 11.1

```
>> num = [0 0 0 1 2];
>> den = [1 7 17 15 0];
>> rlocus(num,den,'k')
>> title('Root Locus Plot of Example in Sub-section 11.1.2')
```

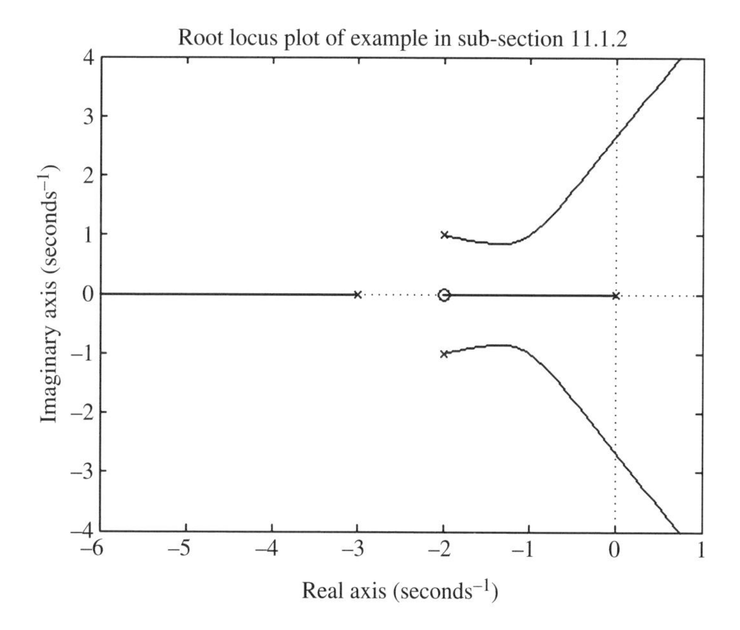

Figure 11E1 Root locus of a fourth-order system

Example 2

By employing MATLAB, perform the root locus plot of the following control system whose open-loop transfer function

$$G(s)H(s) = KG_o(s) = \frac{K(5s+10)}{s^2(s+1)}$$

Solution:

Expanding the given open-loop transfer function, one has

$$G_o(s) = \frac{5s+10}{s^3+s^2}$$

The input to and output from MATLAB are included in Program Listing 11.2 and Figure 11E2.

Program Listing 11.2

```
>> num = [0 0 5 10];
>> den = [1 1 0 0];
>> rlocus(num,den,'k')
>> title('Root Locus Plot of Example 2 Section 11.8')
```

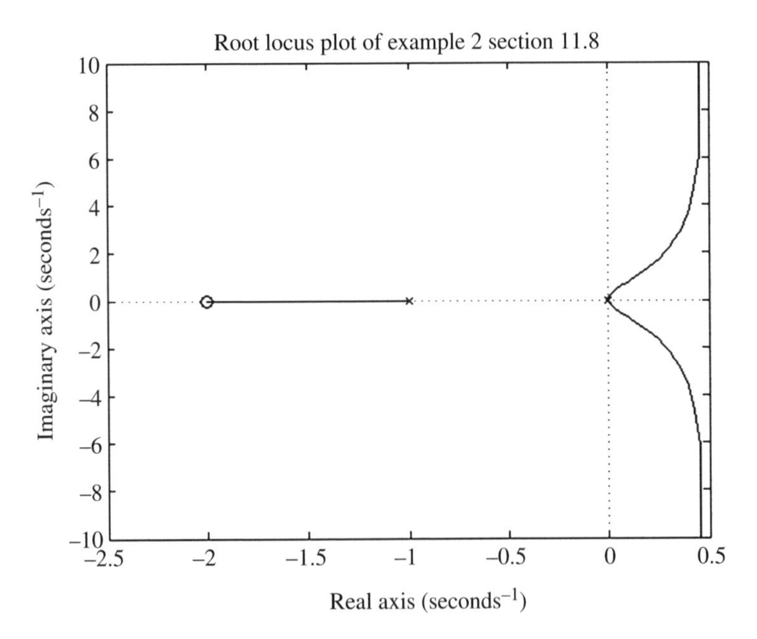

Figure 11E2 Root locus of a third-order system

Example 3

By employing MATLAB, perform the root locus plot of the following control system whose open-loop transfer function

$$G(s)H(s) = KG_o(s) = \frac{K(s+1)}{s(s-1)(s^2+4s+16)}.$$

Solution:

A simple expansion and simplification gives

$$G_o(s) = \frac{s+1}{s^4+3s^2+12s^2-16s}.$$

Therefore, the input to and output from MATLAB are included in Program Listing 11.3a and Figure 11E3a.

Note that in Figure 11E3a the loci on the right-hand side of the figure have two small parts on the left-hand half of the complex plane, meaning the system with the gains of these small parts are stable. In order to provide a better view of these two small parts, a statement is added to Program Listing 11.3a. The new program is presented as Program Listing 11.3b and the output is included in Figure 11E3b.

Program Listing 11.3a

```
>> num = [0 0 0 1 1];
>> den = [1 3 12 -16 0];
>> rlocus(num,den,'k')
>> title('Root Locus Plot of Example 3 Section 11.8')
```

Program Listing 11.3b

```
>> num = [0 0 0 1 1];
>> den = [1 3 12 -16 0];
>> rlocus(num,den,'k')
>> axis([-5 5 -10 10]);
>> title('Root Locus Plot of Example 3 Section 11.8')
```

Example 4

By employing MATLAB, perform the root locus plot of the following control system whose open-loop transfer function

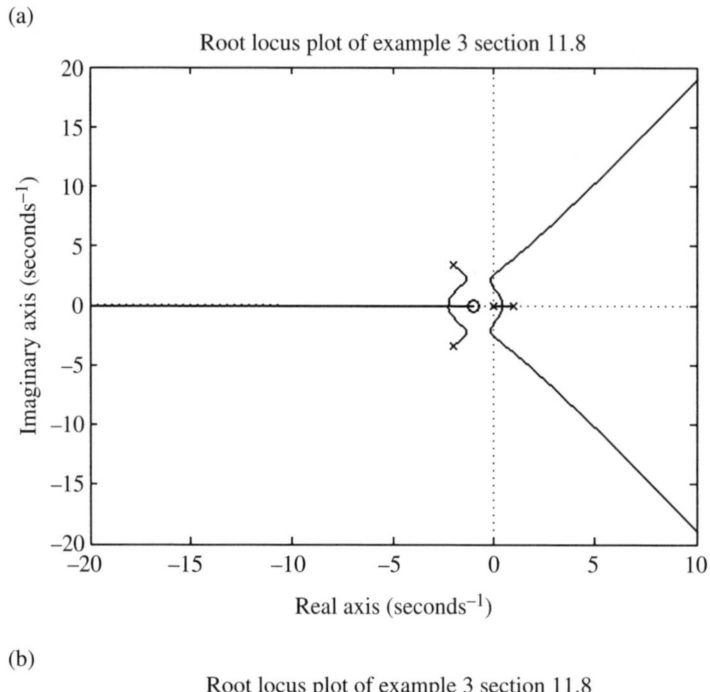

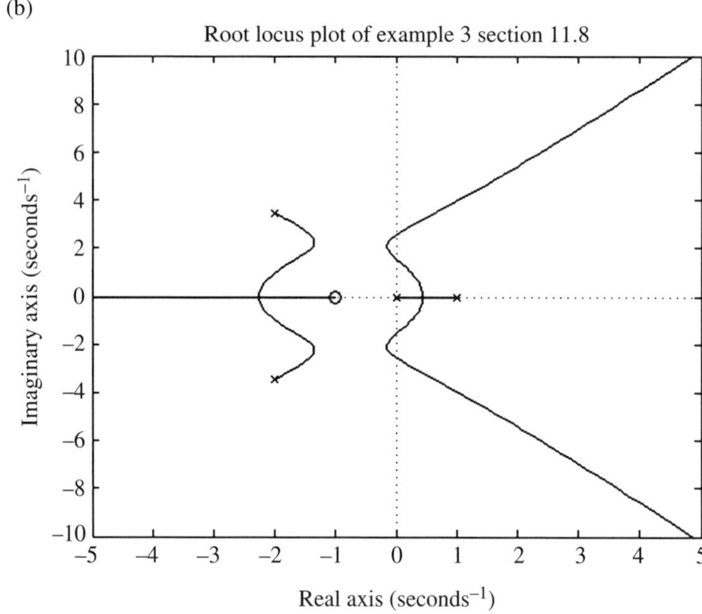

Figure 11E3 (a) Root locus of a fourth-order system; (b) root locus of a fourth-order system with controlled axes for plotting

$$G(s)H(s) = K\,G_o(s) = \frac{K(2s+5)(10s+7)}{(s^2+5s+40)(100s^2+3s+6)}.$$

Solution:

Expanding and simplifying the given open-loop transfer function, one can show that

$$G_o(s) = \frac{20s^2+64s+35}{100s^4+503s^3+4021s^2+150s+240}$$

which is applied to the root locus construction.

The input to and output from MATLAB are included in Program Listing 11.4a and Figure 11E4a. As can be observed the loci near and on the real axis are too small to distinguish their features a statement is added to Program Listing 11.4a in order to provide a better view of the plot. This new program is included in Program Listing 11.4b. The output from MATLAB is presented in Figure 11E4b.

Program Listing 11.4a

```
>> num = [0 0 20 64 35];
>> den = [100 503 4021 150 240];
>> rlocus(num,den,'k')
>> title('Root Locus Plot of Example 4 Section 11.8')
```

Program Listing 11.4b

```
>> num = [0 0 20 64 35];
>> den = [100 503 4021 150 240];
>> rlocus(num,den,'k')
>> axis([-1.5 0.5 -3 3]);
>> title('Root Locus Plot of Example 4 Section 11.8')
```

Example 5

By employing MATLAB, perform the root locus plot of the following control system whose open-loop transfer function

$$G(s)H(s) = K\,G_o(s) = \frac{K(s^2+1)}{s(s^2+100)}$$

Solution:

The input to and output from MATLAB are included in Program Listing 11.5 and Figure 11E5, respectively.

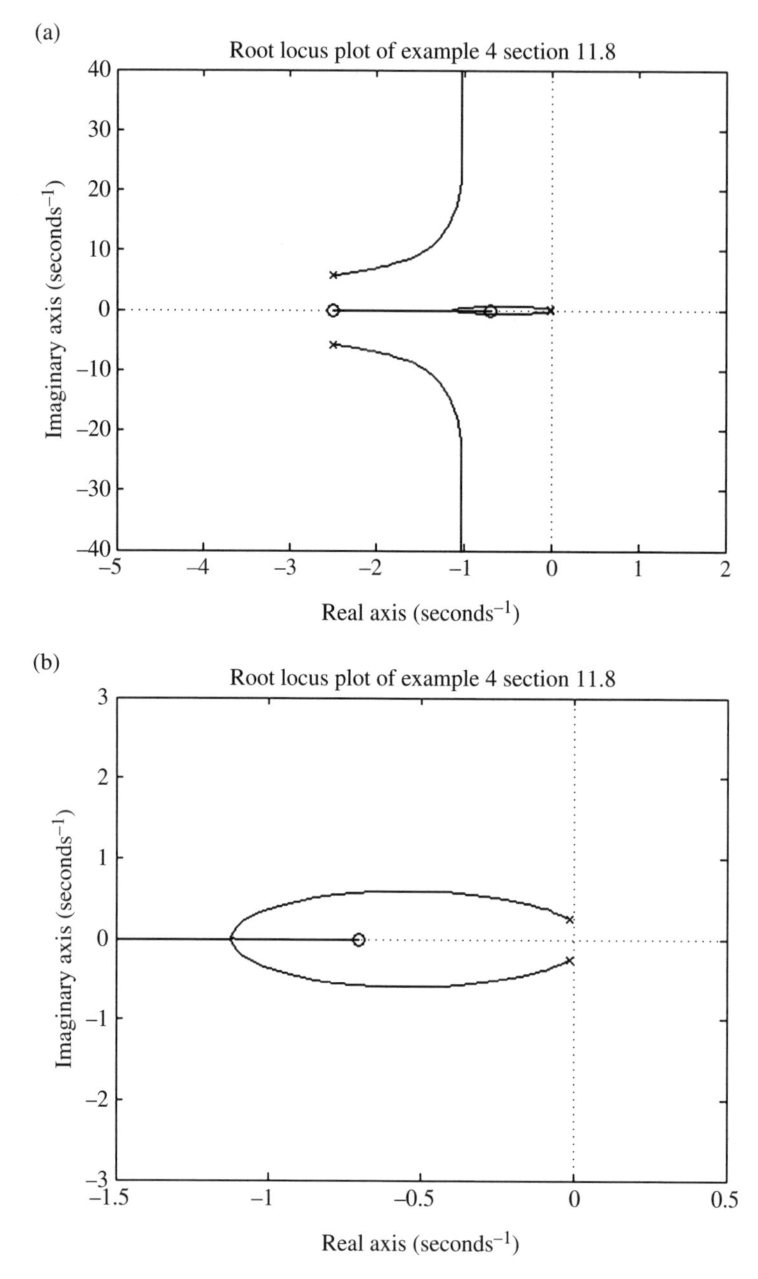

Figure 11E4 (a) Root locus of a fourth-order system; (b) root locus of a fourth-order system with controlled axes for plotting

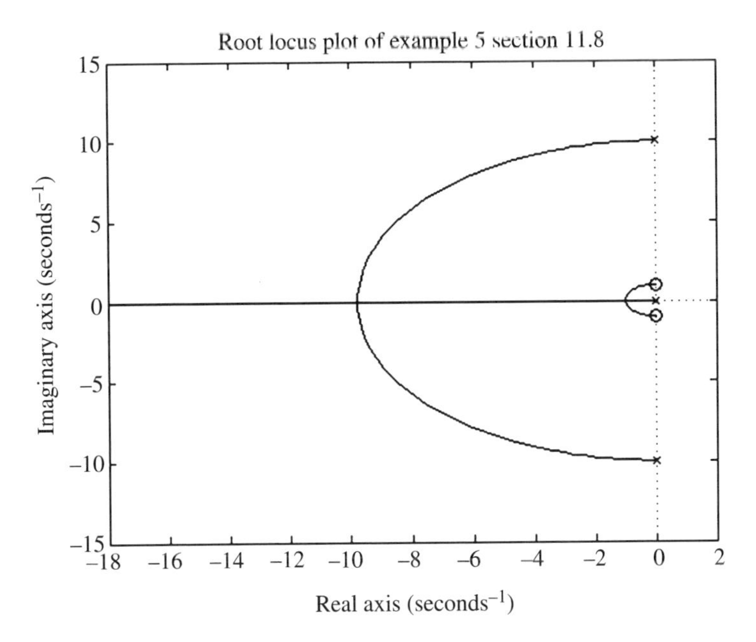

Figure 11E5 Root locus of a third-order system

Program Listing 11.5

```
>> num = [0 1 0 1];
>> den = [1 0 100 0];
>> rlocus(num,den,'k')
>> title('Root Locus Plot of Example 5 Section 11.8')
```

11.8.2 Bode Plots

In this subsection, six systems are considered. The first two are those studied in Section 11.2, and therefore the results obtained by MATLAB are applied to verify the asymptotical constructions.

Example 6

By employing MATLAB, perform the Bode plots of magnitude and phase for the simple first-order system whose transfer function is given by

$$G(s) = \frac{1}{s+1}$$

This is the simple lag system considered in Section 11.2.

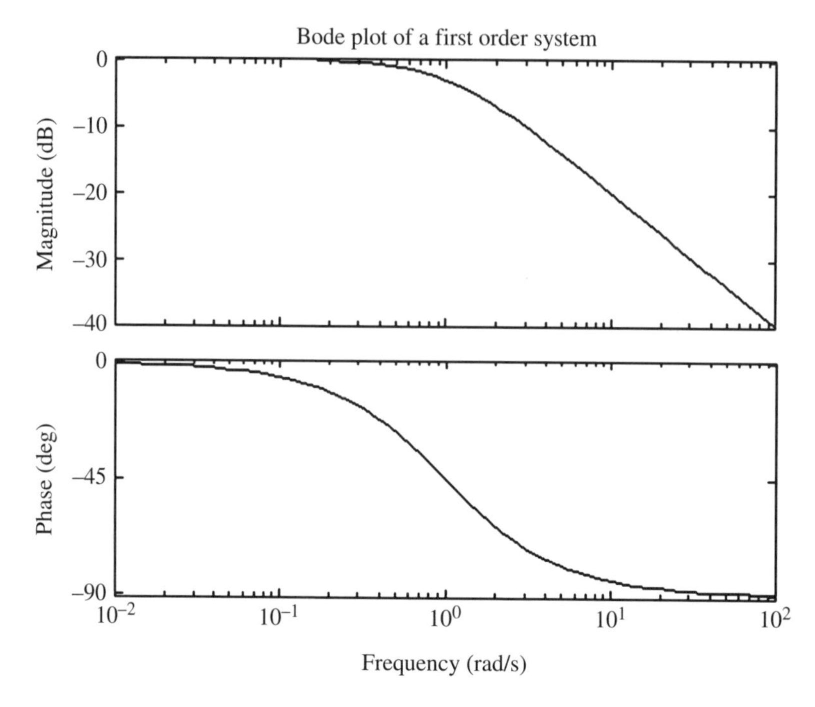

Figure 11E6 Bode plot of a first-order system

Solution:
The input to and output from MATLAB are included in Program Listing 11.6 and Figure 11E6.

Program Listing 11.6

```
>> num = [0 1];
>> den = [1 1];
>> bode(num,den)
>> title('Bode Plot of a First Order System')
```

It is clear that the Bode plots in Figure 11E6 compare very well with those presented in Figure 11.3 of Section 11.2.

Example 7
By applying MATLAB, perform the Bode plots of magnitude and phase for the third-order system whose transfer function is given by Equation (11.9) as

$$G(s) = \frac{5}{10s^3 + 31s^2 + 253s + 25}$$

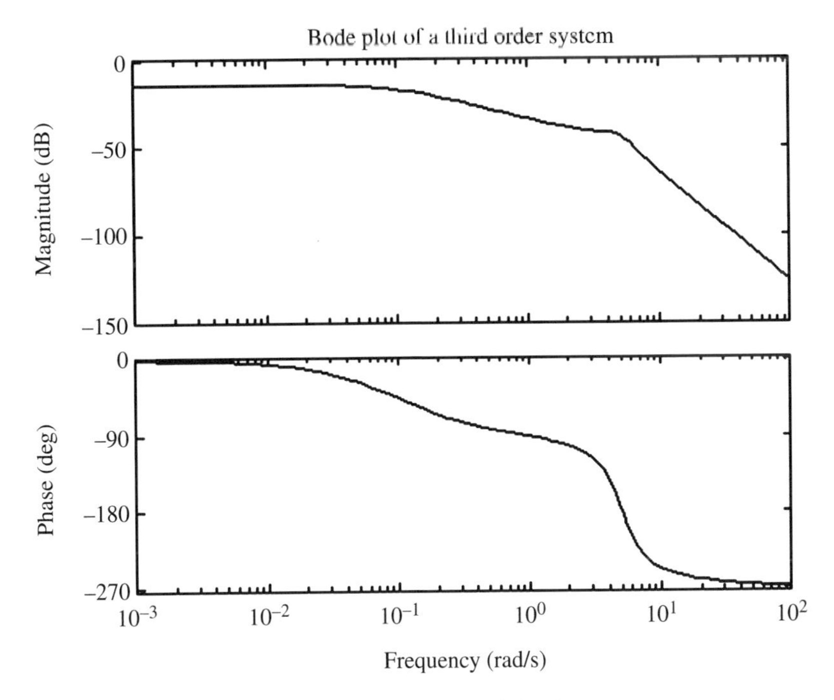

Figure 11E7 Bode plot of a third-order system

Solution:
The input to and output from MATLAB are included in Program Listing 11.7 and Figure 11E7.

Program Listing 11.7

```
>> num = [0 0 0 5];
>> den = [10 31 253 25];
>> bode(num,den)
>> title('Bode Plot of a Third Order System')
```

The Bode plots in Figure 11E7 compare very well with the asymptotic Bode plots in Figure 11.4 in Section 11.2.

Example 8
By employing MATLAB, perform the Bode plots of magnitude and phase of the system in Example 2 of Section 11.8.1, which gives the transfer function as

$$G(s) = \frac{5s + 10}{s^3 + s^2}$$

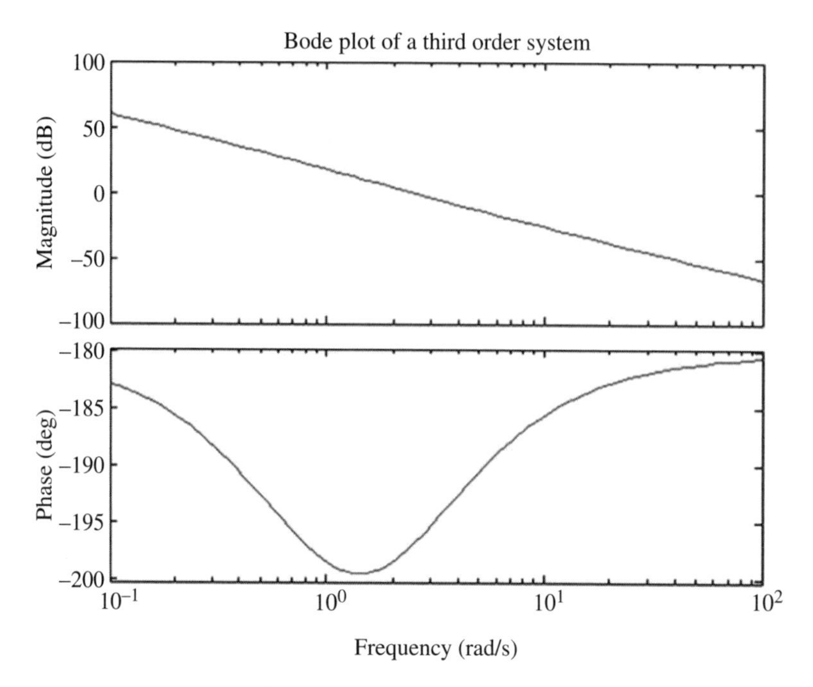

Figure 11E8 Bode plot of a third-order system

Solution:
The input to and output from MATLAB are included in Program Listing 11.8 and Figure 11E8.

Program Listing 11.8

```
>> num = [0 0 5 10];
>> den = [1 1 0 0];
>> bode(num,den)
>> title('Bode Plot of a third order system')
```

Example 9
By employing MATLAB, perform the Bode plots of magnitude and phase of the control system in Example 3 of Section 11.8.1. The transfer function is

$$G(s) = \frac{s+1}{s^4 + 3s^2 + 12s^2 - 16s}$$

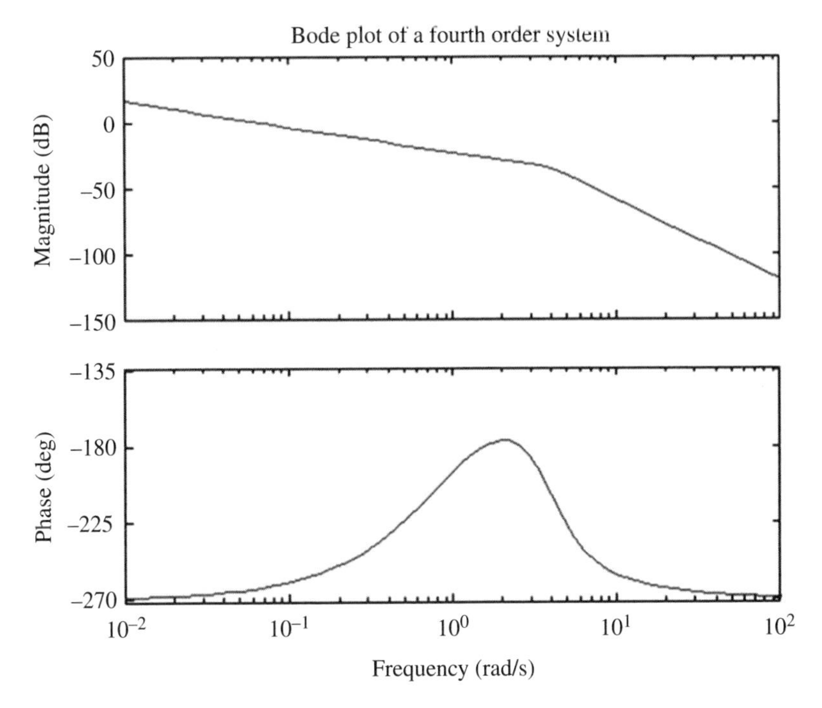

Figure 11E9 Bode plot of a fourth-order system

Solution:
The input to and output from MATLAB are included in Program Listing 11.9 and Figure 11E9.

Program Listing 11.9

```
>> num = [0 0 0 1 1];
>> den = [1 3 12 -16 0];
>> bode(num,den)
>> title('Bode Plot of a fourth order system')
```

Example 10
By employing MATLAB, perform the Bode plots of magnitude and phase of the control system in Example 4 of Section 11.8.1. The transfer function is

$$G(s) = \frac{20s^2 + 64s + 35}{100s^4 + 503s^3 + 4021s^2 + 150s + 240}$$

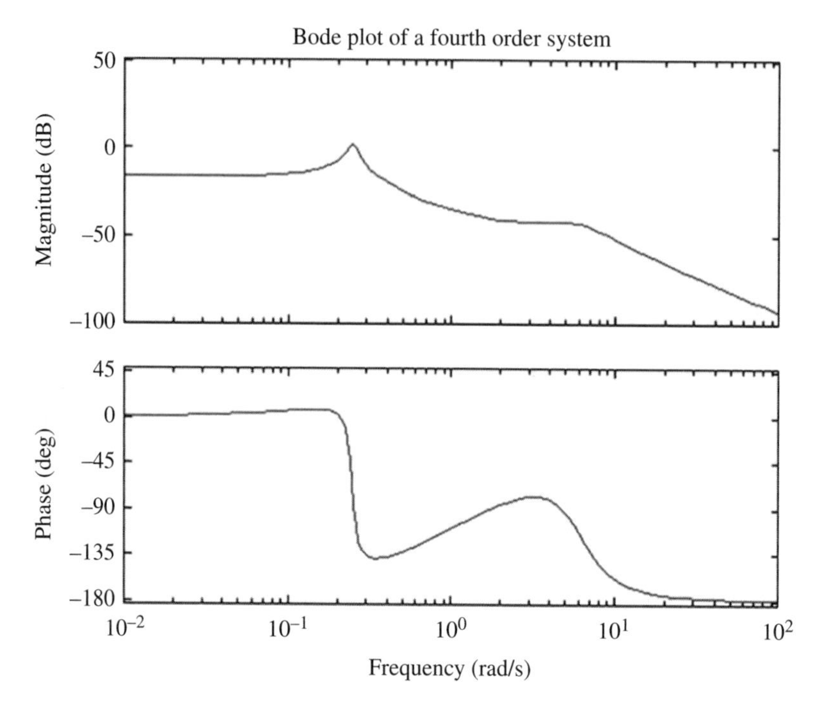

Figure 11E10 Bode plot of a fourth order system

Solution:
The input to and output from MATLAB are included in Program Listing 11.10 and Figure 11E10.

Program Listing 11.10

```
>> num = [0 0 20 64 35];
>> den = [100 503 4021 150 240];
>> bode(num,den)
>> title('Bode Plot of a fourth order system')
```

Example 11
By employing MATLAB, perform the Bode plots of magnitude and phase of the control system in Example 5 of Section 11.8.1. The transfer function is

$$G(s) = \frac{s^2 + 1}{s(s^2 + 100)}$$

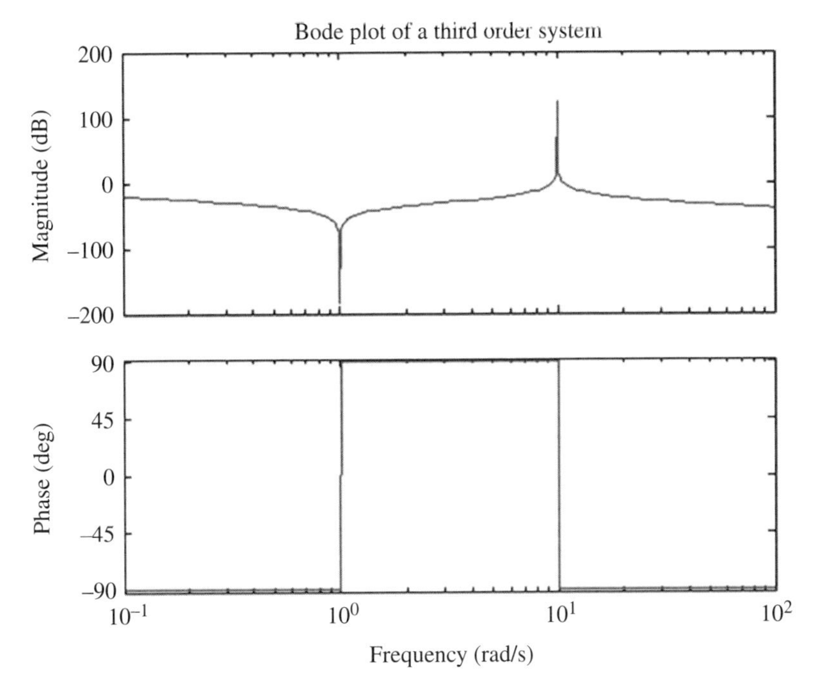

Figure 11E11 Bode plot of a third-order system

Solution:
The input to and output from MATLAB are included in Program Listing 11.11 and Figure 11E11, respectively.

Program Listing 11.11

```
>> num = [0 1 0 1];
>> den = [1 0 100 0];
>> bode(num,den)
>> title('Bode Plot of a third order system')
```

11.8.3 Nyquist Plots

In this subsection, Nyquist plots are presented. The first example is concerned with the system defined by Equation (11.9).

Example 12

By applying MATLAB, perform the Nyquist plots in the complex plane of the third-order system whose transfer function is given by Equation (11.9), that is:

$$G(s) = \frac{5}{10s^3 + 31s^2 + 253s + 25}.$$

Solution:

The input to and output from MATLAB are included in Program Listing 11.12a and Figure 11E12a, respectively.

Program Listing 11.12a

```
>> num = [0 0 0 5];
>> den = [10 31 253 25];
>> nyquist(num,den)
>> title('Nyquist Diagram of a Third Order System')
```

In order to provide a better or larger view of the plot close to the origin of the complex plane in Figure 11E12a, one would add a statement in the above MATLAB so that a closer inspection may be made. The input to and output from MATLAB are presented in Program Listing 11.12b and Figure 11E12b, respectively.

Program Listing 11.12b

```
>> num = [0 0 0 5];
>> den = [10 31 253 25];
>> nyquist(num,den)
>> axis([-0.02 0.02 -0.03 0.03]);
>> title('Nyquist Diagram of a Third Order System')
```

Example 13

By applying MATLAB, perform the Bode and Nyquist plots in the complex plane of the third-order system whose transfer function is given by

$$G(s)H(s) = KG_o(s) = \frac{K}{s(s+4)(s+5)} = \frac{K}{s^3 + 9s^2 + 20s}$$

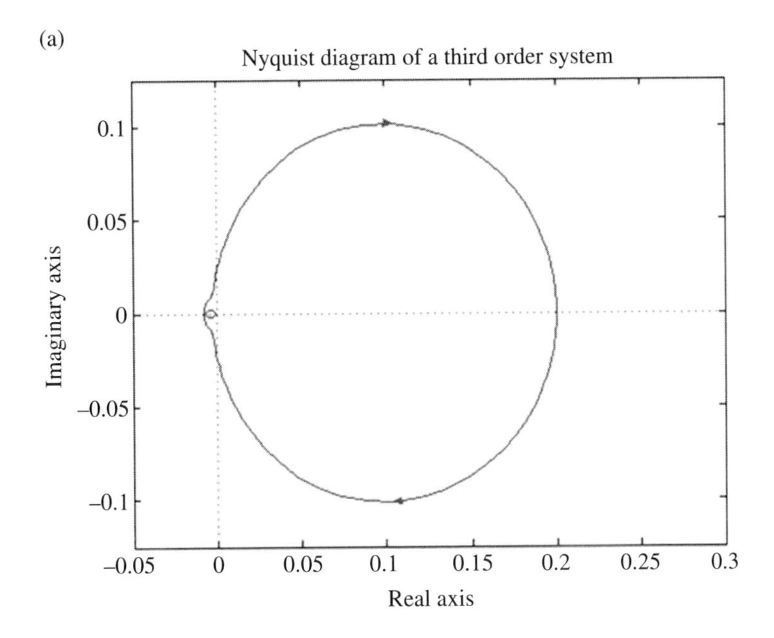

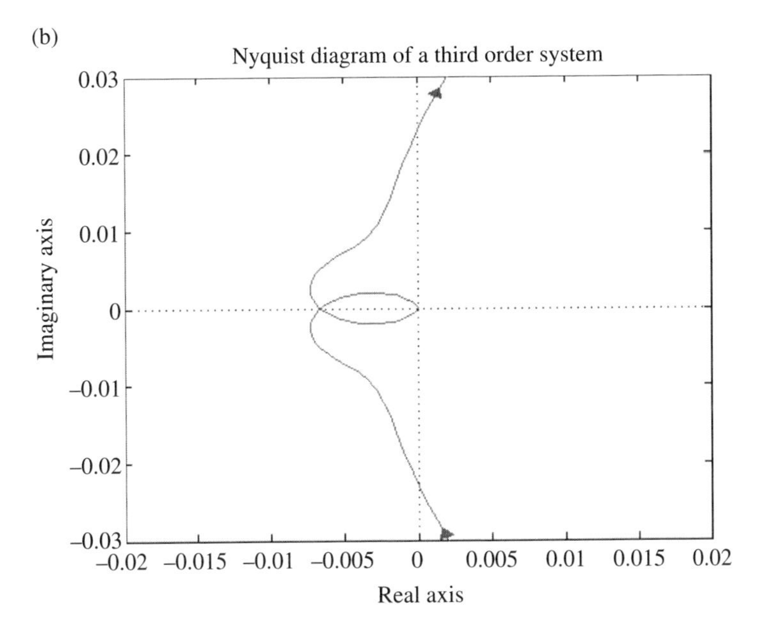

Figure 11E12 (a) Nyquist plot of a third-order system. (b) Nyquist plot of a third-order system with imaginary and real axes controlled for plotting

Solution:

The input to and output from MATLAB are included in Program Listing 11.13 and Figure 11E13.

Program Listing 11.13

```
>> num = [0 0 0 1];
>> den = [1 9 20 0];
>> bode(num,den)
>> title('Bode Plot of a Third Order System')
>> nyquist(num,den)
>> title('Nyquist Plot of a Third Order System')
```

Example 14

By applying MATLAB, perform the Bode and Nyquist plots in the complex plane of the third-order system whose transfer function is given by

$$G(s)H(s) = KG_o(s) = \frac{K}{s^3}$$

Solution:

The input to and output from MATLAB are included in Program Listing 11.14 and Figure 11E14.

Program Listing 11.14

```
>> num = [0 0 0 1];
>> den = [1 0 0 0];
>> bode(num,den)
>> title('Bode Plot of a Third Order System')
>> nyquist(num,den)
>> title('Nyquist Plot of a Third Order System')
```

Before considering the next example, it may be appropriate to observe that the Bode plots in Figures 11E13a and 11E14a are consistent with the theory. For example, in Figure 11E14a the magnitude plot gives a negative gradient of 60 dB per decade. However, the Nyquist plots in Figures 11E13b and 11E14b do not reveal any observable meaning. Computational experiments with MATLAB by increasing and decreasing the ranges of both the real and imaginary axes were unable to provide any meaningful information. This reflects the fact that there are limitations to employing MATLAB for the construction of Nyquist plots.

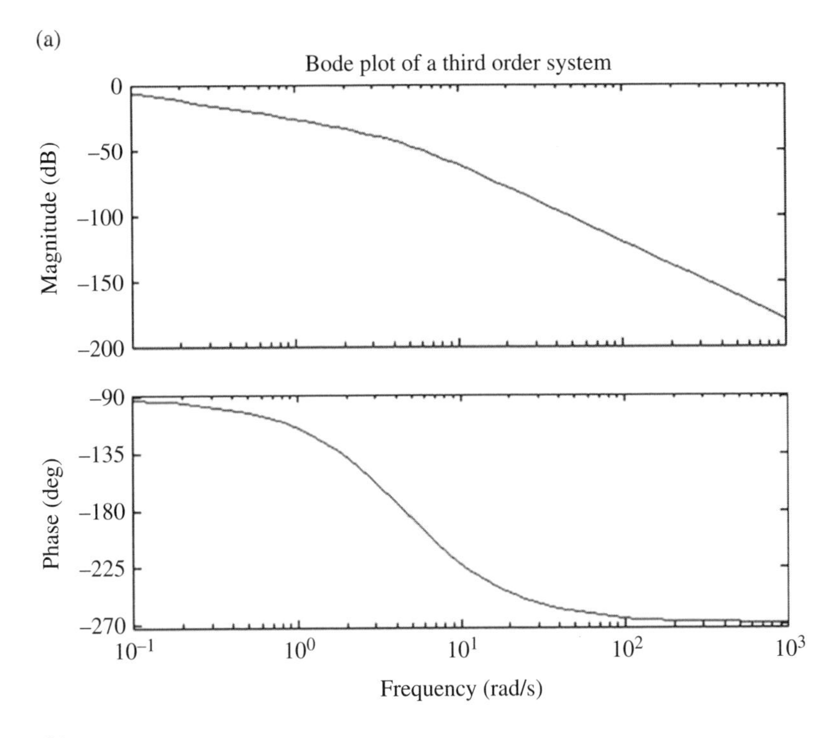

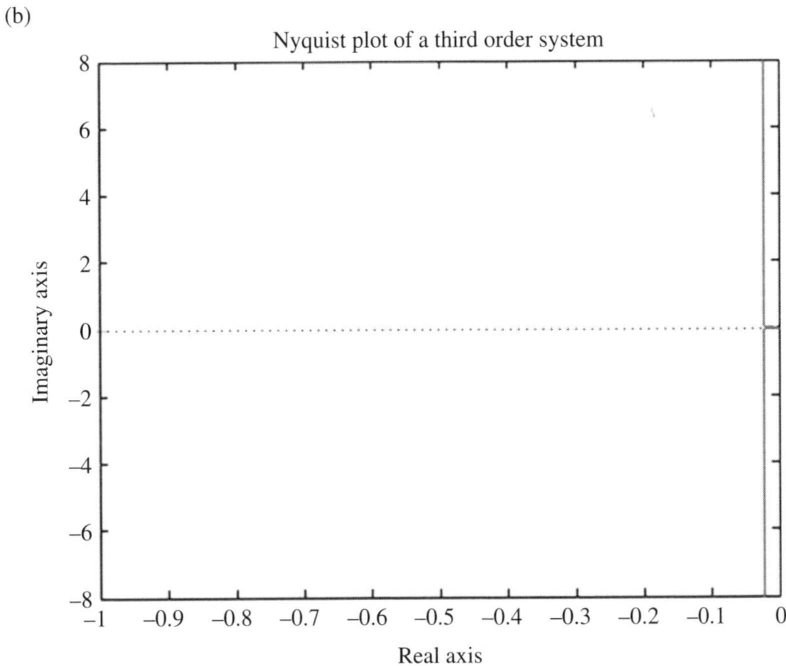

Figure 11E13 (a) Bode plot of a third order system $G_o(s) = \dfrac{K}{s^3 + 9s^2 + 20s}$. (b) Nyquist plot of the same third-order system

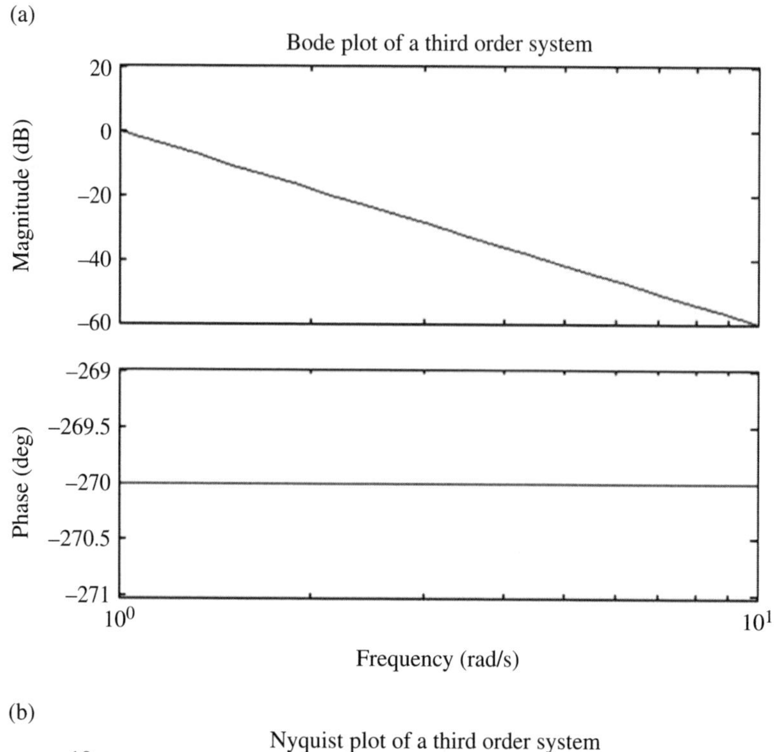

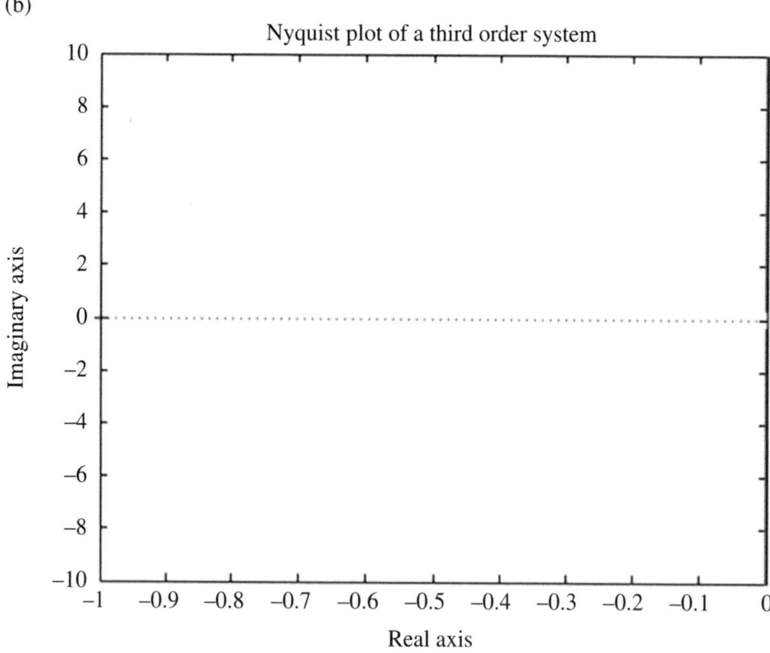

Figure 11E14 (a) Bode plot of a third-order system $G_o(s) = \frac{K}{s^3}$. (b) Nyquist plot of the same third-order system

Example 15

By applying MATLAB and the Nyquist criterion, determine the stability of the system whose open-loop transfer function is given by

$$G(s)H(s) = KG_o(s) = \frac{K}{s(s+1)(s+2)} = \frac{K}{s^3 + 3s^2 + 2s}$$

Solution:

The input to and output from MATLAB are included in Program Listing 11.15 and Figure 11E15.

Program Listing 11.15

```
>> num = [0 0 0 1];
>> den = [1 3 2 0];
>> nyquist(num,den)
>> title('Nyquist Plot of K/[s(s + 1)(s+2)] ')
```

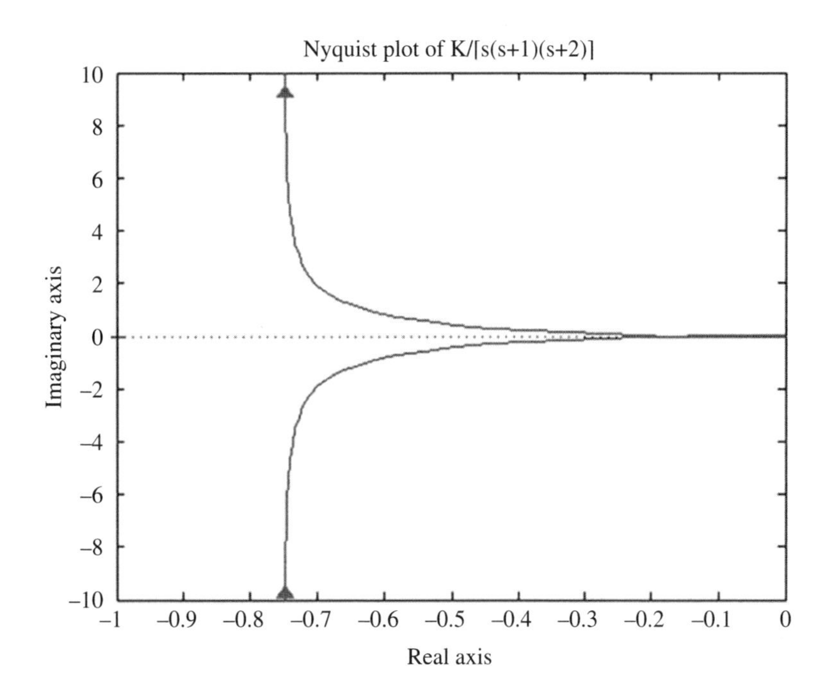

Figure 11E15 Nyquist plot of a third-order system $G_o(s) = \frac{K}{s^3 + 3s^2 + 2s}$.

With reference to Figure 11E15, one can observe that the system is stable since the contours do not enclose the (−1, 0) point and are on the left-hand side of this point. In addition, the c.e. of the system is

$$1+G(s)H(s)=1+\frac{K}{s^3+3s^2+2s}=0$$

Thus, $s^3+3s^2+2s+K=0$.

By the Routh array one can show that the system is stable within the range $0<K<6$.

Example 16

By applying MATLAB and the Nyquist criterion, determine the stability of the system whose open-loop transfer function is given by

$$G(s)H(s)=KG_o(s)=\frac{K}{s^2(s+1)(s+2)}$$

Solution:

The open-loop transfer function is

$$G(s)H(s)=\frac{K}{s^2(s+1)(s+2)}=\frac{K}{s^4+3s^3+2s^2}.$$

The input to and output from MATLAB are included in Program Listing 11.16 and Figure 11E16, respectively.

With reference to the Nyquist criterion, one can observe that the system is unstable since the contours do enclose the (−1, 0) point.

Program Listing 11.16

```
>> num = [0 0 0 0 1];
>> den = [1 3 2 0 0];
>> nyquist(num,den)
>> title('Nyquist Plot of K / [s^2(s+1)(s+2)]')
```

Example 17

By applying MATLAB and the Nyquist criterion, determine the stability of the system whose open-loop transfer function is given by

$$G(s)H(s)=KG_o(s)=\frac{K}{s^2+s+1}$$

Compare this result with that of the Bode plot.

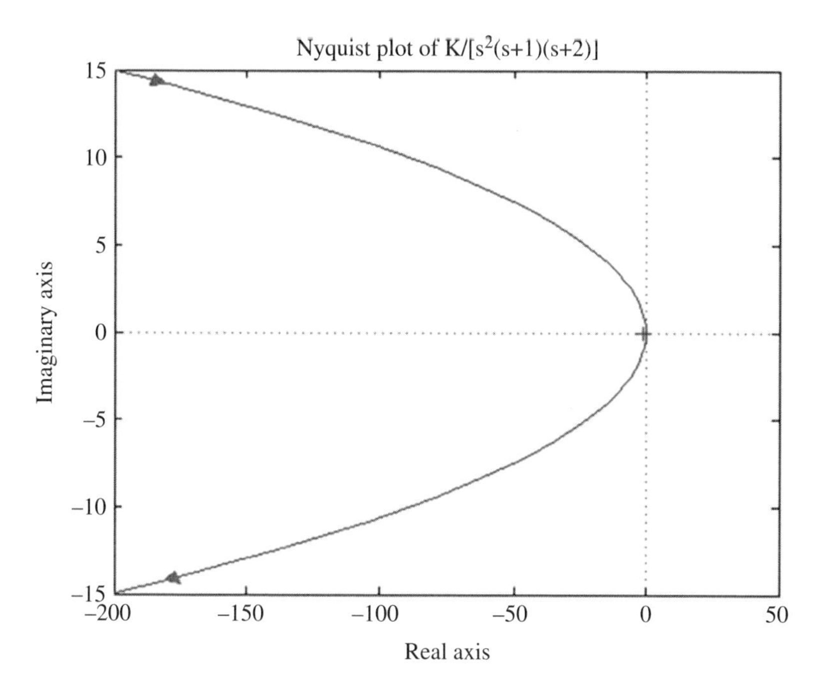

Figure 11E16 Nyquist plot of a fourth-order system $G_o(s) = \dfrac{K}{s^4 + 3s^3 + 2s^2}$.

Solution:

The input to and output from MATLAB are included in Program Listing 11.17 and Figure 11E17.

With reference to Figure 11E17, the system is stable for all positive K.

Program Listing 11.17

```
>> num = [0 0 1];
>> den = [1 1 1];
>> nyquist(num,den)
>> title('Nyquist Plot of K/(s^2+s+1)')
>> bode(num,den)
>> title('Bode Plot of K/(s^2+s+1)')
```

Exercise Questions

Q1. A unity feedback control system has its closed-loop transfer function defined by $\dfrac{C(s)}{R(s)} = \dfrac{K}{s(s+1)(s^2+s+1)+K}$ such that its corresponding open-loop transfer function becomes $G(s)H(s) = \dfrac{K}{s(s+1)(s^2+s+1)}$. Sketch the root locus for this closed-loop control system for K = −∞ to K = +∞. Find the range of K that will keep this system stable.

(a)

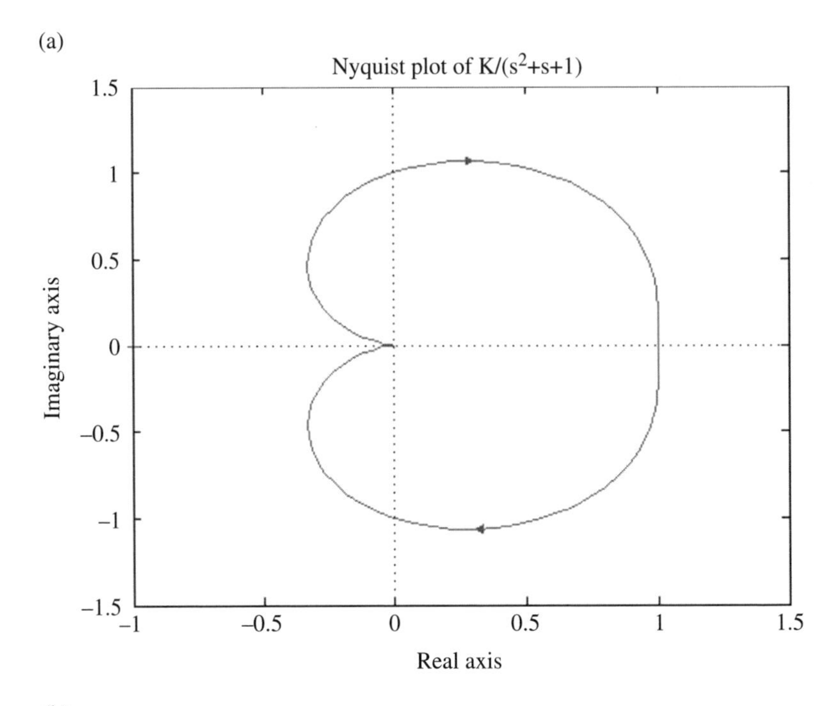

(b)

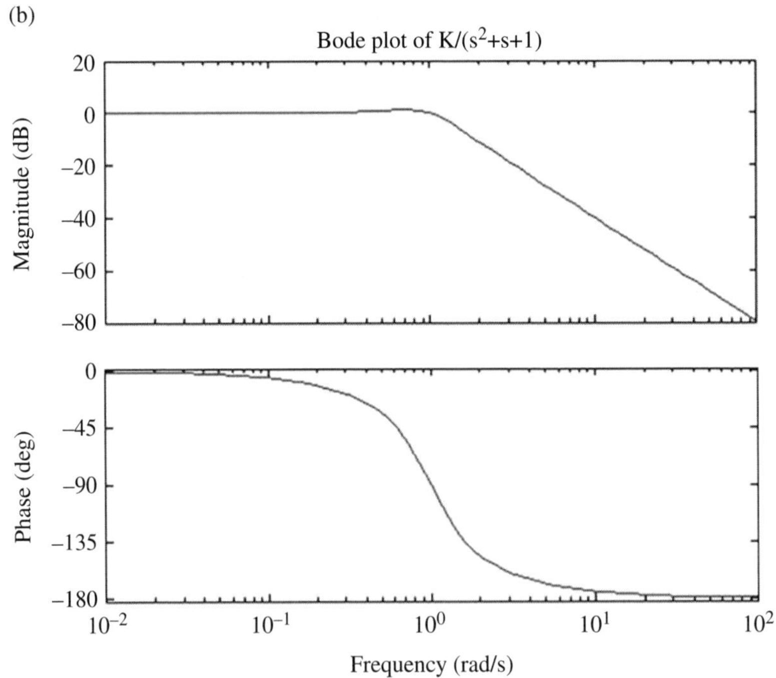

Figure 11E17 (a) Nyquist plot of a second-order system $G_o(s) = \dfrac{K}{s^2 + s + 1}$. (b) Bode plot of the same second-order system

Q2. Construct the root locus for the unity feedback control system whose open-loop transfer function is given by

$$G(s) = \frac{K}{s(s+4)(s+5)}$$

Q3. Construct the asymptotic approximations of a so-called quadratic lag whose transfer function is given by

$$G(s) = \frac{25}{s^2 + s + 25}$$

Q4. Draw the Nyquist diagram for the control system with an open-loop transfer function

$$G(s) = \frac{K}{(s+4)(s+5)}$$

By making use of the Nyquist diagram, determine whether or not the closed-loop feedback system is stable.

Q5. Consider a closed-loop control system whose open-loop transfer function is given by

$$G(s)H(s) = \frac{K}{s^3 + 6s^2 + 11s + 6}.$$

Determine the gain margin and the maximum value of K for a stable response.

Q6. By using MATLAB, construct the root locus for the unity feedback control system whose open-loop transfer function is given by:

$$G(s) = \frac{K(s+3)}{s(s+4)(s+5)}.$$

Q7. By using MATLAB, construct the Bode plots for the quadratic lag:

$$G(s) = \frac{25}{s^2 + s + 25}.$$

Q8. By applying MATLAB, draw the Nyquist diagram for the control system with an open-loop transfer function:

$$G(s) = \frac{K}{(s+4)(s+5)}.$$

Q9. By applying MATLAB, draw the Nyquist diagram for the control system with an open-loop transfer function

$$G(s) = \frac{25}{s^2 + s + 25}.$$

Q10. By applying MATLAB, draw the Nyquist diagram for the control system with an open-loop transfer function:

$$G(s)=\frac{Ks(s+3)}{(s+4)(s+5)}.$$

References

[1] The Math Works, Inc. (2014). *MATLAB R2014a*. The Math Works, Inc., Natick, MA.

[2] Shahian, B., and Hassul, M. (1993). *Control System Design Using MATLAB*. Prentice-Hall, Englewood Cliffs, NJ.

[3] Evans, W.R. (1948). Graphical analysis of control systems. *Transactions of the American Institute of Electrical Engineers*, **67**, 547–551.

[4] Bode, H.W. (1940). Feedback amplifier design. *Bell System Technical Journal*, **19**, 44.

[5] Nyquist, H. (1932). Regeneration theory. *Bell System Technical Journal,* **11**, 1, 126–147.

[6] Anand, D.K. (1984). *Introduction to Control Systems*, 2nd edn. Pergamon Press, New York.

[7] Churchill, R.V., and Brown, J.W. (1984). *Complex Variables and Applications*, 4th edn. McGraw-Hill, New York.

[8] James, H.M., Nichols, N.B., and Phillips, R.S. (1947). *Theory of Servomechanisms*. McGraw-Hill, New York.

[9] Ogata, K. (1998). *System Dynamics*, 3rd edn. Prentice-Hall, Englewood Cliffs, NJ.

[10] Kuo, B.C., and Golnaraghi, F. (2003). *Automatic Control Systems*, 8th edn. John Wiley & Sons, Inc., Hoboken, NJ.

[11] Lobontiu, N. (2010). *System Dynamics for Engineering Students: Concepts and Applications*. Academic Press, New York.

12

Modern Control System Analysis

Modern control system analysis is known as analysis of control systems by applying the *state space* or *vector space* method. Among many advantages of the state space method, the following three are the main important ones. First, the matrix theory can be explored and exploited. Second, it is not limited to linear systems, having time-invariant system parameters. Third, it is not limited to applications for single input and single output (SISO) control systems.

In the following section the state space method is introduced. Section 12.2 deals with derivation of the state transition matrix (STM). Section 12.3 is concerned with the development of the relationship between the Laplace transformed state equation and transfer function. Stability based on eigenvalues of the coefficient matrix of a system is discussed in Section 12.4. The concepts of controllability and observability are presented in detail in Section 12.5, while the concepts of stabilizability and detectability are defined in Section 12.6. Application of MATLAB [1] for solutions of systems to step and impulse inputs are included in Section 12.7.

12.1 State Space Method

Consider a simple single dof oscillator, which is shown in Figure 12.1. The equation of motion for this system is given by

$$\begin{gathered} m\ddot{x} + c\dot{x} + kx = f \quad \text{or} \\ \ddot{x} + 2\zeta\omega_n\dot{x} + \omega_n^2 x = r \end{gathered} \tag{12.1}$$

where the symbols have their usual meaning.

Introduction to Dynamics and Control in Mechanical Engineering Systems, First Edition. Cho W. S. To.

Companion website: www.wiley.com/go/to/dynamics

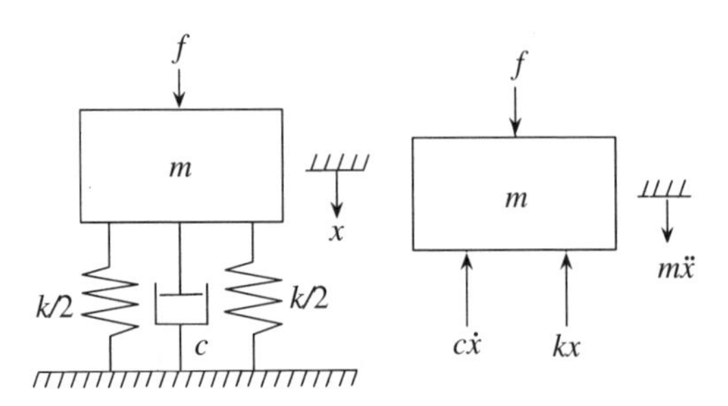

Figure 12.1 The single dof oscillator and its free body diagram

The state space method essentially consists of transforming the second-order differential equation into two first-order differential equations. Thus, one defines the two state variables

$$x_1 = x, \quad x_2 = \dot{x}$$

so that Equation (12.1) may be written as

$$\begin{aligned}\dot{x}_1 &= x_2\\ \dot{x}_2 &= -2\zeta\omega_n x_2 - \omega_n^2 x_1 + r.\end{aligned}$$

These two first-order differential equations may be written in matrix form as

$$\begin{pmatrix}\dot{x}_1\\ \dot{x}_2\end{pmatrix} = \begin{bmatrix}0 & 1\\ -\omega_n^2 & -2\zeta\omega_n\end{bmatrix}\begin{pmatrix}x_1\\ x_2\end{pmatrix} + \begin{pmatrix}0\\ r\end{pmatrix}$$

or in more concise form

$$\dot{y} = Ay + Br, \tag{12.2}$$

where the vectors and matrix are defined as

$$\dot{y} = \begin{pmatrix}\dot{x}_1\\ \dot{x}_2\end{pmatrix}, \quad A = \begin{bmatrix}0 & 1\\ -\omega_n^2 & -2\zeta\omega_n\end{bmatrix}, \quad y = \begin{pmatrix}x_1\\ x_2\end{pmatrix}, \quad B = \begin{pmatrix}0\\ 1\end{pmatrix}.$$

In Equation (12.2), A is known as the *coefficient* or *amplification* matrix of the system, whereas Equation (12.2) is called the *state equation* of the system.

For any m, not to be confused with the mass of the single dof system in Equation (12.1), dof system whose state equation is identical in form to Equation (12.2), the vector y now is of order $2m \times 1$ and r is scalar input. On the other hand, matrix A is of order $2m \times 2m$ and vector B is of order $2m \times 1$.

12.2 State Transition Matrix

The STM is defined as one that satisfies the homogeneous state equation. With reference to Equation (12.2), the homogeneous state equation is

$$\dot{y} = Ay. \tag{12.3}$$

Suppose Φ is the STM. Then it must satisfy the following equations:

$$\frac{d\Phi}{dt} = A\Phi, \tag{12.4}$$

and

$$y(t) = \Phi(t)y(0), \tag{12.5}$$

which is the solution of Equation (12.3) for $t \geq 0$.

The STM Φ can be obtained as in the following. One takes the Laplace transform of Equation (12.3) so that

$$sY(s) - y(0) = AY(s), \tag{12.6}$$

where $Y(s)$ is the Laplace transform of the state vector $y(t)$.

Re-arranging Equation (12.6), one has

$$Y(s) = (sI - A)^{-1}y(0). \tag{12.7}$$

Taking the inverse Laplace transform of Equation (12.7) results in

$$y(t) = \mathcal{L}^{-1}\left[(sI - A)^{-1}\right]y(0). \tag{12.8}$$

Comparing Equation (12.8) with Equation (12.5), the STM is

$$\Phi(t) = \mathcal{L}^{-1}\left[(sI - A)^{-1}\right]. \tag{12.9}$$

By the Maclaurin series expansion as shown in Appendix 12B,

$$\Phi(t) = e^{At} = I + At + \frac{1}{2!}A^2t^2 + \ldots \tag{12.10}$$

Remark 12.2.1 The STM represents the free or transient response of the system. In other words, it is the response governed by the initial conditions.

Remark 12.2.2 The inverse of the STM exists for all time t, since it is not singular.

Remark 12.2.3 The STM has the following important properties:

$$\Phi(t_0 - t_0) = I, \quad \Phi(t_{i+2} - t_i) = \Phi(t_{i+2} - t_{i+1})\Phi(t_{i+1} - t_i),$$
$$[\Phi(t)]^k = \Phi(kt), \quad [\Phi(t)]^{-1} = \Phi(-t).$$

The matrix in the last property is known as *symplectic* whereas i and k are integers. These properties are very useful in numerical integration and many other applications.

12.3 Relationship between Laplace Transformed State Equation and Transfer Function

Consider a general linear time-invariant system of m dof with feedback control. Such a system may be described by

$$\frac{dy(t)}{dt} = Ay(t) + Br(t), \tag{12.11a}$$

$$c(t) = \tilde{C}y(t) + Dr(t), \tag{12.11b}$$

where

$y(t)$ is the state vector of order $2m \times 1$,
$r(t)$ is the input vector of order $p \times 1$,
$c(t)$ is the output vector of order $q \times 1$,
B is now a matrix of $2m \times p$,
$\tilde{C}$ is the constant output gain or coefficient matrix of $q \times 2m$, and
D is a matrix of $q \times p$.

For simplicity, one assumes that $D = 0$ in Equation (12.11b) such that upon taking the Laplace transform of Equation (12.11a) it gives

$$Y(s) = (sI - A)^{-1}y(0) + (sI - A)^{-1}BR(s) \tag{12.12}$$

Taking the Laplace transform of Equation (12.11b) results in

$$C(s) = \tilde{C}Y(s). \tag{12.13}$$

Substituting Equation (12.12) into Equation (12.13), one obtains

$$C(s) = \tilde{C}(sI - A)^{-1}y(0) + \tilde{C}(sI - A)^{-1}BR(s).$$

The definition of a transfer function requires that $y(0) = 0$, therefore the Laplace transformed output becomes

$$C(s) = \tilde{C}(sI - A)^{-1}BR(s). \tag{12.14}$$

By definition, the transfer function of a SISO system is the ratio of the Laplace transformed output to that of the input, and therefore by analogy, the transfer function of a multiple input and multiple output (MIMO) system is defined as

$$G(s)=\widetilde{C}(sI-A)^{-1}B. \tag{12.15}$$

Before leaving this section it should be pointed out that the output in the time domain $c(t)$ can be obtained by taking the inverse Laplace transform of Equation (12.14). That is,

$$\mathcal{L}^{-1}[C(s)]=\mathcal{L}^{-1}\left[\widetilde{C}(sI-A)^{-1}\right]y(0)+\mathcal{L}^{-1}\left[\widetilde{C}(sI-A)^{-1}BR(s)\right].$$

By application of Equation (12.9) and the convolution theorem defined by Equation (2.12), the last equation becomes

$$c(t)=\widetilde{C}\Phi(t)y(0)+\widetilde{C}\int_0^t \Phi(t-\tau)Br(\tau)d\tau. \tag{12.16a}$$

With reference to Equations (12.11b) and (12.10), one can show that

$$y(t)=e^{A(t-t_0)}y(t_0)+\int_0^t e^{A(t-\tau)}Br(\tau)d\tau. \tag{12.16b}$$

The first term on the rhs of Equation (12.16b) is the so-called transient solution while the second term is the forced response of the system.

In Appendix 12A, another approach, generally known as the method of diagonalization, for the determination of a solution to the state equation is introduced. It is based on the concept of coordinate transformation. This approach is an alternative to the foregoing steps for obtaining Equation (12.16b).

Example

A system has the following state equation

$$\begin{pmatrix}\dot{x}_1\\ \dot{x}_2\end{pmatrix}=\begin{bmatrix}0 & 1\\ -6 & -5\end{bmatrix}\begin{pmatrix}x_1\\ x_2\end{pmatrix}+\begin{pmatrix}0\\ 1\end{pmatrix}r(t)$$

in which the states are $x_1=x, x_2=\dot{x}$, and the scalar input $r(t)$ is a unit step function.

a. Determine the STM $\Phi(t)$, and
b. when the system starts from rest, find the forced responses of the system.

Solution:

a. *State transition matrix*

   With reference to the given state equation,

$$A=\begin{bmatrix}0 & 1\\ -6 & -5\end{bmatrix},\ B=\begin{pmatrix}0\\ 1\end{pmatrix}.$$

Therefore, $sI-A=\begin{bmatrix} s & 0 \\ 0 & s \end{bmatrix}-\begin{bmatrix} 0 & 1 \\ -6 & -5 \end{bmatrix}=\begin{bmatrix} s & -1 \\ 6 & s+5 \end{bmatrix}$.

$$(sI-A)^{-1}=\frac{\begin{bmatrix} s+5 & 1 \\ -6 & s \end{bmatrix}}{s(s+5)+6}=\frac{\begin{bmatrix} s+5 & 1 \\ -6 & s \end{bmatrix}}{(s+2)(s+3)}.$$

By Equation (12.9), the STM

$$\Phi(t)=\mathcal{L}^{-1}\left[(sI-A)^{-1}\right].$$

That is, $\Phi_{11}(t)=\mathcal{L}^{-1}\left[\frac{s+5}{(s+2)(s+3)}\right]=\mathcal{L}^{-1}\left[\frac{s+2}{(s+2)(s+3)}+\frac{3}{(s+2)(s+3)}\right]$.

$$\Phi_{11}(t)=\mathcal{L}^{-1}\left[\frac{1}{(s+3)}\right]+\mathcal{L}^{-1}\left[\frac{3}{(s+2)(s+3)}\right].$$

Applying first the partial fraction method to the second term on the rhs and disregarding the factor 3 for the time being, one has

$$\frac{A_1}{(s+2)}+\frac{A_2}{(s+3)}=\frac{A_1(s+3)+A_2(s+2)}{(s+2)(s+3)}=\frac{(A_1+A_2)s+3A_1+2A_2}{(s+2)(s+3)}.$$

Equating the coefficients on the lhs to those on the rhs, one finds $A_1=1$, $A_2=-1$. Therefore,

$$\Phi_{11}(t)=\mathcal{L}^{-1}\left[\frac{1}{(s+3)}\right]+3\mathcal{L}^{-1}\left[\frac{1}{(s+2)}-\frac{1}{(s+3)}\right].$$

$$\Phi_{11}(t)=\mathcal{L}^{-1}\left[\frac{3}{(s+2)}-\frac{2}{(s+3)}\right].$$

Thus, after taking the inverse Laplace transform on the rhs, one obtains

$$\Phi_{11}(t)=3e^{-2t}-2e^{-3t}.$$

Similarly,

$$\Phi_{12}(t)=\mathcal{L}^{-1}\left[\frac{1}{(s+2)(s+3)}\right]=\mathcal{L}^{-1}\left[\frac{1}{(s+2)}-\frac{1}{(s+3)}\right].$$

Therefore, $\Phi_{12}(t)=e^{-2t}-e^{-3t}$.

$$\Phi_{21}(t)=\mathcal{L}^{-1}\left[\frac{-6}{(s+2)(s+3)}\right]=\mathcal{L}^{-1}\left[\frac{6}{(s+3)}-\frac{6}{(s+2)}\right].$$

$$\Phi_{21}(t)=6e^{-3t}-6e^{-2t}.$$

Also, $\Phi_{22}(t) = \mathcal{L}^{-1}\left[\frac{s}{(s+2)(s+3)}\right] = \mathcal{L}^{-1}\left[\frac{1}{(s+3)} - \frac{1}{(s+2)}\right]$.

Therefore, $\Phi_{22}(t) = e^{-3t} - e^{-2t}$.

Collecting all the elements, the STM becomes

$$\Phi(t) = \begin{bmatrix} 3e^{-2t} - 2e^{-3t} & e^{-2t} - e^{-3t} \\ 6e^{-3t} - 6e^{-2t} & e^{-3t} - e^{-2t} \end{bmatrix}.$$

b. *Forced responses*

Applying Equation (12.16a), one obtains

$$\begin{pmatrix} x_1 \\ x_2 \end{pmatrix} = \Phi(t)\begin{pmatrix} x_1(0) \\ x_2(0) \end{pmatrix} + \int_0^t \Phi(t-\tau)B\, r(\tau)d\tau.$$

Since the system starts from rest, implying that $x_1(0) = 0$, and $x_2(0) = 0$ while $r(t) = 1$, the first term on the rhs disappears and the responses become

$$\begin{pmatrix} x_1 \\ x_2 \end{pmatrix} = \int_0^t \Phi(t-\tau)\begin{pmatrix} 0 \\ 1 \end{pmatrix} 1 d\tau = \int_0^t \begin{pmatrix} \Phi_{12}(t-\tau) \\ \Phi_{22}(t-\tau) \end{pmatrix} d\tau.$$

Therefore,

$$x_1 = \int_0^t \Phi_{12}(t-\tau)d\tau = \int_0^t \left[e^{-2(t-\tau)} - e^{-3(t-\tau)}\right] d\tau.$$

$$x_1 = e^{-2t}\int_0^t e^{2\tau}d\tau - e^{-3t}\int_0^t e^{3\tau}d\tau = e^{-2t}\left[\frac{e^{2t}}{2} - \frac{e^0}{2}\right] - e^{-3t}\left[\frac{e^{3t}}{3} - \frac{e^0}{3}\right].$$

$$x_1 = \frac{1}{6} - \frac{e^{-2t}}{2} + \frac{e^{-3t}}{3}.$$

Similarly,

$$x_2 = \int_0^t \Phi_{22}(t-\tau)d\tau = -\int_0^t \Phi_{12}(t-\tau)d\tau.$$

Therefore,

$$x_2 = -x_1 = -\frac{1}{6} + \frac{e^{-2t}}{2} - \frac{e^{-3t}}{3}.$$

12.4 Stability Based on Eigenvalues of the Coefficient Matrix

From Equation (12.15) it is apparent that the c.e. of the system can be formed by making use of the determinant of matrix $(sI - A)$ since matrices D and B are constant while the inverse of matrix $(sI - A)$ is equal to the ratio of the adjoint of $(sI - A)$ to the determinant of $(sI - A)$. In other words, the c.e. is

$$|sI - A| = 0.$$

Recall that for a system to be stable, the roots of the c.e. or eigenvalues of the system have no positive real parts nor repeated roots on the imaginary axis. In other words, when any or all real parts of the eigenvalues are positive, the responses of the system grow with time, rendering the system unstable.

To provide concrete illustrations, the following two examples are considered in detail. The first one is a lightly damped single dof oscillator, while the second is the 2-dof system of a simplified model of a rotating rotor. In this second example, as will be observed later, the governing equation of motion contains a damping matrix which is skew-symmetric. At first glance and without knowledge of advanced matrix theory, it is not clear whether or not this system is stable. It is therefore of interest to study this issue by determining the eigenvalues of the system.

Example 1

Consider the simple, lightly damped single dof oscillator defined by Equation (12.1). Determine the eigenvalues of the system.

Solution:

The eigenvalues of this system are given by

$$|sI - A| = 0$$

in which the coefficient matrix A is defined in Equation (12.2). Therefore,

$$\left| s\begin{bmatrix} 1 & 0 \\ 0 & 1 \end{bmatrix} - \begin{bmatrix} 0 & 1 \\ -\omega_n^2 & -2\zeta\omega_n \end{bmatrix} \right| = 0 \text{ giving } \begin{vmatrix} s & -1 \\ \omega_n^2 & (2\zeta\omega_n + s) \end{vmatrix} = 0.$$

Operating the determinant, one has

$$s^2 + (2\zeta\omega_n)s + \omega_n^2 = 0.$$

This is a quadratic equation that gives the two complex roots as

$$s_{1,2} = -\zeta\omega_n \pm i\omega_d,$$

where the damped natural frequency is given by $\omega_d = \omega_n\sqrt{1-\zeta^2}$. These two roots are identical to those defined in Equation (4.3). If ζ is negative such that the real parts of the complex roots are positive, then the corresponding system is unstable.

Example 2

Consider the following equation of motion for a rotating rotor:

$$M\ddot{x} + \underline{C}\dot{x} + Kx = \mathbf{0}$$

in which the mass, damping, stiffness matrices, and displacement vector are given, respectively as

$$M=\begin{bmatrix} m & 0 \\ 0 & m \end{bmatrix},\quad \underline{C}=\begin{bmatrix} 0 & -2m\Omega \\ 2m\Omega & 0 \end{bmatrix},\quad K=\begin{bmatrix} k-m\Omega^2 & 0 \\ 0 & k-m\Omega^2 \end{bmatrix},$$

and $x=[x_1\ \ x_2]^T$. The damping matrix $\underline{C}$ is skew-symmetric and is known as the gyroscopic matrix. With m = 10 kg, k = 20 000 N/m, and the angular speed of the rotor Ω = 10 rad/s, solve

a. for the eigenvalues of the system and determine whether or not the system is stable, and
b. for the eigenvectors or mode shapes.

Solution:

a. Let $Z=[z_1\ \ z_2]^T$ where $z_1=[x_1\ \ x_2]^T$ and $z_2=[\dot{x}_1\ \ \dot{x}_2]^T$ such that

$$\dot{z}_1=z_2, \dot{z}_2=-M^{-1}\underline{C}z_2-M^{-1}Kz_1.$$

Therefore, the two second-order equations are transformed into the following first-order differential equations in matrix form

$$\begin{pmatrix} \dot{z}_1 \\ \dot{z}_2 \end{pmatrix}=\begin{bmatrix} 0 & I \\ -M^{-1}K & -M^{-1}\underline{C} \end{bmatrix}\begin{pmatrix} z_1 \\ z_2 \end{pmatrix} \quad \text{or} \quad \dot{Z}=AZ, \tag{i}$$

where the coefficient or amplification matrix is given by

$$A=\begin{bmatrix} 0 & I \\ -M^{-1}K & -M^{-1}\underline{C} \end{bmatrix}.$$

Now, consider M^{-1}, which is

$$M^{-1}=\begin{bmatrix} 10 & 0 \\ 0 & 10 \end{bmatrix}^{-1}=\begin{bmatrix} 0.1 & 0 \\ 0 & 0.1 \end{bmatrix}.$$

Therefore,

$$M^{-1}K=\begin{bmatrix} 0.1 & 0 \\ 0 & 0.1 \end{bmatrix}\begin{bmatrix} 20{,}000-1000 & 0 \\ 0 & 20{,}000-1000 \end{bmatrix}$$
$$=\begin{bmatrix} 1900 & 0 \\ 0 & 1900 \end{bmatrix}.$$

Similarly,

$$M^{-1}\underline{C}=\begin{bmatrix} 0.1 & 0 \\ 0 & 0.1 \end{bmatrix}\begin{bmatrix} 0 & -200 \\ 200 & 0 \end{bmatrix}=\begin{bmatrix} 0 & -20 \\ 20 & 0 \end{bmatrix}.$$

By substituting all the above, one has

$$A=\begin{bmatrix} 0 & 0 & 1 & 0 \\ 0 & 0 & 0 & 1 \\ -1900 & 0 & 0 & 20 \\ 0 & -1900 & -20 & 0 \end{bmatrix}.$$

To obtain the eigenvalues of the system, one applies

$$|sI-A|=0 \text{ or } |A-sI|=0 \text{ or}$$

$$\begin{vmatrix} -s & 0 & 1 & 0 \\ 0 & -s & 0 & 1 \\ -1900 & 0 & -s & 20 \\ 0 & -1900 & -20 & -s \end{vmatrix}=0. \quad \text{(ii)}$$

Operating on Equation (ii), one obtains

$$-s\begin{vmatrix} -s & 0 & 1 \\ 0 & -s & 20 \\ -1900 & -20 & -s \end{vmatrix}+(1)\begin{vmatrix} 0 & -s & 1 \\ -1900 & 0 & 20 \\ 0 & -1900 & -s \end{vmatrix}=0$$

$$s^4+1900\, s^2+400\, s^2+1900^2+1900\, s^2=0, \text{ or}$$

$$s^4+4200\, s^2+3610000=0.$$

This is a quadratic equation in s^2. Thus, if one writes $u = s^2$, the last equation becomes

$$u^2+4200\, u+3610000=0. \quad \text{(iii)}$$

Solving this quadratic equation and, in turn, taking the square roots of the solutions, the eigenvalues of the system are

$$s_1=-10(2\sqrt{5}-1)i=-34.7214i, \quad s_2=10(2\sqrt{5}-1)i=34.7214i,$$
$$s_3=-10(2\sqrt{5}+1)i=-54.7214i, \quad s_4=10(2\sqrt{5}+1)i=54.7214i.$$

Clearly, there are no real parts in all four eigenvalues that are not repeated roots. Therefore, the system is stable.

b. For the eigenvectors or mode shapes, one makes use of Equation (ii) so that one has

$$(A-sI)X=0 \text{ or}$$

$$\begin{bmatrix} -s & 0 & 1 & 0 \\ 0 & -s & 0 & 1 \\ -1900 & 0 & -s & 20 \\ 0 & -1900 & -20 & -s \end{bmatrix}\begin{pmatrix} X_1 \\ X_2 \\ X_3 \\ X_4 \end{pmatrix}=0, \quad \text{(iv)}$$

where $X_3=\dot{X}_1, X_4=\dot{X}_2$.

Substituting $s = s_1 = -10(2\sqrt{5}-1)i$ in Equation (iv), one finds

$$\begin{pmatrix} X_1 \\ X_2 \\ X_3 \\ X_4 \end{pmatrix} = \begin{pmatrix} 1 \\ i \\ -34.7214i \\ 34.7214 \end{pmatrix} X_1.$$

Assume $X_1 = 1$ so this eigenvector can be written as

$$\begin{pmatrix} X_1 \\ X_2 \\ X_3 \\ X_4 \end{pmatrix}_{(1)} = \begin{pmatrix} 1 \\ i \\ -34.7214i \\ 34.7214 \end{pmatrix}. \tag{v}$$

The subscript (1) on the lhs of Equation (v) denotes the eigenvector associated with s_1. Similarly, the remaining eigenvectors can be obtained as

$$\begin{pmatrix} X_1 \\ X_2 \\ X_3 \\ X_4 \end{pmatrix}_{(2)} = \begin{pmatrix} 1 \\ -i \\ 34.7214\,i \\ 34.7214 \end{pmatrix}, \tag{vi}$$

$$\begin{pmatrix} X_1 \\ X_2 \\ X_3 \\ X_4 \end{pmatrix}_{(3)} = \begin{pmatrix} 1 \\ -i \\ -54.7214\,i \\ -54.7214 \end{pmatrix}, \tag{vii}$$

$$\begin{pmatrix} X_1 \\ X_2 \\ X_3 \\ X_4 \end{pmatrix}_{(4)} = \begin{pmatrix} 1 \\ i \\ 54.7214i \\ -54.7214 \end{pmatrix}. \tag{viii}$$

Thus, the required eigenvectors are given by Equations (v)–(viii).

12.5 Controllability and Observability

These concepts are important in the design of MIMO feedback control systems. They were introduced by Kalman [2] in 1960. First, controllability is introduced generally in the following.

Controllability: *In order for the designer to do whatever s/he wants with the given dynamic system under control input, the system must be controllable.*

More specifically, controllability is defined as an ability to transfer the dynamic system from any initial state $y(t_0) = y(0) = y_0$ to any desired final state $y(t_f) = y_f$ in a finite time. Thus, the question to be answered is the following: Can one find a control input $r(t)$ that will transfer the dynamic system from the initial state to the final state?

For the purpose of investigating controllability of a m dof linear system, one can start by considering Equation (12.11a). With a scalar input, Equation (12.11a) is given by:

$$\frac{dy(t)}{dt} = Ay(t) + br(t), \quad y(t_0) = y(0) = y_0. \tag{12.17}$$

It may be appropriate to note that a time shift in the discrete time corresponds to a derivative in the continuous time, and therefore derivatives of Equation (12.17) are required in studying the concept of controllability.

Taking the derivative of this state equation with respect to time t, one has

$$\frac{d^2y(t)}{dt^2} = A\frac{dy(t)}{dt} + b\frac{dr(t)}{dt} \tag{12.18}$$

Substituting Equation (12.17) into the first term on the rhs of Equation (12.18) leads to

$$\frac{d^2y(t)}{dt^2} = A^2y(t) + Abr(t) + b\frac{dr(t)}{dt} \tag{12.19}$$

If one repeatedly performs the above steps up to the $n = 2m$ derivative and then substitutes the $(n-1)$ differentiated equation into the first term on the rhs of the n differentiated equation, one has

$$\begin{aligned} \frac{dy(t)}{dt} &= Ay(t) + br(t) \\ \frac{d^2y(t)}{dt^2} &= A^2y(t) + Abr(t) + b\frac{dr(t)}{dt} \\ &\vdots \\ \frac{d^ny(t)}{dt^n} &= A^ny(t) + A^{n-1}br(t) + A^{n-2}b\frac{dr(t)}{dt} + \cdots + b\frac{d^{n-1}r(t)}{dt^{n-1}}. \end{aligned} \tag{12.20}$$

Moving the first term on the rhs of this equation to the lhs gives

$$\frac{d^ny(t)}{dt^n} - A^ny(t) = A^{n-1}br(t) + A^{n-2}b\frac{dr(t)}{dt} + \cdots + b\frac{d^{n-1}r(t)}{dt^{n-1}}.$$

This equation can further be written as

$$\frac{d^ny(t)}{dt^n} - A^ny(t) = \begin{bmatrix} b & Ab & \dots & A^{n-1}b \end{bmatrix} \begin{pmatrix} \frac{d^{n-1}r(t)}{dt^{n-1}} \\ \frac{d^{n-2}r(t)}{dt^{n-2}} \\ \vdots \\ \frac{dr(t)}{dt} \\ r(t) \end{pmatrix}. \tag{12.21}$$

When the control input $r(t)$ is a vector, then Equation (12.21) becomes

$$\frac{d^n y(t)}{dt^n} - A^n y(t) = \begin{bmatrix} B & AB & \dots & A^{n-1}B \end{bmatrix} \begin{pmatrix} \frac{d^{n-1} r(t)}{dt^{n-1}} \\ \frac{d^{n-2} r(t)}{dt^{n-2}} \\ \vdots \\ \frac{dr(t)}{dt} \\ r(t) \end{pmatrix}. \quad (12.22)$$

Equations (12.21) and (12.22) are valid for any time t between t_0 and t_f, with t_f being finite. The matrix $\mathbb{C} = [B\ AB\ \dots\ A^{n-1}B]$ is a square matrix and is called the *controllability matrix*. If $\mathbb{C}$ is non-singular, Equation (12.22) provides the unique solution to the generalized input vector given by

$$\begin{pmatrix} \frac{d^{n-1} r(t)}{dt^{n-1}} \\ \frac{d^{n-2} r(t)}{dt^{n-2}} \\ \vdots \\ \frac{dr(t)}{dt} \\ r(t) \end{pmatrix} = \mathbb{C}^{-1} \left[\frac{d^n y(t)}{dt^n} - A^n y(t) \right]. \quad (12.23)$$

It follows that the controllability condition is equivalent to non-singularity of the controllability matrix $\mathbb{C}$. Thus, a solution of Equation (11.23) exists if and only if (iff) the rank of $\mathbb{C}$ is equal to n.

Theorem 12.1 *The linear dynamic system is controllable iff the controllability matrix $\mathbb{C} = [B\ AB\ \dots\ A^{n-1}B]$ has full rank. That is, rank of $\mathbb{C}$ is equal to n.*

Another path to derive the controllability matrix $\mathbb{C}$ is to make use of the concept of controllability, which is an ability to transfer the dynamic system from any initial state $y(t_0) = y(0) = y_0$ to any desired final state $y(t_f) = y_f$ in a finite time. Thus, starting with Equations (12.16b) and (12.10), one has

$$y(t) = e^{A(t-t_0)} y(t_0) + \int_{t_0}^{t} e^{A(t-\tau)} Br(\tau) d\tau.$$

At the final time t_f the above equation becomes

$$y(t_f) = e^{A(t_f - t_0)} y(t_0) + \int_{t_0}^{t_f} e^{A(t_f - \tau)} Br(\tau) d\tau.$$

Pre-multiplying both sides by e^{-At_f} and re-arranging, one obtains

$$e^{-At_f}y(t_f) - e^{-At_0}y(t_0) = \int_{t_0}^{t_f} e^{-A\tau}Br(\tau)d\tau. \tag{12.24a}$$

By Maclaurin series in Appendix 12B,

$$e^{-A\tau} = I - A\tau + \frac{1}{2!}A^2\tau^2 - \frac{1}{3!}A^3\tau^3 + \cdots$$

Substituting this into the last equation, one has

$$\int_{t_0}^{t_f} e^{-A\tau}Br(\tau)d\tau = \int_{t_0}^{t_f}\left(I - A\tau + \frac{1}{2!}A^2\tau^2 - \frac{1}{3!}A^3\tau^3 + \cdots\right)Br(\tau)d\tau$$

$$= \left[B \;\; AB \;\; \ldots \;\; A^{n-1}B\right]\begin{pmatrix} \int_{t_0}^{t_f} r(\tau)d\tau \\ \int_{t_0}^{t_f} -\tau r(\tau)d\tau \\ \vdots \\ \int_{t_0}^{t_f} \frac{(-\tau)^{n-1}}{(n-1)!} r(\tau)d\tau \end{pmatrix}.$$

Writing in a more concise form,

$$\int_{t_0}^{t_f} e^{-A\tau}Br(\tau)d\tau = \mathbb{C}\begin{pmatrix} f_1[r(\tau)] \\ f_2[r(\tau)] \\ \vdots \\ f_{n-1}[r(\tau)] \end{pmatrix},$$

where $f_{n-1}[r(\tau)] = \int_{t_0}^{t_f} \frac{(-\tau)^{n-1}}{(n-1)!} r(\tau)d\tau$. Therefore, by making use of Equation (12.24a), it can be written as

$$\begin{pmatrix} f_1[r(\tau)] \\ f_2[r(\tau)] \\ \vdots \\ f_{n-1}[r(\tau)] \end{pmatrix} = \mathbb{C}^{-1}\left[e^{-At_f}y(t_f) - e^{-At_0}y(t_0)\right]. \tag{12.24b}$$

A solution of this equation exists iff the rank of $\mathbb{C}$ is equal to n.

Having dealt with the concept of controllability, now the concept of observability is presented below.

Observability: *In order to investigate the inside of a dynamic system under observation, the system must be observable.*

In order to investigate the observability of a dynamic m dof linear system, one can consider an input-free or unperturbed system

$$\frac{dy(t)}{dt} = Ay(t), \quad y(t_0) = y(0) = y_0 \tag{12.25a}$$

where y_0 is unknown.

With the corresponding outputs or measurements,

$$c(t) = \tilde{C}y(t). \tag{12.25b}$$

Note that the dimensions of the matrices and vectors are those in Equation (12.11). Following the same reasoning as in the study of controllability in the foregoing, one can conclude that the knowledge of y_0 is sufficient to determine $y(t)$ at any time t, since from Equation (12.25a) the solution of the state vector $y(t)$, by applying Equations (12.5) and (12.10), is given by

$$y(t) = e^{A(t-t_0)}y(t_0). \tag{12.26}$$

The problem one encounters is to find $y(t_0)$ from the available measurements or outputs in Equation (12.25b). Similar to the steps for the controllability of a system, one takes the derivatives of the measurements in Equation (12.25b) such that

$$\begin{aligned}
c(t_0) &= \tilde{C}y(t_0). \\
\frac{dc(t_0)}{dt} &= \tilde{C}\frac{dy(t_0)}{dt} = \tilde{C}Ay(t_0) \\
\frac{d^2c(t_0)}{dt^2} &= \tilde{C}\frac{d^2y(t_0)}{dt^2} = \tilde{C}A^2y(t_0) \\
&\vdots \\
\frac{d^{n-1}c(t_0)}{dt^{n-1}} &= \tilde{C}\frac{d^{n-1}y(t_0)}{dt^{n-1}} = \tilde{C}A^{n-1}y(t_0).
\end{aligned} \tag{12.27}$$

Equation (12.27) can be cast in matrix form as

$$\begin{pmatrix} c(t_0) \\ \frac{dc(t_0)}{dt} \\ \frac{d^2c(t_0)}{dt^2} \\ \vdots \\ \frac{d^{n-1}c(t_0)}{dt^{n-1}} \end{pmatrix} = \begin{pmatrix} \tilde{C} \\ \tilde{C}A \\ \tilde{C}A^2 \\ \vdots \\ \tilde{C}A^{n-1} \end{pmatrix} y(t_0) = \tilde{O}y(t_0), \tag{12.28}$$

where $\tilde{O} = \begin{bmatrix} \tilde{C} & \tilde{C}A & \tilde{C}A^2 & \dots & \tilde{C}A^{n-1} \end{bmatrix}^T$ is the observability matrix. Therefore, the initial condition $y(t_0)$ can be uniquely obtained from Equation (12.28) iff the observability matrix has full rank. That is, rank of $\tilde{O}$ is equal to n.

Theorem 12.2 *The linear dynamic system with output or measurements is observable iff the observability matrix $\tilde{O}$ has full rank.*

Example

The state and output equations of a system are given by

$$\begin{pmatrix} \dot{x}_1 \\ \dot{x}_2 \\ \dot{x}_3 \end{pmatrix} = \begin{bmatrix} 0 & -1 & 0 \\ 1 & 0 & 1 \\ 0 & -2 & -3 \end{bmatrix} \begin{pmatrix} x_1 \\ x_2 \\ x_3 \end{pmatrix} + \begin{bmatrix} 1 \\ 0 \\ 0 \end{bmatrix} r(t), \text{ and}$$

$$c(t) = [1 \ 0 \ 0] \begin{pmatrix} x_1 \\ x_2 \\ x_3 \end{pmatrix}.$$

Determine whether or not this system is controllable and/or observable.

Solution:

Consider the controllability condition. One applies

$$B = \begin{bmatrix} 1 \\ 0 \\ 0 \end{bmatrix}, \quad AB = \begin{bmatrix} 0 & -1 & 0 \\ 1 & 0 & 1 \\ 0 & -2 & -3 \end{bmatrix} \begin{bmatrix} 1 \\ 0 \\ 0 \end{bmatrix} = \begin{bmatrix} 0 \\ 1 \\ 0 \end{bmatrix},$$

$$A^2B = \begin{bmatrix} -1 & 0 & -1 \\ 0 & -3 & -3 \\ -2 & 6 & 7 \end{bmatrix} \begin{bmatrix} 1 \\ 0 \\ 0 \end{bmatrix} = \begin{bmatrix} -1 \\ 0 \\ -2 \end{bmatrix}.$$

Therefore, the controllability matrix becomes

$$\mathbb{C} = \begin{bmatrix} B & AB & \ldots & A^{n-1}B \end{bmatrix} = \begin{bmatrix} 1 & 0 & -1 \\ 0 & 1 & 0 \\ 0 & 0 & -2 \end{bmatrix}.$$

The determinant of the controllability matrix is $\det \mathbb{C} = |\mathbb{C}| = -2 \neq 0$. Thus, $\mathbb{C}$ has a rank of 3 such that the system is controllable.

Now, consider the observability matrix. One has

$$\widetilde{C} = [1 \ 0 \ 0], \quad \widetilde{C}A = [1 \ 0 \ 0] \begin{bmatrix} 0 & -1 & 0 \\ 1 & 0 & 1 \\ 0 & -2 & -3 \end{bmatrix} = [0 \ -1 \ 0],$$

$$\widetilde{C}A^2 = [1 \ 0 \ 0] \begin{bmatrix} -1 & 0 & -1 \\ 0 & -3 & -3 \\ -2 & 6 & 7 \end{bmatrix} = [-1 \ 0 \ -1].$$

Therefore, the observability matrix becomes

$$\widetilde{O}=\begin{pmatrix}\widetilde{C}\\ \widetilde{C}A\\ \widetilde{C}A^2\end{pmatrix}=\begin{bmatrix}1 & 0 & 0\\ 0 & -1 & 0\\ -1 & 0 & -1\end{bmatrix}.$$

The determinant of the observability matrix is det $\widetilde{O}=\left|\widetilde{O}\right|=1\neq 0$. Thus, $\tilde{O}$ has a rank of 3 such that the system is observable.

12.6 Stabilizability and Detectability

In the last section the concepts of controllability and observability have been introduced and studied. It is shown that the system is controllable (observable) if all elements of the state vector are controllable (observable). Naturally, the logical question to be asked is the following: does one really need to control and observe all state variables? In some situations, only the unstable components of the state vector have to be considered. This leads to the concepts of stabilizability and detectability. These latter concepts play very important roles in optimal control theory. However, they are beyond the scope of the present book and will not be studied in detail except to provide their definitions and meanings in the following.

Stabilizability: *A linear dynamic system is stabilizable if all unstable modes are controllable, or if it is uncontrollable but the uncontrollable modes are stable.*

Thus, if a system is unstable due to the fact that its eigenvalues contain positive real parts in the s-plane, and ***Theorem*** 12.1 shows that it is uncontrollable, the unstable modes can be checked one at a time to determine whether the system is stabilizable.

Detectability: *A linear dynamic system is detectable if all unstable modes are observable.*

12.7 Applications of MATLAB

While the responses of the systems considered in this section are confined to those under inputs that are available in MATLAB [1], they are selected to illustrate the power and ease in using this computational tool. The latter can also be applied to verify the analytical results for a particular system.

Example 1

Consider a system having the following state and output equations

$$\begin{pmatrix}\dot{x}_1\\ \dot{x}_2\end{pmatrix}=\begin{bmatrix}0 & 1\\ -\dfrac{4}{3} & -\dfrac{5}{3}\end{bmatrix}\begin{pmatrix}x_1\\ x_2\end{pmatrix}+\begin{bmatrix}1 & 0\\ 0 & 1\end{bmatrix}\begin{pmatrix}r_1\\ r_2\end{pmatrix},\quad \text{and}$$

$$c(t)=\begin{bmatrix}1 & 0\\ 0 & 1\end{bmatrix}\begin{pmatrix}x_1\\ x_2\end{pmatrix}+\begin{bmatrix}0 & 0\\ 0 & 0\end{bmatrix}\begin{pmatrix}r_1\\ r_2\end{pmatrix},$$

in which $r_1=1, r_2=1,\ x_1=x,\ x_2=\dot{x}$. Obtain the responses of this system by applying MATLAB.

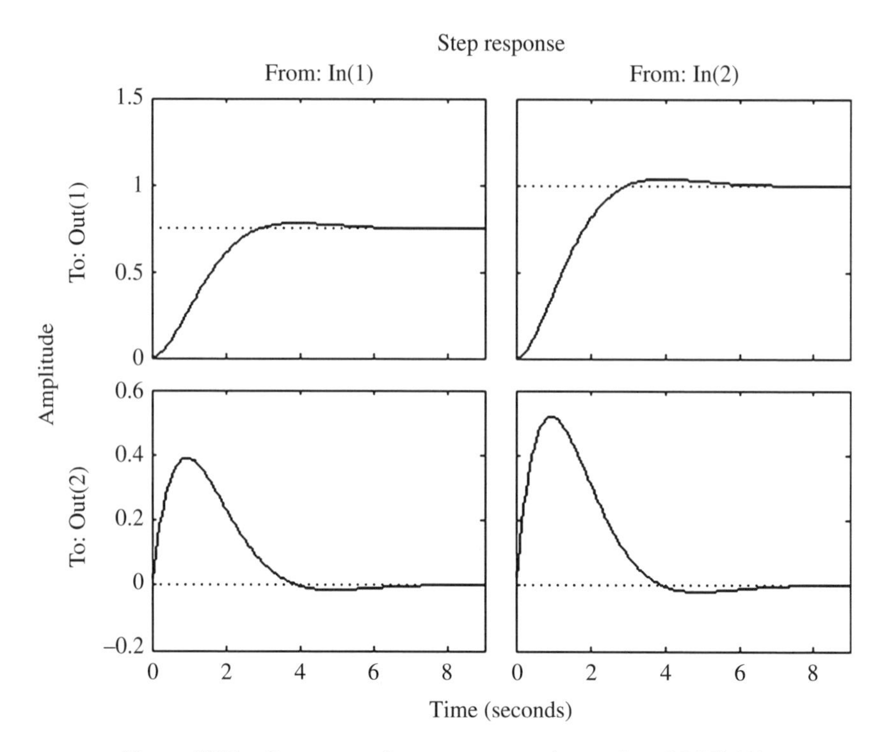

Figure 12E1 Responses of a system to step inputs from MATLAB

Solution:

The input to and output from MATLAB are presented in Program Listing 12.1 and Figure 12E1, respectively. It should be noted that the output coefficient matrix C in the program listing is $\widetilde{C}$ in the given output equation. Henceforth, this applies to the remaining part of this chapter.

Furthermore, the system has two inputs, $r_1 = 1$ and $r_2 = 1$, and two outputs, x_1 and x_2. The results from MATLAB are those with $r_1 = 1$ input into the system when $r_2 = 0$, and with $r_2 = 1$ input into the system when $r_1 = 0$. With reference to Figure 12E1, the top left-hand plot, for example, is the response x_1 due to the input $r_1 = 1$, while the top right-hand plot is the response x_2 due to the input $r_1 = 1$.

Program Listing 12.1

```
>> A = [0  1;-1.3333  -1.6667];
>> B = [0  0;1  1.3333];
>> C = [1  0;0  1];
>> D = [0  0;0  0];
>> step(A,B,C,D)
```

Example 2

A system has the following state and output equations

$$\begin{pmatrix} \dot{x}_1 \\ \dot{x}_2 \end{pmatrix} = \begin{bmatrix} -1 & -1 \\ 9 & 0 \end{bmatrix} \begin{pmatrix} x_1 \\ x_2 \end{pmatrix} + \begin{bmatrix} 1 & 0 \\ 0 & 1 \end{bmatrix} \begin{pmatrix} r_1 \\ r_2 \end{pmatrix}, \text{ and}$$

$$c(t) = \begin{bmatrix} 1 & 0 \\ 0 & 1 \end{bmatrix} \begin{pmatrix} x_1 \\ x_2 \end{pmatrix} + \begin{bmatrix} 0 & 0 \\ 0 & 0 \end{bmatrix} \begin{pmatrix} r_1 \\ r_2 \end{pmatrix},$$

in which $r_1 = 1$, $r_2 = 1$, $x_1 = x$, $x_2 = \dot{x}$.

a. Obtain the responses of the system by applying MATLAB.
b. Instead of the step inputs, obtain the responses of the system due to impulses.
c. Find the free vibration responses of the system under initial conditions

$$x_1(0) = 5, \quad x_2(0) = 1.$$

Solution:

a. The input to and output from MATLAB are presented in Program Listing 12.2a and Figure 12E2, respectively.

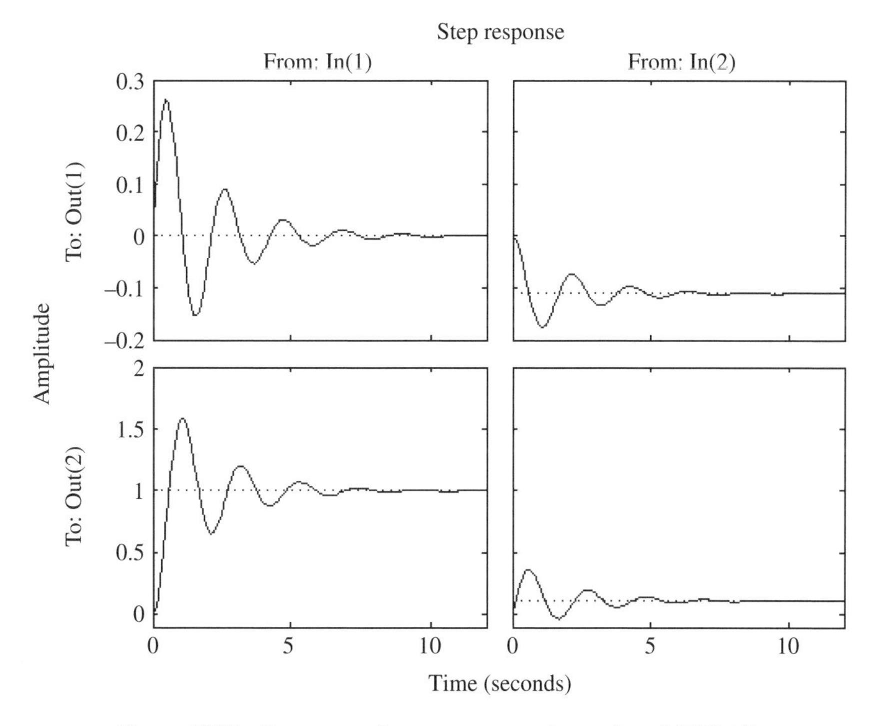

Figure 12E2 Responses of a system to step inputs from MATLAB

Furthermore, the results from MATLAB are those with $r_1 = 1$ input into the system when $r_2 = 0$, and with $r_2 = 1$ input into the system when $r_1 = 0$.

Program Listing 12.2a

```
>> A = [-1  -1;9  0];
>> B = [1  0;0  1];
>> C = [1  0;0  1];
>> D = [0  0;0  0];
>> step(A,B,C,D)
```

b. The input to MATLAB is identical to that in Program Listing 12.2a except that the last statement is changed. For completeness, it is included in Program Listing 12.2b, while the output from MATLAB is presented in Figure 12E3.

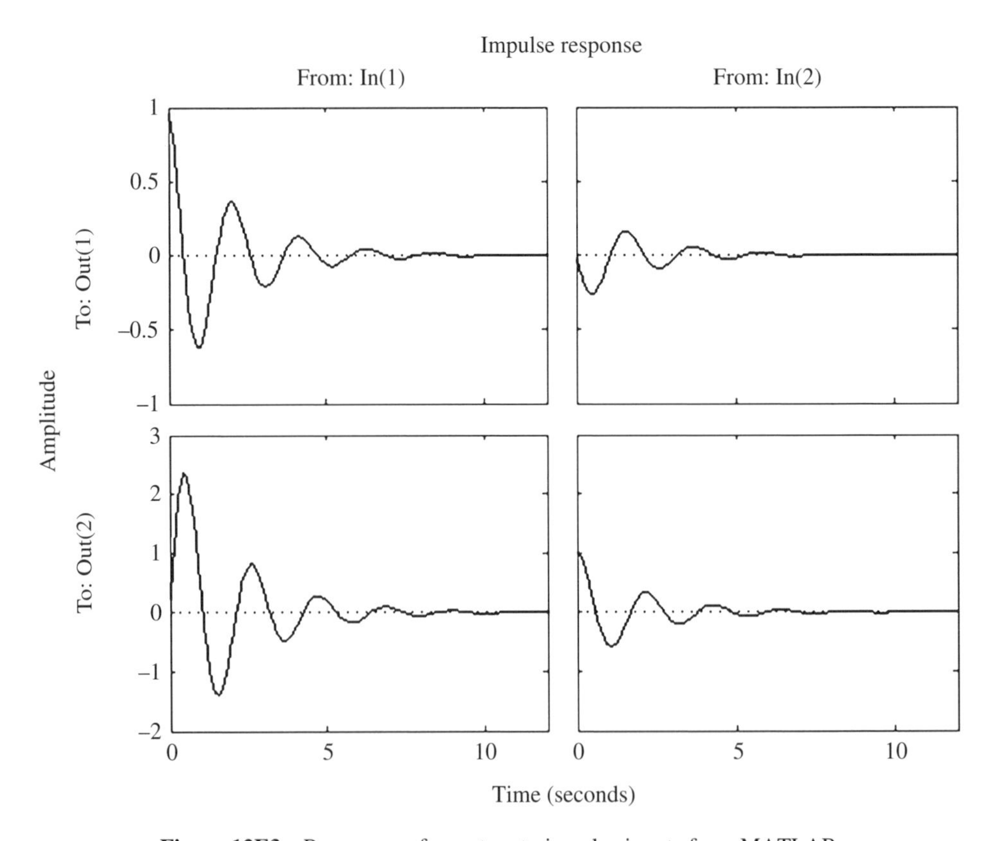

Figure 12E3 Responses of a system to impulse inputs from MATLAB

Program Listing 12.2b

```
>> A = [-1 -1;9 0];
>> B = [1 0;0 1];
>> C = [1 0;0 1];
>> D = [0 0;0 0];
>> impulse(A,B,C,D)
```

c. For this part of the question, the input to and output from MATLAB are presented in Program Listing 12.2c and Figure 12E4, respectively.

Program Listing 12.2c

```
>> A = [-1 -1;9 0];
>> B = [1 0;0 1];
>> C = [1 0;0 1];
>> D = [0 0;0 0];
>> x0 = [5 ; 1];
>> sys = ss (A , B , C , D);
>> initial (sys, x0,'k');
```

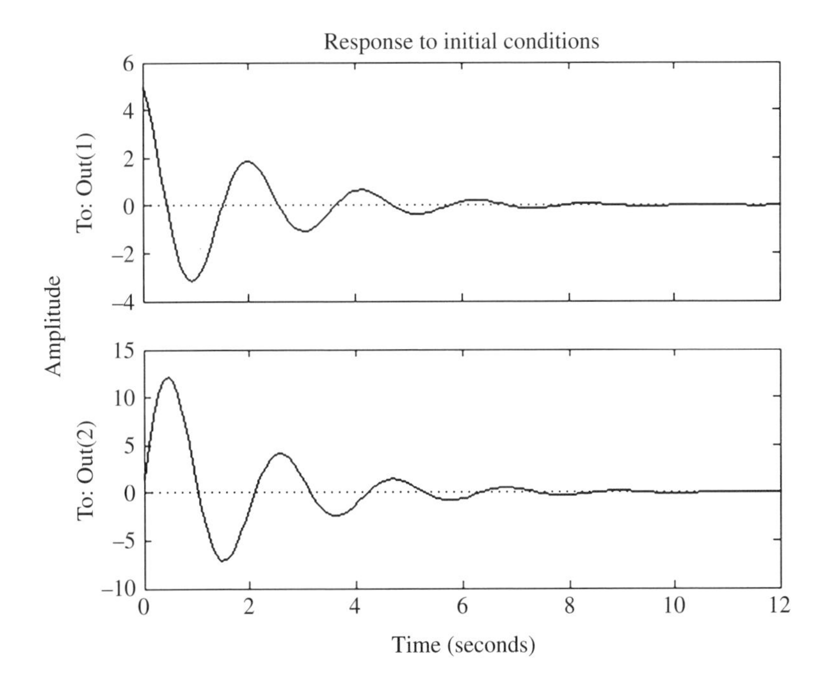

Figure 12E4 Free response of non-zero initial conditions

Example 3
A system has the following state and output equations

$$\begin{pmatrix} \dot{x}_1 \\ \dot{x}_2 \end{pmatrix} = \begin{bmatrix} -1 & -1 \\ 9 & 0 \end{bmatrix} \begin{pmatrix} x_1 \\ x_2 \end{pmatrix} + \begin{pmatrix} 1 \\ 1 \end{pmatrix} r(t), \text{ and}$$

$$c(t) = \begin{bmatrix} 1 & 0 \\ 0 & 1 \end{bmatrix} \begin{pmatrix} x_1 \\ x_2 \end{pmatrix} + \begin{pmatrix} 0 \\ 0 \end{pmatrix} r(t)$$

in which $x_1 = x, \ x_2 = \dot{x}$.

Determine the responses of the system under inputs $r(t) = 0.02 \sin(4t)$, and initial conditions $x_1(0) = 0$, $x_2(0) = 10^{-3}$.

Solution:
The input to MATLAB is included in Program Listing 12.3, while the output from MATLAB is presented in Figure 12E5.

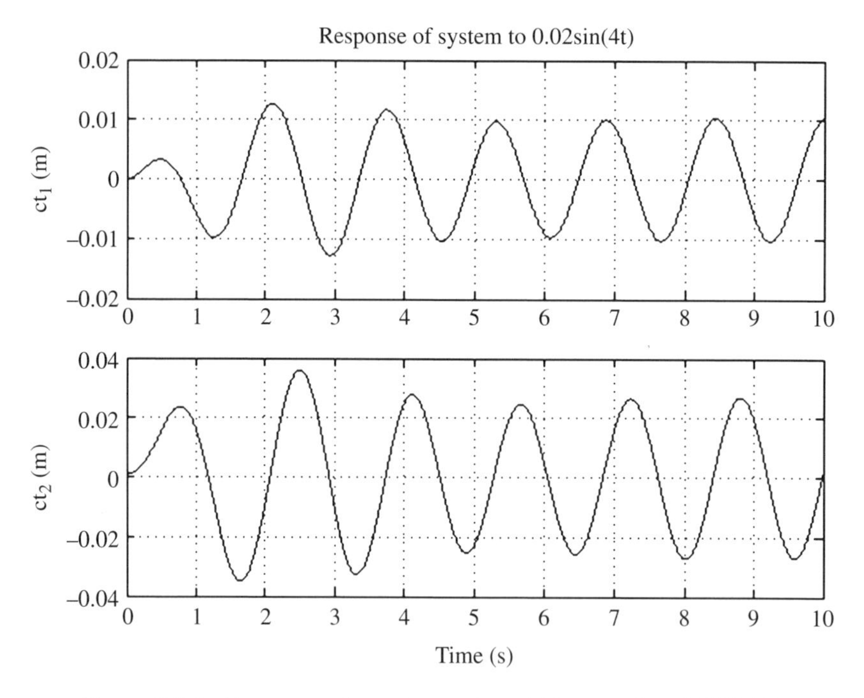

Figure 12E5 Responses of a system to sinusoidal inputs from MATLAB

Program Listing 12.3

```
>> t = 0:0.0001:10;
>> A = [-1  -1;9  0];
>> B = [1; 1];
>> C = [1 0; 0 1];
>> D = [0;0];
>> r = 0.02*sin(4*t);
>> sys = ss(A,B,C,D);
>> %
>> % individual response cf to inputs
>> %
>> [cf , t , x] = lsim(sys, r , t);
>> %
>> % individual response cic to initial conditions
>> %
>> x0 = [0 ; 0.001];
>> [cic, t , x] = initial (sys, x0, t);
>> %
>> % total response ct as superposition of cf and cic
>> %
>> ct = cf + cic;
>> ct1= ct;
>> %
>> ct1(: , 2) = [ ];  % deletes the second column and keeps
>> % the first column of ct1
>> %
>> ct2 = ct;
>> %
>> ct2(: , 1) = [ ];  % deletes the first column and keeps
>> % the second column of ct2
>> %
>> subplot(211);
>> plot(t , ct1,'k');
>> ylabel ( 'ct_1 (m)')
>> grid on
>> subplot(212);
>> plot(t , ct2,'k');
>> ylabel ('ct_2 (m)')
>> xlabel ('Time (s)')
>> grid on
```

Example 4

Consider the 2-dof system in the second example in Section 12.4. Now, the equation of motion for a forced rotating rotor is

$$M\ddot{x} + \underline{C}\dot{x} + Kx = F$$

in which the mass, damping, stiffness matrices, and displacement vector are given, respectively as

$$M = \begin{bmatrix} m & 0 \\ 0 & m \end{bmatrix}, \quad \underline{C} = \begin{bmatrix} 0 & -2m\Omega \\ 2m\Omega & 0 \end{bmatrix}, \quad K = \begin{bmatrix} k - m\Omega^2 & 0 \\ 0 & k - m\Omega^2 \end{bmatrix},$$

and $x = [x_1 \;\; x_2]^T$. The force vector F is defined as $F = [u(t) \;\; 0]^T$ where $u(t) = r(t)$. Meanwhile, the coefficient matrix A was obtained in Section 12.4 as

$$A = \begin{bmatrix} 0 & I \\ -M^{-1}K & -M^{-1}\underline{C} \end{bmatrix} = \begin{bmatrix} 0 & 0 & 1 & 0 \\ 0 & 0 & 0 & 1 \\ -1900 & 0 & 0 & 20 \\ 0 & -1900 & -20 & 0 \end{bmatrix}.$$

The force vector pre-multiplied by the inverse of the mass matrix becomes

$$M^{-1}F = \begin{bmatrix} 0.1 & 0 \\ 0 & 0.1 \end{bmatrix} \begin{pmatrix} u(t) \\ 0 \end{pmatrix} = \begin{pmatrix} 1 \\ 0 \end{pmatrix} r(t).$$

This system is excited by a step function so that its state and output equations are

$$\begin{pmatrix} \dot{y}_1 \\ \dot{y}_2 \\ \dot{y}_3 \\ \dot{y}_4 \end{pmatrix} = \begin{bmatrix} 0 & 0 & 1 & 0 \\ 0 & 0 & 0 & 1 \\ -1900 & 0 & 0 & 20 \\ 0 & -1900 & -20 & 0 \end{bmatrix} \begin{pmatrix} y_1 \\ y_2 \\ y_3 \\ y_4 \end{pmatrix} + \begin{pmatrix} 0 \\ 0 \\ 1 \\ 0 \end{pmatrix} r(t) \tag{i}$$

$$\begin{pmatrix} c_1 \\ c_2 \end{pmatrix} = \begin{bmatrix} 1 & 0 & 0 & 0 \\ 0 & 1 & 0 & 0 \end{bmatrix} \begin{pmatrix} y_1 \\ y_2 \\ y_3 \\ y_4 \end{pmatrix} + \begin{pmatrix} 0 \\ 0 \end{pmatrix} r(t) \tag{ii}$$

in which $r(t) = 1$, and the displacement vector is identified as

$$\begin{pmatrix} y_1 \\ y_2 \\ y_3 \\ y_4 \end{pmatrix} = \begin{pmatrix} x_1 \\ x_2 \\ \dot{x}_1 \\ \dot{x}_2 \end{pmatrix}. \tag{iii}$$

By making use of MATLAB and assuming the system starts from rest such that

$$\begin{pmatrix} y_1(0) \\ y_2(0) \\ y_3(0) \\ y_4(0) \end{pmatrix} = \begin{pmatrix} 0 \\ 0 \\ 0 \\ 0 \end{pmatrix},$$

plot the outputs c_1 and c_2.

Solution:
The input to and output from MATLAB are presented in Program Listing 12.4 and Figure 12E6, respectively.

Program Listing 12.4

```
>> t = 0:0.0001:1;
>> A =[0 0 1 0; 0 0 0 1;-1900 0 0 20;0 -1900 -20 0];
>> B = [0;0;1;0];
>> C = [1 0 0 0;0 1 0 0];
>> D = [0;0];
>> sys = ss(A,B,C,D);
>> %
>> % individual response cf to inputs
>> %
>> [cf , t , x] = step(sys,t);
>> %
>> % individual response cic to initial conditions
>> %
>> x0 = [0;0;0;0];
>> [cic, t , x] = initial (sys, x0, t);
>> %
>> % total response ct as superposition of cf and cic
>> %
>> ct = cf + cic;
>> ct1= ct;
>> ct1(: , 2) = [ ];  % deletes the second column and keeps
>> % the first column of ct1 = ct
>> %
>> ct2 = ct;
>> ct2(: , 1) = [ ];  % deletes the first column and keeps
>> % the second column of ct2 = ct
>> %
>> subplot(211);
>> plot(t , ct1,'k');
>> title('Responses of rotating rotor to unit step input');
>> ylabel ( 'ct_1 (m)')
>> grid on
>> subplot(212);
>> plot(t , ct2,'k');
>> ylabel ('ct_2 (m)')
>> xlabel ('Time (s)')
>> grid on
```

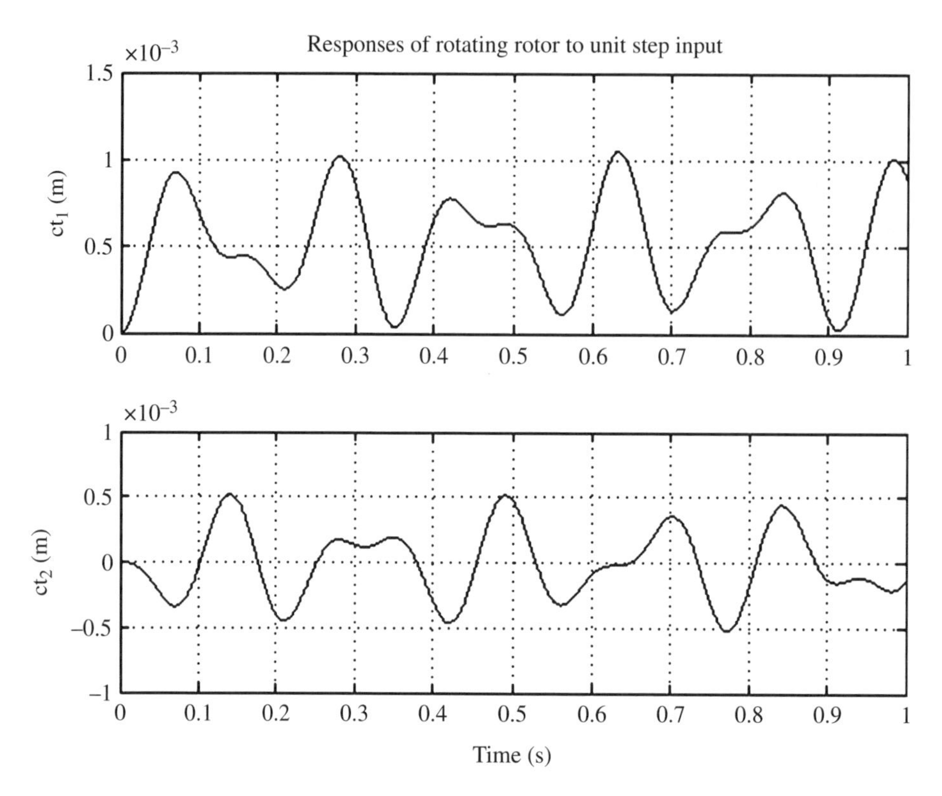

Figure 12E6 Responses of forced rotating rotor to unit step input

Appendix 12A: Solution of System of First-Order Differential Equations

A system has the following state equation

$$\begin{pmatrix} \dot{x}_1 \\ \dot{x}_2 \\ \dot{x}_3 \end{pmatrix} = \begin{bmatrix} -2 & 2 & 2 \\ 0 & -1 & 0 \\ -2 & 4 & 3 \end{bmatrix} \begin{pmatrix} x_1 \\ x_2 \\ x_3 \end{pmatrix} + \begin{pmatrix} \sin t \\ 3 \\ 0 \end{pmatrix}$$

which may be written more concisely as

$$\dot{y} = Ay + g, \tag{A1}$$

where the coefficient or amplification matrix A and vector g are given by

$$A = \begin{bmatrix} -2 & 2 & 2 \\ 0 & -1 & 0 \\ -2 & 4 & 3 \end{bmatrix}, \text{ and } g = \begin{pmatrix} \sin t \\ 3 \\ 0 \end{pmatrix}.$$

a. Solve the eigenvalue problem of this system of equations, and

b. solve for $\begin{pmatrix} x_1 \\ x_2 \\ x_3 \end{pmatrix}$ by the method of diagonalization if the initial conditions are

$$[x_1(0)\ \ x_2(0)\ \ x_3(0)]^T = [0\ \ 0\ \ 0]^T.$$

Solution:

The method of diagonalization for responses of systems of first-order differential equations consists of three steps as follows. Before the first step, one may write

$$y = Rz,$$

where R is known as the transformation matrix whose columns, as to be shown later, are the eigenvectors of the coefficient matrix A, and

$$\widetilde{D} = R^{-1}AR,$$

a diagonal matrix whose diagonal elements are the eigenvalues of A. The diagonal matrix $\widetilde{D}$ should not be confused with the output or measurement matrices in Equations (12.11b) and (12.25b).

Step 1: Eigenvalue problem of A

Since $A = \begin{bmatrix} -2 & 2 & 2 \\ 0 & -1 & 0 \\ -2 & 4 & 3 \end{bmatrix}$, the eigenvector $\boldsymbol{X}$ is given by the equation

$$(A - \lambda I)X = 0.$$

Thus, the frequency equation becomes

$$|A - \lambda I| = \begin{vmatrix} -2-\lambda & 2 & 2 \\ 0 & -1-\lambda & 0 \\ -2 & 4 & 3-\lambda \end{vmatrix} = 0.$$

Operating on the determinant, one has

$$(-2-\lambda)(-1-\lambda)(3-\lambda) - 4(1+\lambda) = 0.$$

Expanding, one has

$$\lambda^3 - 3\lambda - 2 = 0$$

which can be factored as $(\lambda + 1)^2(\lambda - 2) = 0$.

The eigenvalues are $\lambda_1 = \lambda_2 = -1$, $\lambda_2 = 2$.

Note that the first two eigenvalues are identical (repeated eigenvalues). To obtain the eigenvectors of A, one applies

$$(A-\lambda I)X=0 \tag{A2}$$

where $X=\begin{bmatrix}x_1 & x_2 & x_3\end{bmatrix}^T$.

When $\lambda = \lambda_1 = -1$, substitute this eigenvalue into Equation (A2) to give

$$\begin{aligned} -x_1+2x_2+2x_3&=0 \\ 0&=0 \\ -2x_1+4x_2+4x_3&=0. \end{aligned} \tag{A3}$$

Since the eigenvalues $\lambda_1=\lambda_2=-1$ (that is, an eigenvalue of multiplicity 2) and there is only one equation for the three unknowns, therefore, one assumes $x_1 = 0$ and $x_2 = 1$. Substituting these values into the last or first equation above, one has $x_3 = -1$. Therefore, the first eigenvector becomes

$$\begin{pmatrix} x_1 \\ x_2 \\ x_3 \end{pmatrix}_1 = \begin{pmatrix} 0 \\ 1 \\ -1 \end{pmatrix},$$

where the subscript of the eigenvector denotes its association with the first eigenvalue.

Since the first eigenvalue is of multiplicity 2, one therefore has to assume a set of new values for x_1 and x_2. To this end, one can let $x_1 = 1$ and $x_2 = 0$, so that upon substituting these values into Equation (A3) one obtains $x_3 = ½$. Therefore, the second eigenvector becomes

$$\begin{pmatrix} x_1 \\ x_2 \\ x_3 \end{pmatrix}_2 = \begin{pmatrix} 1 \\ 0 \\ 1/2 \end{pmatrix}.$$

When $\lambda = \lambda_3 = 2$, substitute this eigenvalue into Equation (A2) to give

$$\begin{aligned} -4x_1+2x_2+2x_3&=0 \\ 3x_2&=0 \\ -2x_1+4x_2+x_3&=0. \end{aligned} \tag{A4}$$

Clearly, the second equation above gives $x_2 = 0$, and one can assume $x_1 = 1$, so that upon substituting $x_1 = 1$ and $x_2 = 0$ into Equation (A4), one finds $x_3 = 2$. Thus, the third eigenvector becomes

$$\begin{pmatrix} x_1 \\ x_2 \\ x_3 \end{pmatrix}_3 = \begin{pmatrix} 1 \\ 0 \\ 2 \end{pmatrix}.$$

Therefore, the transformation matrix

$$R=\begin{bmatrix} 0 & 1 & 1 \\ 1 & 0 & 0 \\ -1 & 1/2 & 2 \end{bmatrix}. \tag{A5}$$

Step 2: Diagonalization of A

The diagonal matrix $\widetilde{D}$ can be constructed by applying the fact that its diagonal elements are the eigenvalues; that is,

$$\widetilde{D}=\begin{bmatrix} -1 & 0 & 0 \\ 0 & -1 & 0 \\ 0 & 0 & 2 \end{bmatrix}. \tag{A6}$$

One can verify Equation (A6) by performing the following matrix operation, since

$$\widetilde{D}=R^{-1}AR=\begin{bmatrix} 0 & 1 & 0 \\ 4/3 & -2/3 & -2/3 \\ -1/3 & 2/3 & 2/3 \end{bmatrix}\begin{bmatrix} -2 & 2 & 2 \\ 0 & -1 & 0 \\ -2 & 4 & 3 \end{bmatrix}\begin{bmatrix} 0 & 1 & 1 \\ 1 & 0 & 0 \\ -1 & 1/2 & 2 \end{bmatrix}.$$

Upon performing the matrix multiplication, one has

$$\widetilde{D}=\begin{bmatrix} -1 & 0 & 0 \\ 0 & -1 & 0 \\ 0 & 0 & 2 \end{bmatrix}.$$

Thus, Equation (A6) is verified.

Step 3: Solution of state equation

With the transformation matrix R one can write Equation (A1) as

$$R\dot{z}=ARz+g, \tag{A7}$$

Pre-multiplying this equation by the inverse of the transformation matrix R,

$$\begin{gathered} R^{-1}R\dot{z}=R^{-1}ARz+R^{-1}g, \\ \text{or}\quad \dot{z}=\widetilde{D}z+h, \end{gathered} \tag{A8}$$

in which

$$h=R^{-1}g=\begin{bmatrix} 0 & 1 & 0 \\ \frac{4}{3} & -\frac{2}{3} & -\frac{2}{3} \\ -\frac{1}{3} & \frac{2}{3} & \frac{2}{3} \end{bmatrix}\begin{pmatrix} \sin t \\ 3 \\ 0 \end{pmatrix}=\begin{pmatrix} 3 \\ \frac{4}{3}\sin t-2 \\ -\frac{1}{3}\sin t+2 \end{pmatrix}. \tag{A9}$$

Therefore, upon integrating Equation (A8) and remembering that the initial conditions for y in Equation (A1) are not the same as for z in Equation (A8), one obtains

$$\begin{aligned} z_1 &= a_1 e^{\lambda_1 t} + e^{\lambda_1 t}\int_0^t e^{-\lambda_1 \tau} h_1(\tau)\mathrm{d}\tau \\ &= a_1 e^{-t} + e^{-t}\int_0^t e^{\tau} 3\mathrm{d}\tau = a_1 e^{-t} + 3. \end{aligned} \tag{A10}$$

Similarly,

$$\begin{aligned} z_2 &= a_2 e^{\lambda_2 t} + e^{\lambda_2 t}\int_0^t e^{-\lambda_2 \tau} h_2(\tau)\mathrm{d}\tau \\ &= a_2 e^{-t} + e^{-t}\int_0^t e^{\tau}\left(\frac{4}{3}\sin\tau - 2\right)\mathrm{d}\tau \\ &= a_2 e^{-t} + \frac{4}{3}e^{-t}\int_0^t e^{\tau}\left(\sin\tau - \frac{3}{2}\right)\mathrm{d}\tau \\ &= a_2 e^{-t} + \left(\frac{2}{3}\right)(\sin t - \cos t) - 2. \end{aligned} \tag{A11}$$

Also,

$$\begin{aligned} z_3 &= a_3 e^{\lambda_3 t} + e^{\lambda_3 t}\int_0^t e^{-\lambda_3 \tau} h_3(\tau)\mathrm{d}\tau \\ &= a_3 e^{2t} + e^{2t}\int_0^t e^{-2\tau}\left(-\frac{1}{3}\sin\tau + 2\right)\mathrm{d}\tau \\ &= a_3 e^{2t} + \left(\frac{2}{15}\right)\sin t + \left(\frac{1}{15}\right)\cos t - 1. \end{aligned} \tag{A12}$$

In the foregoing, a_1, a_2 and a_3 are the arbitrary constants that have to be determined by applying the initial conditions in y.

Transforming back to the original coordinate system,

$$y = \begin{pmatrix} x_1 \\ x_2 \\ x_3 \end{pmatrix} = Rz = \begin{bmatrix} 0 & 1 & 1 \\ 1 & 0 & 0 \\ -1 & 1/2 & 2 \end{bmatrix}\begin{pmatrix} z_1 \\ z_2 \\ z_3 \end{pmatrix} = \begin{pmatrix} z_2 + z_3 \\ z_1 \\ -z_1 + \frac{z_2}{2} + 2z_3 \end{pmatrix}.$$

Substituting for Equations (12.10)–(12.12), one has

$$\begin{pmatrix} x_1 \\ x_2 \\ x_3 \end{pmatrix} = \begin{pmatrix} a_2 e^{-t} + a_3 e^{2t} + \frac{4}{5}\sin t - \frac{3}{5}\cos t - 3 \\ a_1 e^{-t} + 3 \\ -a_1 e^{-t} + \frac{a_2}{2}e^{-t} + a_3 e^{2t} + \frac{3}{5}\sin t - \frac{1}{5}\cos t - 6 \end{pmatrix}. \tag{A13}$$

Now, one applies the initial conditions $y(0)=[0\ \ 0\ \ 0]^T$ so that the second equation in Equation (A13) gives

$$x_2(0)=0=a_1+3 \text{ giving } a_1=-3. \tag{A14a}$$

$$\text{For } x_1(0)=0=a_2+a_3-\frac{3}{5}-3,\ \ a_2+a_3=\frac{18}{5}. \tag{A14b}$$

$$\text{For } x_3(0)=0=-a_1+\frac{a_2}{2}+2a_3-\frac{31}{5},\ \ a_2+4a_3=\frac{32}{5}. \tag{A14c}$$

Subtracting Equation (A14b) from Equation (A14c), one has $a_3=\frac{14}{15}$.

Substituting this value into Equation (A14b), one obtains $a_2=\frac{8}{3}$.

Substituting all these constants into Equation (A13), it becomes

$$\begin{pmatrix} x_1 \\ x_2 \\ x_3 \end{pmatrix}=\begin{pmatrix} \frac{8}{3}e^{-t}+\frac{14}{15}e^{2t}+\frac{4}{5}\sin t-\frac{3}{5}\cos t-3 \\ -3e^{-t}+3 \\ \frac{13}{3}e^{-t}+\frac{28}{15}e^{2t}+\frac{3}{5}\sin t-\frac{1}{5}\cos t-6 \end{pmatrix}. \tag{A15}$$

Note that in the foregoing it has been shown that, even with repeated eigenvalues, the method of diagonalization can be applied for the solution of the system of differential equations, since in this problem

$$|R|=\begin{vmatrix} 0 & 1 & 1 \\ 1 & 0 & 0 \\ -1 & 1/2 & 2 \end{vmatrix}\neq 0$$

and therefore its inverse exists. This enables one to operate on the diagonal matrix

$$\widetilde{D}=R^{-1}AR.$$

At this stage, a comment about the present approach is in order. While it is true that it is not necessarily more efficient than other methods, such as the method of Laplace transform, for the above problem, the fact that matrix analysis is employed makes it possible and easier to automate the above steps with a computer program. When a relatively large number of equations is involved, many efficient eigenvalue solution subroutines for the coefficient or amplification matrix can be selected such that the solution of responses of the system may be obtained much more efficiently.

Appendix 12B: Maclaurin's Series

This appendix is concerned with the derivation of Taylor's formula and Maclaurin's series for square matrices. The steps in the derivation follow closely those presented by Pipes [3]. In the derivation, the first step is to consider the integration

$$\int_{x_o}^{x_o+h} f'(x)dx=f(x_o+h)-f(x_o). \tag{B1}$$

where the prime denotes the differentiation with respect to x; that is, $f'(x)=\dfrac{df(x)}{dx}$.

Now, change the variable of integration from x to X so that

$$x=(x_o+h)-X \quad \text{or} \quad X=(x_o+h)-x.$$

Substituting this new variable into Equation (B1), one can obtain

$$\begin{aligned}\int_{x_o}^{x_o+h} f'(x)dx &= -\int_h^0 f'[(x_o+h)-X]dX \\ &= \int_0^h f'(x_o+h-X)dX.\end{aligned} \tag{B2}$$

Integrating Equation (B2) by parts and letting

$$u=f'(x_o+h-X),\quad du=-f''(x_o+h-X)dX,$$
$$dv=dX,\quad v=X,$$

with the formula $\int udv=uv-\int vdu$, one has

$$\begin{aligned}\int_0^h f'(x_o+h-X)dX &= Xf'(x_o+h-X)|_0^h+\int_0^h Xf''(x_o+h-X)dX \\ &= hf'(x_o)+\int_0^h Xf''(x_o+h-X)dX.\end{aligned} \tag{B3}$$

Integrating Equation (B3) by parts again, but this time letting

$$u=f''(x_o+h-X),\quad du=-f'''(x_o+h-X)dX,$$
$$dv=XdX,\quad v=\frac{X^2}{2},$$

then the second term on the rhs of Equation (B3) becomes

$$\begin{aligned}\int_0^h Xf''(x_o+h-X)dX &= \frac{X^2}{2!}f''(x_o+h-X)|_0^h \\ &\quad +\int_0^h \frac{X^2}{2!}f'''(x_o+h-X)dX \\ &= \frac{h^2}{2!}f''(x_o)+\int_0^h \frac{X^2}{2!}f'''(x_o+h-X)dX.\end{aligned} \tag{B4}$$

After n integrations by parts, one has

$$\int_{x_o}^{x_o+h} f'(x)dx = hf'(x_o) + \frac{h^2}{2!}f''(x_o) + \frac{h^3}{3!}f'''(x_o) + \cdots + \frac{h^n}{n!}f^{(n)}(x_o) + \int_0^h \frac{X^n}{n!}f^{(n+1)}(x_o + h - X)dX = f(x_o+h) - f(x_o), \quad \text{(B5)}$$

in which $f^{(n)}(x_o) = \left.\frac{d^n f(x)}{dx^n}\right|_{x = x_0}$. The last integral on the rhs of Equation (B5) may be written as

$$I_{n+1} = \int_0^h \frac{X^n}{n!}f^{(n+1)}(x_o + h - X)dX = \frac{1}{n!}\int_0^h X^n f^{(n+1)}(x_o + h - X)dX.$$

The integral I_{n+1} may be interpreted as representing the area under the curve $X^n f^{(n+1)}(x_o + h - X)$ from the point $X = 0$ to $X = h$, so that if $f^{(n+1)}(x_o + h - X)$ is a continuous function of X, there will be some point X_o such that $0 < X_o < h$, for which we have

$$I_{n+1} = \frac{1}{n!}\int_0^h X^n f^{(n+1)}(x_o + h - X)dX = \frac{f^{(n+1)}(x_o + h - X_o)}{n!}\int_0^h X^n dX.$$

Simplifying the rhs, one has

$$I_{n+1} = \frac{h^{n+1}}{(n+1)!}\, f^{(n+1)}(x_o + \alpha h), \quad 0 < \alpha < 1$$

where $\alpha h = h - X_o$. Thus, Equation (B5) may be written as

$$f(x_o + h) = f(x_o) + hf'(x_o) + \frac{h^2}{2!}f''(x_o) + \cdots + \frac{h^n}{n!}f^{(n)}(x_o) + I_{n+1}. \quad \text{(B6)}$$

This is known as Taylor's formula with the Lagrangian form of the remainder. It should be mentioned that in the above derivation it was assumed that $f(x)$ has a continuous nth derivative such that I_{n+1} is called the remainder after $n + 1$ terms. Further,

$$\lim_{n\to\infty} I_{n+1} = \lim_{n\to\infty} \frac{h^{n+1}}{(n+1)!}\, f^{(n+1)}(x_o + \alpha h) = 0.$$

Therefore, Equation (B6) becomes a convergent infinite series

$$f(x_o + h) = f(x_o) + hf'(x_o) + \frac{h^2}{2!}f''(x_o) + \cdots + \frac{h^n}{n!}f^{(n)}(x_o) + \cdots \quad \text{(B7)}$$

If one let $x_o = 0$ and $h = x = At$, where A is a square matrix, one obtains

$$f(At) = f(0) + At f'(0) + \frac{A^2t^2}{2!} f''(0) + \cdots + \frac{A^n t^n}{n!} f^{(n)}(0) + \cdots \tag{B8}$$

When the matrix A becomes a scalar quantity, Equation (B8) is known as the Maclaurin's series. Specifically, if the function $f(At) = e^{At}$, then by Equation (B8) it gives

$$e^{At} = I + At + \frac{A^2t^2}{2!} + \cdots + \frac{A^n t^n}{n!} + \cdots \tag{B9}$$

since $f(0) = I$, $f'(0) = I, \ldots, f^{(n)}(0) = I$, which is the unit matrix with the same order as matrix A.

Appendix 12C: Rank of *A* Matrix

A zero matrix is said to have rank zero. A non-zero matrix A is said to have rank k if at least one of the k-square minors of A is not equal to zero, whereas every $(k + 1)$-square minor, if any, is zero.

Example

Find the rank of matrix $A = \begin{bmatrix} 1 & 2 & 3 \\ 0 & 1 & -1 \\ 1 & 0 & 1 \end{bmatrix}$.

Solution:
The determinant of A is

$$|A| = \begin{vmatrix} 1 & 2 & 3 \\ 0 & 1 & -1 \\ 1 & 0 & 1 \end{vmatrix} = 1 - 2 - 3 = -4 \neq 0.$$

Therefore, the rank of matrix A is 3.

Exercise Questions

Q1. The state equation of a system is given by

$$\begin{pmatrix} \dot{x}_1 \\ \dot{x}_2 \end{pmatrix} = \begin{bmatrix} 0 & 1 \\ -\frac{4}{3} & -\frac{5}{3} \end{bmatrix} \begin{pmatrix} x_1 \\ x_2 \end{pmatrix} + \begin{bmatrix} 0 & 0 \\ 1 & \frac{4}{3} \end{bmatrix} \begin{pmatrix} r_1 \\ r_2 \end{pmatrix}$$

in which $r_1 = r,\ r_2 = \dot{r},\ x_1 = x,\ x_2 = \dot{x}$,

$$A = \begin{bmatrix} 0 & 1 \\ -\frac{4}{3} & -\frac{5}{3} \end{bmatrix}, \quad \text{and} \quad B = \begin{bmatrix} 0 & 0 \\ 1 & \frac{4}{3} \end{bmatrix}.$$

a. Find the state transition matrix of this system.
b. Evaluate the forced responses of the system if $r_1 = r = 1$.

Q2. The state equation of an undamped dynamic system is given by

$$\begin{pmatrix} \dot{x}_1 \\ \dot{x}_2 \end{pmatrix} = \begin{bmatrix} 0 & 1 \\ -\omega_n^2 & 0 \end{bmatrix} \begin{pmatrix} x_1 \\ x_2 \end{pmatrix} + \begin{bmatrix} 1 & 0 \\ 0 & \frac{1}{m} \end{bmatrix} \begin{pmatrix} r_1 \\ r_2 \end{pmatrix}$$

in which $r_1 = 0,\ r_2 = f(t),\ x_1 = x$, and $x_2 = \dot{x}$. Thus,

$$A = \begin{bmatrix} 0 & 1 \\ -\omega_n^2 & 0 \end{bmatrix}, \quad \text{and} \quad B = \begin{bmatrix} 1 & 0 \\ 0 & \frac{1}{m} \end{bmatrix}.$$

The STM for this system, by Equation (12.10), is

$$\Phi(t) = e^{At}.$$

Show that

$$e^{At} = \begin{bmatrix} \cos \omega_n t & \dfrac{\sin \omega_n t}{\omega_n} \\ -\omega_n \sin \omega_n t & \cos \omega_n t \end{bmatrix}.$$

Q3. The state and output equations of a system are given by

$$\begin{pmatrix} \dot{x}_1 \\ \dot{x}_2 \end{pmatrix} = \begin{bmatrix} -2 & -3 \\ 1 & 0 \end{bmatrix} \begin{pmatrix} x_1 \\ x_2 \end{pmatrix} + \begin{bmatrix} 1 \\ 1 \end{bmatrix} r(t), \quad \text{and}$$

$$c(t) = \begin{bmatrix} 1 & 1 \end{bmatrix} \begin{pmatrix} x_1 \\ x_2 \end{pmatrix},$$

in which

$$A = \begin{bmatrix} -2 & -3 \\ 1 & 0 \end{bmatrix}, \quad B = \begin{bmatrix} 1 \\ 1 \end{bmatrix}, \quad \tilde{C} = \begin{bmatrix} 1 & 1 \end{bmatrix}.$$

Determine whether or not this system is controllable and/or observable.

Q4. The state and output equations of a system are given by

$$\begin{pmatrix} \dot{x}_1 \\ \dot{x}_2 \\ \dot{x}_3 \end{pmatrix} = \begin{bmatrix} 0 & 1 & 0 \\ 0 & -1 & -1 \\ 0 & 0 & -3 \end{bmatrix} \begin{pmatrix} x_1 \\ x_2 \\ x_3 \end{pmatrix} + \begin{bmatrix} 1 \\ 1 \\ 1 \end{bmatrix} r(t), \quad \text{and}$$

$$c(t) = \begin{bmatrix} 1 & 1 & 0 \end{bmatrix} \begin{pmatrix} x_1 \\ x_2 \\ x_3 \end{pmatrix}$$

such that

$$A = \begin{bmatrix} 0 & 1 & 0 \\ 0 & -1 & -1 \\ 0 & 0 & -3 \end{bmatrix}, \quad B = \begin{bmatrix} 1 \\ 1 \\ 1 \end{bmatrix},$$

$$\widetilde{C} = \begin{bmatrix} 1 & 1 & 0 \end{bmatrix}.$$

Determine whether or not this system is controllable and/or observable.

Q5. A system has the following state and output equations

$$\begin{pmatrix} \dot{x}_1 \\ \dot{x}_2 \\ \dot{x}_3 \end{pmatrix} = \begin{bmatrix} -2 & 2 & 2 \\ 0 & -1 & 0 \\ -2 & 4 & 3 \end{bmatrix} \begin{pmatrix} x_1 \\ x_2 \\ x_3 \end{pmatrix} + \begin{bmatrix} 1 \\ 0 \\ 1 \end{bmatrix} \sin \omega t, \quad \text{and}$$

$$c(t) = \begin{bmatrix} 1 & 1 & 1 \end{bmatrix} \begin{pmatrix} x_1 \\ x_2 \\ x_3 \end{pmatrix}$$

where the coefficient matrices are given by

$$A = \begin{bmatrix} -2 & 2 & 2 \\ 0 & -1 & 0 \\ -2 & 4 & 3 \end{bmatrix}, \quad B = \begin{bmatrix} 1 \\ 0 \\ 1 \end{bmatrix},$$

$$\widetilde{C} = \begin{bmatrix} 1 & 1 & 1 \end{bmatrix}.$$

Determine whether or not this system is controllable and/or observable.

References

[1] The Math Works, Inc. (2014). *MATLAB R2014a*. The Math Works, Inc., Natick, MA.

[2] Kalman, R.E. (1963). Mathematical description of linear dynamical systems. *SIAM Journal of Control*, **1**, 152–192.

[3] Pipes, L.A. (1958). *Applied Mathematics for Engineers and Physicists*, 2nd edn. McGraw-Hill, New York.

Index

Introduction to Dynamics and Control in Mechanical Engineering Systems, First Edition. Cho W. S. To.

Companion website: www.wiley.com/go/to/dynamics